T0325873

# Sensors for Mechatronics

Sensor for Mechatronics

# Sensors for Mechatronics

*Paul P.L. Regtien*

Hengelo
The Netherlands

AMSTERDAM • BOSTON • HEIDELBERG • LONDON • NEW YORK • OXFORD
PARIS • SAN DIEGO • SAN FRANCISCO • SINGAPORE • SYDNEY • TOKYO

Elsevier
32 Jamestown Road, London NW1 7BY
225 Wyman Street, Waltham, MA 02451, USA

First edition 2012

Copyright © 2012 Elsevier Inc. All rights reserved

No part of this publication may be reproduced or transmitted in any form or by any means,
electronic or mechanical, including photocopying, recording, or any information storage and
retrieval system, without permission in writing from the publisher. Details on how to seek
permission, further information about the Publisher's permissions policies and our arrangement
with organizations such as the Copyright Clearance Center and the Copyright Licensing Agency,
can be found at our website: www.elsevier.com/permissions

This book and the individual contributions contained in it are protected under copyright by the
Publisher (other than as may be noted herein).

**Notices**

Knowledge and best practice in this field are constantly changing. As new research and experience
broaden our understanding, changes in research methods, professional practices, or medical
treatment may become necessary.

Practitioners and researchers must always rely on their own experience and knowledge in
evaluating and using any information, methods, compounds, or experiments described herein.
In using such information or methods they should be mindful of their own safety and the safety
of others, including parties for whom they have a professional responsibility.

To the fullest extent of the law, neither the Publisher nor the authors, contributors, or editors,
assume any liability for any injury and/or damage to persons or property as a matter of products
liability, negligence or otherwise, or from any use or operation of any methods, products,
instructions, or ideas contained in the material herein.

**British Library Cataloguing-in-Publication Data**
A catalogue record for this book is available from the British Library

**Library of Congress Cataloguing-in-Publication Data**
A catalogue record for this book is available from the Library of Congress

ISBN: 978-0-12-391497-2

For information on all Elsevier publications
visit our website at elsevierdirect.com

This book has been manufactured using Print On Demand technology. Each copy is produced
to order and is limited to black ink. The online version of this book will show color
figures where appropriate.

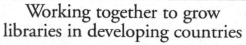

## Working together to grow
## libraries in developing countries

www.elsevier.com | www.bookaid.org | www.sabre.org

ELSEVIER    BOOK AID
            International    Sabre Foundation

# Contents

# Preface

This book offers an overview of various sensors and sensor systems as required and applied in mechatronics. Emphasis lies on the physical background of the operating principles, illustrated with examples of commercially available sensors and of recent and future developments. The work is adapted from a course book on sensors for mechatronic systems, used by students in Electrical and Mechanical Engineering of the University of Twente, The Netherlands. It tries to give an answer to the following questions:

- What is the best sensing principle for a specified mechatronic system or robot task?
- Are commercially available sensors adequate for these tasks?
- How to make the most of these sensors?
- What are their basic limitations?
- How to design a sensor system based on available sensing devices?
- Which improvements could be expected in the near future, based on current sensor research?

A mechatronic system uses information from its internal state as well as from its environment, in order to operate properly. This information is obtained using sensors. In many situations turnkey sensing systems are available on the sensor market. For many other, newly defined applications, such a solution is not feasible, and a sensing system should be assembled from obtainable or specifically designed subsystems. This book offers a help in achieving the best solution to various kinds of sensor problems encountered in mechatronics, by reviewing the major types of transducers, a characterization of the state-of-the-art in sensing technology and a view on current sensor research.

Sensors operate in at least two different physical domains. Therefore, the first chapter discusses general aspects of sensors, with a special section on quantities, notations and relations concerning these various domains. Moreover, it includes a section devoted to sensor errors and error minimization that apply to most of the sensors discussed in this book. Each of the subsequent chapters deals with one class of sensors, pursuing a classification according to physical principles rather than to measurands. These categories are resistive, capacitive, inductive and magnetic, optical, piezoelectric and acoustic sensors. Each of these chapters starts with a brief introduction to the physical background, from which follow the basic properties and limitations of such sensors. In subsequent sections, different types and constructions are given, together with the typical specifications and maximal ratings for devices available on the sensor market. For each category of sensors, a number of applications is given. Where appropriate, a section is added on the interfacing of the sensor.

This book is written for people involved in designing mechatronic systems and robots but who are not specialist in sensor technology. However, the reader is assumed to have some background knowledge on physics and electronics. Nevertheless, some basic principles are briefly reviewed as a recollection, to facilitate reading. Moreover, readers who are not familiar with certain parts of physics are briefly introduced into the subject. Finally, each chapter ends with a section on literature (books and journal papers) as a reference for readers where to find further information on the topics discussed in that chapter. No references have been made to web pages because this information source appears to be too volatile for citation in a book. At the back of the book, appendices are found providing elementary electronic interface circuits, together with expressions describing their performance.

The author welcomes all kind of comments and suggestions that will result in an increased quality of the work.

Paul P.L. Regtien  
Hengelo  
The Netherlands  
November 2011

# 1 Introduction

Worldwide, sensor development is a fast growing discipline. Today's sensor market offers thousands of sensor types, for almost every measurable quantity, for a broad area of applications, and with a wide diversity in quality. Many research groups are active in the sensor field, exploring new technologies, investigating new principles and structures, aiming at reduced size and price, at the same or even better performance.

System engineers have to select the proper sensors for their design, from an overwhelming volume of sensor devices and associated equipment. A well motivated choice requires thorough knowledge of what is available on the market, and a good insight in current sensor research to be able to anticipate forthcoming sensor solutions.

This introductory chapter gives a general view on sensors — their functionality, the nomenclature and global properties — as a prelude to a more in-depth discussion about sensor performance and operation principles.

## 1.1 Sensors in Mechatronics

### 1.1.1 Definitions

A transducer is an essential part of any information processing system that operates in more than one physical domain. These domains are characterized by the type of quantity that provides the carrier of the relevant information. Examples are the optical, electrical, magnetic, thermal and mechanical domains. A transducer is that part of a measurement system that converts information about a measurand from one domain to another, ideally without information loss.

A transducer has at least one input and one output. In measuring instruments, where information processing is performed by electrical signals, either the output or the input is of electrical nature (voltage, current, resistance, capacitance and so on), whereas the other is a non-electrical signal (displacement, temperature, elasticity and so on). A transducer with a non-electrical *input* is an *input transducer*, intended to convert a non-electrical quantity into an electrical signal in order to measure that quantity. A transducer with a non-electrical *output* is called an *output*

**Sensors for Mechatronics.** DOI: 10.1016/B978-0-12-391497-2.00001-7
© 2012 Elsevier Inc. All rights reserved.

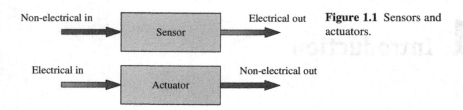

**Figure 1.1** Sensors and actuators.

*transducer*, intended to convert an electrical signal into a non-electrical quantity in order to control that quantity. So, a more explicit definition of a transducer is *an electrical device that converts one form of energy into another*, with the intention of preserving information.

According to common terminology, these transducers are also called *sensor* and *actuator*, respectively (Figure 1.1). So, a sensor is an input transducer and an actuator is an output transducer. It should be noted, however, that this terminology is not standardized. In literature other definitions are found. Some authors make an explicit difference between a *sensor* and a (input) *transducer*, stressing a distinction between the element that performs the physical conversion and the complete device − for instance, a strain gauge (the transducer) and a load cell (the sensor) with one or more strain gauges and an elastic element.

Attempts to standardize terminology in the field of metrology have resulted in the *Vocabulaire International de Métrologie* (VIM) [1]. According to this document a *transducer* is a device, used in measurement, that provides an output quantity having a specified relation to the input quantity. The same document defines a *sensor* as the element of a measuring system that is directly affected by a phenomenon, body or substance carrying a quantity to be measured.

Modern sensors not only contain the converting element but also part of the signal processing (analogue processing such as amplification and filtering, AD conversion and even some digital electronics). Many of such sensors have the electronics integrated with the transducer part onto a single chip. Present-day sensors may have a bus-compatible output, implying full signal conditioning on board. There is a trend to also include transmission electronics within the device, for instance for biomedical applications.

Signal conditioning may be included:

- to protect the sensor from being loaded or to reduce loading errors;
- to match the sensor output range to the input range of the ADC;
- to enhance the S/N (signal-to-noise ratio) prior to further signal processing;
- to generate a digital, bus-compatible electrical output; or
- to transmit measurement data for wireless applications.

In conclusion, the boundaries between sensor and transducer as proclaimed in many sensor textbooks are disappearing or losing their usefulness: the user buys and applies the sensor system as a single device, with a non-electrical input and an electrical (e.g. analogue, digital and bus compatible) output.

## 1.1.2  Sensor Development

Sensors provide the essential information about the state of a (mechatronic) system and its environment. This information is used to execute prescribed tasks, to adapt the system properties or operation to the (changing) environment or to increase the accuracy of the actions to be performed.

Sensors play an important role not only in mechatronics but also in many other areas. They are widely applied nowadays in all kind of industrial products and systems. A few examples are as follows:

• Consumer electronics
• Household products
• Public transport, automotive
• Process industry
• Manufacturing, production
• Agriculture and breeding industry
• Medical instruments

and many other areas where the introduction of sensors has increased dramatically the performance of instruments, machines and products.

The world sensor market is still growing substantially. The worldwide sensor market offers over 100,000 different types of sensors. This figure not only illustrates the wide range of sensor use but also the fact that selecting the right sensor for a particular application is not a trivial task. Reasons for the increasing interest in sensors are as follows:

• *Reduced prices*: the price of sensors not only depends on the technology but also on production volume. Today, the price of a sensor runs from several ten thousands of euros for single pieces down to a few eurocents for a 100 million volume.
• *Miniaturization*: the IC-compatible technology and progress in micromachining technology are responsible for this trend [2–4]. Pressure sensors belong to the first candidates for realization in silicon (early 1960s). Micro-ElectroMechanical Systems (MEMS) are gradually taking over many traditionally designed mechanical sensors [5–7]. Nowadays, solid-state sensors (in silicon or compatible technology) for almost every quantity are available, and there is still room for innovation in this area [8,9].
• *Smart sensing*: the same technology allows the integration of signal processing and sensing functions on a single chip. Special technology permits the processing of both analogue and digital signals ('mixed signals'), resulting in sensor modules with (bus compatible) digital output.

Popular MEMS sensors are accelerometers and gyroscopes. A MEMS accelerometer can be made completely out of silicon, using micromachining technology. The seismic mass is connected to the substrate by thin, flexible beams, acting as a spring. The movement of the mass can be measured by, for instance, integrated piezoresistors positioned on the beam at a location with maximum deformation (Chapter 4) or by a capacitive method (Chapter 5).

In mechatronics, mainly sensors for the measurement of mechanical quantities are encountered. The most frequent sensors are for displacement (position) and force (pressure), but many other sensor types can be found in a mechatronic system.

Many sensors are commercially available and can be added to or integrated into a mechatronic system. This approach is preferred for systems with relatively simple tasks and operating in a well-defined environment, as commonly encountered in industrial applications. However, for more versatile tasks and specific applications, dedicated sensor systems are required, which are often not available. Special designs, further development or even research are needed to fulfil specific requirements, for instance with respect to dimensions, weight, temperature range and radiation hardness.

### 1.1.3    Sensor Nomenclature

In this book, we follow a strict categorization of sensors according to their main physical principle. The reason for this choice is that sensor performance is mainly determined by the physics of the underlying principle of operation. For example, a position sensor can be realized using resistive, capacitive, inductive, acoustic and optical methods. The sensor characteristics are strongly related to the respective physical transduction processes. However, a magnetic sensor of a particular type could be applied as, for instance, a displacement sensor, a velocity sensor or a tactile sensor. For all these applications the performance is limited by the physics of this magnetic sensor.

Apparently, position and movement lead the list of measurement quantities. Common parlance contains many other words for position parameters. Often, transducers are named after these words. Here is a short description of some of these transducers.

| | |
|---|---|
| *Distance sensor* | Measures the length of the straight line between two defined points |
| *Position sensor* | Measures the co-ordinates of a specified point of an object in a specified reference system |
| *Displacement sensor* | Measures the change of position relative to a reference point |
| *Range sensor* | Measures in a 3D space the shortest distance from a reference point (the observer) to various points of object boundaries in order to determine their position and orientation relative to the observer or to get an image of these objects |
| *Proximity sensor* | (a) Determines the sign (positive or negative) of the linear distance between an object point and a fixed reference point; also called a switch<br>(b) A contact-free displacement or distance sensor for short distances (down to zero) |
| *Level sensor* | Measures the distance of the top level of a liquid or granular substance in a container with respect to a specified horizontal reference plane |
| *Angular sensor* | Measures the angle of rotation relative to a reference position |
| *Encoder* | Displacement sensor (linear or angular) containing a binary coded ruler or disk |

| *Tilt sensor* | Measures the angle relative to the earth's normal |
|---|---|
| *Tachometer* | Measures rotational speed |
| *Vibration sensor* | Measures the motion of a vibrating object in terms of displacement, velocity or acceleration |
| *Accelerometer* | Measures acceleration |

Transducers for the measurement of force and related quantities are as follows:

| *Pressure sensor* | Measures pressure difference, relative to either vacuum (absolute pressure), a reference pressure or ambient pressure |
|---|---|
| *Force sensor* | Measures the (normal and/or shear) force exerted on the active point of the transducer |
| *Torque sensor* | Measures torque (moment) |
| *Force−torque sensor* | Measures both forces and torques (up to six components) |
| *Load cell* | Force or pressure sensor, for measuring weight |
| *Strain gauge* | Measures linear relative elongation (positive or negative) of an object, caused by compressive or tensile stress |
| *Touch sensor* | Detects the presence or (combined with a displacement sensor) the position of an object by making mechanical contact |
| *Tactile sensor* | Measures 3D shape of an object by the act of touch, either sequentially using an exploring touch sensor or instantaneously by a matrix of force sensors |

Many transducers have been given names according to their operating principle, construction or a particular property. Examples are as follows:

| *Hall sensor* | Measures magnetic field based on the Hall effect, after the American physicist Edwin Hall (1855−1938) |
|---|---|
| *Coriolis mass flow sensor* | Measures mass flow of a fluid by exploiting the Coriolis force exerted on a rotating or vibrating channel with that fluid; after Gustave-Gaspard de Coriolis, French scientist (1792−1843) |
| *Gyroscope, gyrometer* | A device for measuring angle or angular velocity, based on the gyroscopic effect occurring in rotating or vibrating structures |
| *Eddy current sensor* | Measures short range distances between the sensor front and a conductive object using currents induced in that object due to an applied AC magnetic field; also used for defect detection |
| *LVDT* | or Linear Variable Displacement Transformer, a device that is basically a voltage transformer, with linearly movable core |
| *NTC* | Short for temperature sensor (especially thermistor) with Negative Temperature Coefficient |

Some sensors use a *concatenation* of transduction steps. A displacement sensor, combined with a spring, can act as a force sensor. In combination with a calibrated mass, a displacement sensor can serve as an accelerometer. The performance of such transducers not only depends on the primary sensor but also on the added components: in the examples above the spring compliance and the seismic mass, respectively.

Information about a particular quantity can also be obtained by *calculation* using relations between quantities. The accuracy of the result depends not only on the errors in the quantities that are measured directly but also on the accuracy of the parameters in the model that describes the relation between the quantities involved. For instance, in an acoustic distance measurement the distance is calculated from the measured time-of-flight (ToF; with associated errors) and the sound velocity. An accurate measurement result requires knowledge of the acoustic velocity of the medium at the prevailing temperature.

Some variables can be derived from others by electronic *signal processing*. Speed and acceleration can be measured using a displacement sensor, by differentiating its output signal once or twice, respectively. Conversely, by integrating the output signal of an accelerometer a velocity signal is obtained and, by a second integration, a position signal. Obviously, the performance of the final result depends on the quality of the signal processing. The main problem with differentiation is the increased noise level (in particular in the higher frequency range), and integration may result in large drift due to the integration of offset.

### 1.1.4   Sensors and Information

According to the amount of information a sensor or sensing system offers, three groups of sensors can be distinguished: binary sensors, analogue sensors and image sensors. Binary sensors give only one bit of information but are very useful in mechatronics. They are utilized as end stops, as event detectors and as safety devices. Depending on their output (0 or 1), processes can be started, terminated or interrupted. The binary nature of the output makes them highly insensitive to electrical interference.

Analogue sensors are used for the acquisition of metric information with respect to quantities related to distance (e.g. relative position, linear and angular velocity and acceleration), force (e.g. pressure, gripping force and bending) or others (e.g. thermal, optical, mechanical, electrical or magnetic properties of an object). A wide variety of industrial sensors for these purposes are available.

The third category comprises image sensors, intended for the acquisition of information related to structures and shapes. Depending on the application, the sensor data refer to one-, two- or three-dimensional images. The accuracy requirements are less severe compared to the sensors from the preceding category, but the information content of their output is much larger. As a consequence, the data acquisition and processing for such sensors are more complex and more time consuming.

The next sections present some general aspects of sensors, following the categorization in binary, analogue and image sensors as introduced before. Actually, the section serves as a general overview of the sensors and sensing systems which are discussed in more detail in subsequent chapters. Details on physical background, specifications and typical applications are left for those chapters. Here, the

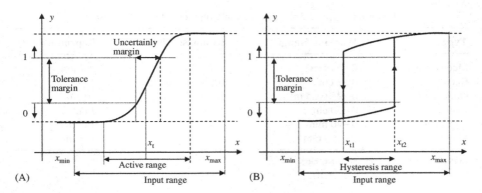

**Figure 1.2** Characteristic of a binary sensor (A) without hysteresis, (B) with imposed hysteresis.

differences in approach are highlighted and their consequences for the applicability in mechatronic systems are emphasized.

### Binary Sensors

A binary sensor has an analogue input and a two-state output (0 or 1). It converts the (analogue) input quantity to an one-bit output signal. These sensors are also referred to as *switches* or *detectors*. They have a fixed or an adjustable threshold level $x_t$ (Figure 1.2A). In fact, there are essentially two levels, marking the hysteresis interval (Figure 1.2B). Any analogue sensor can be converted to a binary sensor by adding a Schmitt trigger (comparator with hysteresis, Appendix C.5). Although hysteresis lowers the accuracy of the threshold detection (down to the hysteresis interval), it may help reduce unwanted bouncing due to noise in the input signal.

Most binary sensors measure position. Binary displacement sensors are also referred to as *proximity* sensors. They react when a system part or a moving object has reached a specified position. Two major types are the mechanically and the magnetically controlled switches.

Mechanically controlled switches are actually touch sensors. They are available in a large variety of sizes and constructions; for special conditions there are waterproof and explosion-proof types; for precision measurements there are switches with an inaccuracy less than $\pm 1\ \mu m$ and a hysteresis interval in the same order, guaranteed over a temperature range from $-20°C$ to $75°C$. Another important parameter of a switch is the reliability, expressed in the minimum number of commutations. Mechanical switches have a reliability of about $10^6$.

A *reed switch* is a magnetically controlled switch: two magnetizable tongues or reeds in a hermetically closed encapsulation filled with an inert gas. The switch is normally off; it can be switched on mechanically by a permanent magnet approaching the sensor. Reed switches have good reliability: over $10^7$ commutations at a switching frequency of 50 Hz. A disadvantage is the *bouncing* effect, the chattering of the contacts during a transition of state. Reed switches are applied in various

**Table 1.1** Typical Specifications of Commercial Binary Sensors

| Type | Working Range | Response Time | Reproducibility |
|------|--------------|---------------|-----------------|
| Mechanical | 0 (contact) | | ±1 μm |
| Reed switch | 0–2 cm | 0.1 ms (on) | |
| Optical | 0–2/10/35 m* | 500 Hz/1 ms | 10 cm |
| Inductive | 0–50 cm | 1 ms | 1 cm |
| Capacitive | 0–40 mm | 1 ms | 1 mm |
| Magnetic | 0–100 mm | | 10 μm |

*Reflection from object/reflector/direct mode.

commercial systems, from cars (monitoring broken lights, level indicators) to electronic organs (playing contacts), to telecommunication devices and testing and
measurement equipments. In mechatronic systems they act as end-of-motion detectors, touch sensors and other safety devices. The technical aspects are described in
Chapter 6 on inductive and magnetic sensors.

The drawbacks of all mechanical switches are a relatively large switch-on time
(for reed switches typically 0.2 ms) and wear. This explains the growing popularity
of electronic switches, such as optically controlled semiconductors and Hall plates.
There is a wide range of binary displacement sensors on the market, for a variety
of distances and performance. Table 1.1 presents a concise overview of
specifications.

All but the mechanical switch operate essentially contact free. Obviously, the
optical types have the widest distance range. The optical, inductive and capacitive
types are essentially analogue sensors, with adjustable threshold levels. The specifications include interface and read-out electronics. In particular, the response time
of the sensor itself may be much better than the value listed in the table. Accuracy
data include hysteresis and apply for the whole temperature range (maximum operating temperature range 70°C typical).

## Analogue Sensors

There is an overwhelming number of analogue sensors on the market, for almost
any physical quantity, and operating according to a diversity of physical principles.
In mechatronics, the major measurement quantities of interest are linear and angular displacement, their time derivatives (velocity and acceleration) and force
(including torque and pressure). These and many other sensors will be discussed in
more detail in later chapters.

## Image Sensors

Imaging is a powerful method to obtain information about geometrical parameters
of objects with a complex shape. The 3D object or a complete scene is transformed
to a set of data points representing the geometrical parameters that describe

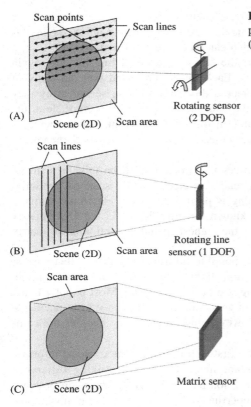

**Figure 1.3** Imaging techniques: (A) 2D point scanning, (B) 1D line scanning and (C) projection on 2D matrix sensor.

particular characteristics of the object, for instance its pose (position and orientation in space), dimensions, shape or identity. An essential condition in imaging is the preservation of the required information. This is certainly not trivial: photographic and camera pictures are 2D representations of a 3D world, and hence much information is lost by the imaging process.

Three basic concepts for image acquisition are depicted schematically in Figure 1.3. In the first method the scene to be imaged is scanned point by point by some mechanical means (e.g. a mirror on a stepping motor) or electronically (for instance with *phased arrays*). Such an imaging system is often referred to as a *range finder*: it yields distance information over an angular range determined by the limits of the scanning mechanism. The output is a sequential data stream containing 3D information about the scene: depth data from the scanning sensor and angular data from the scanning mechanism. Although the data points are three dimensional, information is obtained only about the surface boundary range and only that part of the surface that is connected to the sensor system by a line of direct sight. Therefore, range data are sometimes called 2.5D data. In Figure 1.3A, the sensor and scanning mechanism are presented as a single device. Most scanning systems consist of several parts, for instance a fixed transmitter and receiver and one or more rotating mirrors or reflectors. Sometimes the transmitter, the receiver or both are mounted on the scanning device. Although the scanning method is slow, it requires only a single sensor which can therefore be of high quality.

In the second method (Figure 1.3B) the scene is scanned line by line, again using some mechanical scanning device. Each line is projected onto an array of sensors (in the optical domain for instance a diode array). The sensor array may include electronic scanning to process the data in a proper way. Nevertheless, the mechanical scanning mechanism operates in only one direction, which increases the speed of image formation and lowers construction complexity as compared to point-wise scanning.

The third method (Figure 1.3C) involves the projection of the unknown image on a 2D matrix of point sensors. This matrix is electronically scanned for serial processing of the data. Since all scanning is performed in the electronic domain, the acquisition time is short. The best known imaging device is the CCD matrix camera (Charge Coupled Device). It has the highest spatial resolution of all matrix imagers.

Considering the nature of the various possible information carriers, there are at least three candidates for image acquisition: light, (ultra)sound and contact force. All three are being used in both scanning and projection mode. Most popular is the CCD camera as imager for exploring and analyzing the work space of a mechatronic system or a robot's environment. However, in numerous applications the camera is certainly not the best choice.

The acquisition of an image is just the first step in getting the required information; data processing is another important item. There is a striking difference between (camera based) vision and non-vision data processing. The main problem of the CCD camera is the provision of superfluous data. The first step in image processing is, therefore, to get rid of all irrelevant data in the image. For instance, a mere contour might be sufficient for proper object identification; the point is how to find the right contour. However, most non-vision imagers suffer from a too-low resolution. Here the main problem is the extraction of information from the low-resolution image and − in the case of scanning systems − from other sensors. In all cases, model-driven data processing is required to be able to arrive at proper conclusions about features of the objects or the scene under test.

## Optical Imaging

Most optical imaging systems applied in mechatronics and robotics use a camera (CCD-type or CMOS) and a proper illumination of the scene. The image (or a pair of images or even a sequence when 3D information is required) is analyzed by some image-processing algorithm applied to the intensity and colour distribution in the image. Particular object features are extracted from particular patterns in light intensity in the image. Position information is derived from the position of features in the image, together with camera parameters (position and orientation, focal length).

Specified conditions for getting a proper image must be fulfilled: an illumination that yields adequate contrast and no disturbing shadows and a camera set-up with a full view on the object or the scene and with a camera that has a sufficiently high resolution, so as not to lose relevant details. Obviously, a 2D image shows only a certain prospect of the object, never a complete view (*self-occlusion*). In case of more than one object, some of them could be (partially) hidden behind others (*occlusion*), a situation that makes the identification much more difficult.

Even in the most favourable situation, the image alone does not reveal enough information for the specified task. Besides a proper model of the object, we need a model of the imaging process: position and orientation of the camera(s) and camera parameters like focal length and position of the light source(s) with respect to the object and camera. All of these items determine the quality of the image from which features are to be extracted. The pose of the object in the scene can be derived from the available information and knowledge of the imaging system.

Many algorithms have been developed to extract useful features from an image that is built up of thousands of samples (in space and time) described by colour parameters, grey-tone values or just bits for black and white images. The image is searched for particular combinations of adjacent pixels such as edges, from which region boundaries are derived. Noise in the image may disturb this process, and special algorithms have been developed to reduce its influence. The result is an image that reveals at least some characteristics of the object. For further information on feature extraction the reader is referred to the literature on computer vision and image processing.

## Acoustic Imaging

The interest in acoustic waves for imaging is steadily growing, mainly because of the low cost and simple construction of acoustic transducers. The suitability of acoustic imaging has been proved in medical, geological and submarine applications. Applications in mechatronics have, however, some severe limitations going back to ultrasonic wave propagation in air (where most mechatronic systems operate). Despite these limitations, detailed in Chapter 9, many attempts are being made to improve the accuracy and applicability of acoustic measurement systems, in particular as they are applied to distance measurement and range finding.

The most striking drawback of acoustic imaging is the low spatial resolution, due to the diverging beam of acoustic transducers. The directivity of the transducers can be improved by increasing the ratio between the diameter and the wave length. Even at medium frequencies (i.e. 40 kHz), this results in rather large devices. An alternative method is the use of an array of simultaneously active acoustic elements. Due to interference, the main beam (in the direction of the acoustic axis) is narrowed. Further, the direction of this beam can be electronically controlled by variation of the phase shift or time delay between the elements of the array. This technique, known as *phased arrays*, applies to transmitters as well as receivers.

The recognition of shapes requires a set of distance sensors or scanning with a single sensor, according to one of the principles in Figure 1.3. The shape follows from a series of numerical calculations (see for instance [10,11]).

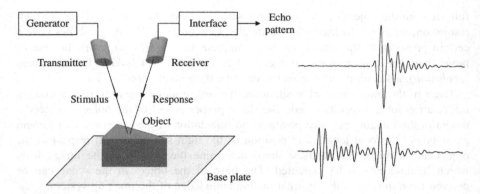

**Figure 1.4** Object recognition using acoustic signature technique; left: system set-up; right top: echo of base plate only; bottom: echo of object on base plate.

Instead of geometric models for use in object recognition, other models may be used. An example of such a different approach is shown schematically in Figure 1.4. An acoustic signal (the stimulus) is transmitted towards the object. The shape of the echo pattern (the response) is determined by the object's shape and orientation. In a learning phase, echo patterns of all possible objects are stored in computer memory. They can be considered acoustic signatures of the objects. The echo pattern from a test object belonging to the trained set is matched to each of the stored signatures. Using a minimum distance criterion reveals the best candidate [12]. Evidently, the test conditions should be the same as during the learning phase: a fixed geometry and a stable stimulus are required.

The comparison process may be performed either in the frequency domain or in the time domain. With this simple technique, it is possible to distinguish between objects whose shapes or orientations (normal versus upside-down position) are quite different. With an adaptive stimulus and a suitable algorithm, even small defects in an object can be detected by ultrasonic techniques. Under certain conditions, very small object differences can be detected, for instance between sides of a coin [13].

## Tactile Imaging

In contrast to optical and acoustic imaging, tactile imaging is performed by mechanical contact between sensor and object. Making contact has advantages as well as disadvantages. Disadvantages are the mechanical load of the object (it may move or be pressed) and the necessity of moving the sensor actively towards the object. Advantages of tactile imaging include the possibility of acquiring force-related information (for instance touching force and torque) and mechanical properties of the object (e.g. elasticity, resilience and surface texture). Another advantage over optical imaging is the insensitivity to environmental conditions. This versatility of a tactile sensor makes it very attractive for control purposes, especially in assembly processes. Moreover, tactile and vision data can be fused, to benefit from both modalities.

In *robotics*, tactile imaging is mostly combined with the gripping action. For in-line control the tactile sensor should be incorporated into the gripper of the robot, allowing simultaneous force distribution and position measurements during the motion of the gripper. This permits continuous force control as well as position correction.

In *inspection* systems (like coordinate measuring machines), the object under test is scanned mechanically by a motion mechanism, with a touch sensor as the end effector. The machine is controlled to follow a path along the object, while keeping the touch force at a constant value. Position data follow from back transformation of the tip (sensor) co-ordinates to world co-ordinates. The scanning is slow but can be very accurate, down to 10 nm in three dimensions.

## 1.2 Selection of Sensors

Choosing a proper sensor is certainly not a trivial task. First of all, the task that is to be supported by one or more sensors needs to be thoroughly analyzed and all possible strategies to be reviewed. Potential sensors should be precisely specified, including environmental conditions and mechanical and electrical constraints. If commercial sensors can be found that satisfy the requirements, purchase is recommended. Special attention should be given to interface electronics (in general available as separate units, but rarely adequate for newly developed mechatronic systems). If the market does not offer the right sensor system, such a system may be assembled from commercial sensor components and electronics. This book gives some physical background for most sensors, to help understand their operation, to assist in making a justified choice, or to provide knowledge for assembling particular sensing systems.

Sensor selection is based on satisfying requirements; however, these requirements are often not known precisely or in detail, in particular when the designer of the system and its user are different persons. The first task of the designer, therefore, is to get as much information as possible about the future applications of the system, all possible conditions of operation, the environmental factors and the specifications, with respect to quality, physical dimensions and costs.

The list of demands should be exhaustive. Even when not all items are relevant, they must be indicated as such. This will leave more room to the designer and minimizes the risk of having to start all over again. The list should be made in a way that enables unambiguous comparison with the final specifications of the designed system. Once the designer has a complete idea about the future use of the system, the phase of the conceptual design can start.

Before thinking about sensors, the *measurement principle* first has to be considered. For the instrumentation of each measurement principle, the designer has a

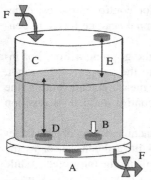

**Figure 1.5** Measuring tank contents.

multitude of *sensing methods* at his disposal. For the realization of a particular sensor method, the designer has to choose the optimal *sensor component* and *sensor type* from a vast collection of sensors offered by numerous sensor manufacturers.

This design process is illustrated by an example of a measurement for just a single, static quantity: the amount of fluid in a container (for instance, a drink dispenser). The first question to be answered is, in what units the amount should be expressed: volume or mass? This may influence the final selection of the sensor. Figure 1.5 shows various *measurement principles* in a schematic way:

A: the tank is placed on a balance, to measure its total weight;
B: a pressure gauge on the bottom of the tank;
C: a gauging-rule from top to bottom with electronic read-out;
D: level detector on the bottom, measuring the column height;
E: level detector from the top of the tank, measuring the height of the empty part;
F: (mass or volume) flow meters at both inlet and outlet.

Obviously, many more principles can be found to measure a quantity that is related to the amount of fluid in the reservoir.

In the conceptive phase of the design as many principles as possible should be considered, even unconventional ones. Based on the list of demands it should be possible to find a proper candidate principle from this list, or at least to delete many of the principles, on an argued base. For instance, if the tank contains a corrosive fluid, a non-contact measurement principle is preferred, putting principles B, C and D on a lower position in the list.

Further, for very large tanks, method A can possibly be eliminated because of high costs. The conceptual design ends up with a set of principles with pros and cons, ranked according to the prospects of success.

After having specified a list of candidate principles, the next step is to find a suitable sensing method for each of them. In the example of Figure 1.5 we will further investigate principle E, a level detector placed at the top of the tank. It should be noted that from level alone the amount of liquid cannot be determined: the shape of the container should also be taken into account. Again, a list of the various possible *sensor methods* is made, as follows:

E1: a float, connected to an electronic read-out system;
E2: optical ToF measurement;
E3: optical range measurement;
E4: electromagnetic distance measurement (radar);
E5: acoustic ToF measurement and so on.

As in the conceptual phase, these methods are evaluated using the list of demands, so not only the characteristics of the sensing method but also the properties of the measurement object (e.g. kind of liquid and shape of tank) and the environment should be taken into account. For the tank system, the acoustic ToF method could be an excellent candidate because of its being contact free. In this phase it is also important to consider methods to reduce such environmental factors as temperature. Ultimately, this phase concludes with a list of candidate sensing methods and their merits and demerits with respect to the requirements.

The final step is the selection of the *components* that make up the sensing system. Here a decision must be made between the purchase of a commercially available system and the development of a dedicated system. The major criteria are costs and time: both are often underestimated when development by one's own is considered.

In this phase of the selection process, sensor *specifications* become important. Sensor providers publish specifications in data sheets or on the Internet. However, the accessibility of such data is still poor, making this part of the selection process critical and time consuming, in particular for non-specialists in the sensor field.

Evidently, the example of the level sensor is highly simplified, whereas the selection process is usually not that straightforward. Since the sensor is often just one element in the design of a complex mechatronic system, close and frequent interaction with other design disciplines as well as the customer is recommended.

# References to Cited Literature

[1] International Vocabulary of Metrology – Basic and general concepts and associated terms (VIM); Document produced by Working Group 2 of the Joint Committee for Guides in Metrology (JCGM/WG 2), JCGM 200: 2008.

[2] K.E. Petersen: Silicon as a mechanical material, Proc. IEEE, 70(5) (1982), 420–457.

[3] S. Middelhoek, S.A. Audet: Silicon sensors; Academic Press, London, San Diego, New York, Berkeley, Boston, Sydney, Tokyo, Toronto, 1989; ISBN 0-12-495051-5.

[4] J.W. Gardner: Microsensors – principles and applications; Wiley, New York, Chichester, Weinheim, Brisbane, Singapore, Toronto, 1994; ISBN 0-471-94135-2/94136-0.

[5] R.F. Wolffenbuttel (ed.): Silicon sensors and circuits; on-chip compatibility; Chapman & Hall, London, Glasgow, Weinheim, New York, Tokyo, Melbourne, Madras, 1996; ISBN 0-412-70970-8.

[6]  M.-H. Bao: Micro mechanical transducers − pressure sensors, accelerometers and gyroscopes; Elsevier, Amsterdam, Lausanne, New York, Oxford, Shannon, Singapore, Tokyo, 2000; ISBN 0-444-50558-X.

[7]  M. Elwenspoek, R. Wiegerink: Mechanical microsensors; Springer-Verlag, Berlin, Heidelberg, New York, (Barcelona, Hong Kong, London, Milan, Paris, Singapore, Tokyo), 2001; ISBN 3-540-67582-5.

[8]  Proceedings of various conferences, for instance 'Transducers', 'Eurosensors', 'IEEE Int. Conf. on Micro Electro Mechanical Systems' and many more.

[9]  Various international journals, for instance 'Sens. Actuators A' (Elsevier), 'IEEE Sens. J.' (IEEE).

[10] P. Mattila, J. Siirtola, R. Suoranta: Two-dimensional object detection in air using ultrasonic transducer array and non-linear digital L-filter, Sens. Actuators A, 55 (1996), 107−113.

[11] A.D. Armitage, N.R. Scales, P.J. Hicks, P.A. Payne, Q.X. Chen, J.V. Hatfield: An integrated array transducer receiver for ultrasound imaging, Sens. Actuators A, 46−47 (1995), 542−546.

[12] C. Cai, P.P.L. Regtien: A smart sonar object recognition system for robots, Meas. Sci. Technol., 4 (1993), 95−100.

[13] J.M. Martín Abreu, T. Freire Bastos, L. Calderón: Ultrasonic echoes from complex surfaces: an application to object recognition, Sens. Actuators A, 31 (1992), 182−187.

# Literature for Further Reading

## Introductory Books on Sensors and Mechatronics

[1]  P. Ripka, A. Tipek (eds.): Modern sensors handbook; Wiley-ISTE, London; Newport Beach, CA, 2007; ISBN 978-1-905209-66-8.

[2]  T.G. Beckwith, R.D. Marangoni, J.H. Lienhard V: Mechanical measurements; Pearson Prentice Hall, Upper Saddle River, NJ, 2007; ISBN 0-201-84765-5.

[3]  R.S. Figliola, D.E. Beasley: Theory and design for mechanical measurements; Wiley, New York, Chichester, Weinheim, Brisbane, Singapore, Toronto, 2006; ISBN 0-471-44593-2.

[4]  A. Preumont: Mechatronics − dynamics of electromechanical and piezoelectric systems; Springer, Berlin, Heidelberg, New York, (etc.) 2006; ISBN 1-4020-4695-2.

[5]  R.S. Figliola, D.E. Beasley: Theory and design for mechanical measurements; Wiley, New York, Chichester, Weinheim, Brisbane, Singapore, Toronto, 2006; ISBN 0-471-44593-2.

[6]  H.K. Tönshoff, I. Inasaki (eds.): Sensors in manufacturing; Wiley-VCH Verlag GmbH, Weinheim, FRG, 2001; ISBN 3-527-29558-5.

[7]  R. Pallas-Areny, J.G. Webster: Sensors and signal conditioning; 2nd edition, Wiley, New York, Chichester, Weinheim, Brisbane, Singapore, Toronto, 2001; ISBN 0-471-33232-1.

[8]  G. Dudek, M. Jenkin: Computational principles of mobile robotics; Cambridge University Press, Cambridge, 2000; ISBN 0-521-56876-5.

[9]  I.J. Busch-Vishniac: Electromechanical sensors and actuators; Springer-Verlag, Berlin, Heidelberg, New York, (etc.) 1999; ISBN 0-387-98495-X.

[10] P. Hauptmann: Sensors − principles and applications; Hanser, Munich; Prentice Hall, Hemel Hempstead, Engelwood Cliffs NJ, 1993; ISBN 0-13-805-789-3P.

## Books on Semiconductor Sensors

[1]  G.C.M. Meijer (ed.): Smart sensor systems; Wiley, New York, Chichester, Weinheim, Brisbane, Singapore, Toronto, 2008; ISBN 9780470866917.

[2]  S.Y. Yurish, M.T.S.R. Gomes (eds.): Smart sensors and MEMS; Springer-Verlag, Berlin, Heidelberg, New York, (etc.) 2005; ISBN 1-402-02927-6.

[3]  A.J. Wheeler, A.R. Ganji: Introduction to engineering experimentation; Pearson, Upper Saddle River NJ, 2004; ISBN 0-13-065844-8.

[4]  M. Elwenspoek, R. Wiegerink: Mechanical microsensors; Springer-Verlag, Berlin, Heidelberg, New York, (etc.) 2001; ISBN 3-540-67582-5.

[5]  M.-H. Bao: Micro mechanical transducers; Elsevier, Amsterdam, Lausanne, New York, Oxford, Shannon, Singapore, Tokyo, 2000; ISBN 0-444-50558-X.

[6]  R.F. Wolffenbuttel (ed.): Silicon sensors and circuits; on-chip compatibility; Chapman & Hall, London, Glasgow, Weinheim, New York, Tokyo, Melbourne, Madras, 1996; ISBN 0-412-70970-8.

[7]  S.M. Sze (ed.): Semiconductor sensors; Wiley, New York, Chichester, Weinheim, Brisbane, Singapore, Toronto, 1994; ISBN 0-471-54609-7.

[8]  J.W. Gardner: Microsensors − principles and applications; Wiley, New York, Chichester, Weinheim, Brisbane, Singapore, Toronto, 1994; ISBN 0-471-94135-2/ 94136-0.

[9]  L. Ristic: Sensor technology and devices; Artech House Publishers, Boston, London, 1994; ISBN 0-89006-532-2.

[10] S. Middelhoek, S.A. Audet: Silicon sensors; Academic Press, London, San Diego, New York, Berkeley, Boston, Sydney, Tokyo, Toronto, 1989; ISBN 0-12-495051-5.

# 2 Sensor Fundamentals

A sensor performs the exchange of information (hence energy) from one domain to another and as such it operates at the interface between different physical domains. In this chapter we first introduce a notation system for the quantities used in this book. To avoid confusion with notations, we define unambiguous symbols for each quantity. For example, in the electrical domain the symbol $\varepsilon$ usually stands for dielectric constant, whereas it means strain in the mechanical domain. In this book we use $\varepsilon$ for dielectric constant only; strain is denoted by $S$. Several frameworks have been developed for a systematic description of sensors. Various approaches are presented in this chapter. Further, a formal description of relations between quantities, based on energy considerations, is introduced from which particular physical effects are derived serving for specific groups of sensors that are discussed in later chapters.

## 2.1 Physical Quantities

### 2.1.1 Classification of Quantities

Various attempts have been made to set up a consistent framework of quantities. Physical quantities can be divided into subgroups according to various criteria. This leads to subgroups with different characteristics.

- With respect to *direction*:
    - a quantity having a direction is called a *vector* (e.g. velocity);
    - a quantity that does not have a direction is a *scalar* (e.g. temperature).
- With respect to *time behaviour*:
    - a *state variable* describes a static property;
    - a *rate variable* describes a dynamic property.

Within one domain state and rate variables are related as follows:

$$X_{\text{rate}} = \frac{\mathrm{d}}{\mathrm{d}t} X_{\text{state}} = \dot{X}_{\text{state}} \quad \text{or} \quad X_{\text{state}} = \int X_{\text{rate}} \, \mathrm{d}t \qquad (2.1)$$

**Sensors for Mechatronics. DOI: 10.1016/B978-0-12-391497-2.00002-9**
© 2012 Elsevier Inc. All rights reserved.

*Examples*
The electrical domain

$$I = \frac{d}{dt}Q \quad \text{or} \quad Q = \int I dt \tag{2.2}$$

The mechanical (translation) domain

$$v = \frac{d}{dt}x \quad \text{or} \quad x = \int v dt \tag{2.3}$$

- With respect to *energy*:
  a quantity that is associated with an energetic phenomenon: often called a *variable* (e.g. electric current and pressure);
  a quantity that is not associated with energy or only with latent energy: often called a (material) *property* (e.g. length). Sometimes a property is also called a *constant*, but the value of most properties is not constant at all, so we will not use this term.
- With respect to *dependency on mass or size*:
  a quantity which value is independent of the dimensions or the amount of matter is called an *intensive quantity* (e.g. temperature is an intensive variable and resistivity is an intensive property);
  a quantity which value depends on the amount of mass or volume (its extension) is called an *extensive quantity* (e.g. charge is an extensive variable and resistance is an extensive property).

Resistivity $\rho$ ($\Omega$m) is a pure material property, whereas the resistance $R$ ($\Omega$) depends on the material as well as the dimensions of the resistor body. In general, the relation between an intensive and extensive quantity within one domain is given by $A_e = G \cdot A_i$, with $A_e$ and $A_i$, general extensive and intensive quantities, and $G$, a geometrical parameter, representing for instance the dimension of a sensor. In most cases the value of a material property is orientation dependent. This dependency is expressed by subscripts added to the symbols (Appendix A).

Extensive variables are state variables; their time derivatives are rate variables or *flows*. Intensive variables are identical to *efforts*. Flow and effort variables are discussed when conjugated pairs of variables are introduced.

- With respect to the *end points* of a lumped element:
  To explain this classification, we first introduce the term lumped element. A lumped element symbolizes a particular property of a physical component. That property is thought to be concentrated in that element between its two end points or nodes. Exchange of energy or information occurs only through these terminals. In this sense we distinguish:
    an *across-variable*, defined by the difference of its value between the two terminals of a lumped element (e.g. voltage and velocity);
    a *through-variable*, a variable that has the same value at both terminals of the lumped element (e.g. electric current and force).
  Through-variables are also called *generalized I-variables*; across-variables are called *generalized V-variables*. However, this is just a matter of viewpoint. It is perfectly justified to call them generalized forces and displacements. We will use these types of variables in Section 2.1.2 where relations between quantities are discussed.

**Table 2.1** Power Conjugate Variables for Various Domains

| Domain | Effort | Unit | Flow | Unit |
|---|---|---|---|---|
| Mechanical (translation) | Force | N | Velocity | m/s |
| Mechanical (rotation) | Torque | Nm | Angular velocity | rad/s |
| Pneumatic, hydraulic | Pressure | Pa | Volume flow | $m^3/s$ |
| Electrical | Voltage | V | Current | A = C/s |
| Magnetic | Current | A | Voltage | V = Wb/s |
| Thermal | Temperature | K | Entropy flow | J/K/s |
| *Thermal* | *Temperature* | *K* | *Heat flow* | *W* |

- With respect to *cause* and *effect*:
  Output variables are related to input variables according to the physics of the system. Input variables can bring a system into a particular state which is represented by its output variables. So output variables depend on the input variables:
      *independent variables* are applied from an external source to the system;
      *dependent variables* are responses of the system to the input variables.
  Obviously, a variable can be dependent or independent, according to its function in the system. For instance the resistance value of a resistor can be determined by applying a voltage across its terminals and measuring the current through the device or just the other way round. In the former case the voltage is the independent variable, and it is the dependent variable in the latter. The relation between independent and dependent variables is governed by physical effects, by material properties or by a particular system layout. It either acts within one physical domain or crosses domain boundaries. Such relations are the fundamental operation of sensors. This is further discussed in Section 2.1.2.
- With respect to *power conjugation*:
  Within a single energy domain, pairs of variables can be defined in such a way that their product is power. They are called power conjugated variables. The members of such a pair are called *effort variable* and *flow variable*. Table 2.1 lists these variables for various domains.

Note that the dimension of each product is power (W). The magnetic quantity 'current' stems from the definition of magnetic field strength, where the number of ampere-turns (or MMF, magnetomotive force) determines the field strength (see Chapter 6). Its power conjugate variable 'voltage' is actually the rate of change in magnetic flux, with unit Wb/s, but this is equal to the induction voltage. For practical reasons, heat flow (W) is often taken as the thermal flow variable rather than an entropy-related quantity which is not measurable in a straightforward way. The domain is therefore sometimes called pseudothermal (see last row of Table 2.1).

Table 2.2 summarizes various relations between rate, state, effort and flow variables for the mechanical, electrical, magnetic and thermal domains.

- With respect to energy conjugation:
  Another way to define pairs of variables is based on the property that their product equals energy per unit volume $(J/m^3)$. Table 2.3 lists these pairs for the major domains.

**Table 2.2** Summary of Relations Between Types of Variables

| Domain | State/Extensive | Rate/ Flow | Effort/Intensive | Energy (J) | Power (W) |
|---|---|---|---|---|---|
| Mechanical (translation) | Position $x$ (m) | $v = \dot{x}$ | Force $F$ (N) | $F \cdot dx$ | $F \cdot v$ |
| Mechanical (rotation) | Angle $\varphi$ (rad) | $\omega = \dot{\varphi}$ | Torque $T$ (Nm) | $T \cdot d\varphi$ | $T \cdot \omega$ |
| Electrical | Charge $Q$ (C) | $I = \dot{Q}$ | Voltage $V$ (V) | $V \cdot dQ$ | $V \cdot I$ |
| Magnetic | Flux $\Phi$ (Wb) | $V = \dot{\Phi}$ | Current $I$ (A) | $I \cdot d\Phi$ | $I \cdot V$ |
| Thermal | Entropy $\sigma$ (J/K/m$^3$) | $\dot{\sigma}$ | Temperature $\Theta$ (K) | $\Theta \cdot d\sigma$ | $\Theta \cdot \dot{\sigma}$ |

**Table 2.3** Energy Conjugate Variables for Several Domains

| Domain | Effort | Unit | Flow | Unit |
|---|---|---|---|---|
| Mechanical (translation) | Tension | N/m$^2$ | Deformation | — |
| Mechanical (rotation) | Shear tension | N/m$^2$ | Shear angle | — |
| Electrical | Field strength | V/m | Dielectric displacement | C/m$^2$ |
| Magnetic | Magnetic induction | Wb/m$^2$ | Magnetic field strength | A/m |
| Thermal | Temperature | K | Entropy | J/K/m$^3$ |

The most fundamental categorization of quantities is based on thermodynamic laws. The description is in particular useful in the field of material research and optimization of sensor materials. Derived from the thermodynamic approach is the *Bondgraph notation* with a division of variables into effort and flow variables (see Section 2.1.1). This method is not only useful for the description of sensors but also has great significance in the design of all kind of technical systems, irrespective of the domain type.

We repeat the list of pairs of the conjugate variables in Table 2.3, together with their symbols:

- mechanical (translation): tension $T$ (N/m$^2$) and deformation $S$ (−);
- mechanical (rotation): shear tension $\tau$ (N/m$^2$) and shear angle $\gamma$ (−);
- electrical: field strength $E$ (V/m) and dielectric displacement $D$ (C/m$^2$);
- magnetic: magnetic induction $B$ (Wb/m$^2$) and magnetic field strength $H$ (A/m);
- thermal: temperature $\Theta$ (K) and entropy $\sigma$ (J/Km$^3$).

Comparing these pairs with the groups from other categories given previously, we can make the following observations. The quantities $E$, $D$, $B$, $H$, $T$ and $S$ are *vector variables*, whereas $\sigma$ and $\Theta$ are *scalars* (therefore often denoted as $\Delta\sigma$ and $\Delta\Theta$ indicating the difference between two values). Further, in the above groups of quantities, $T$, $E$ and $\Theta$ are *across-variables*. On the other hand, $S$, $D$ and $\sigma$ are *through-variables*. Finally, note that the dimension of the product of each domain pair is always J/m$^3$ (energy per unit volume), whereas the product of the pairs *effort* and *flow* variables in Table 2.1 have the dimension power (W).

### 2.1.2 Relations Between Quantities

The energy content of an infinitely small volume of an elastic dielectric material changes by adding or extracting thermal energy and the work exerted upon it by electrical and mechanical forces. If only through-variables affect the energy content, the change can be written as follows:

$$dU = TdS + EdD + \Theta d\sigma \tag{2.4}$$

where we disregard the magnetic domain (in Appendix B this domain is included).

Obviously, the across-variables in this equation can be expressed as partial derivatives of the energy:

$$
\begin{aligned}
T(S, D, \sigma) &= \left( \frac{\partial U}{\partial S} \right)_{D,\sigma} \\
E(S, D, \sigma) &= \left( \frac{\partial U}{\partial D} \right)_{S,\sigma} \\
\Theta(S, D, \sigma) &= \left( \frac{\partial U}{\partial \sigma} \right)_{S,D}
\end{aligned}
\tag{2.5}
$$

Likewise, if only across-variables effect the energy content, the energy change is written as follows:

$$dG = -SdT - DdE - \sigma d\Theta \tag{2.6}$$

$G$ is called the Gibbs potential (see Appendix B). The through-variables can be written as follows:

$$
\begin{aligned}
S(T, E, \Theta) &= -\left( \frac{\partial G}{\partial T} \right)_{\Theta,E} \\
D(T, E, \Theta) &= -\left( \frac{\partial G}{\partial E} \right)_{T,\Theta} \\
\sigma(T, E, \Theta) &= -\left( \frac{\partial G}{\partial \Theta} \right)_{T,E}
\end{aligned}
\tag{2.7}
$$

From these equations we can derive the various material properties. We will extend these equations only for Eq. (2.7) because the resulting parameters are more in agreement with experimental conditions (constant temperature, electrical field strength and stress), as denoted by the subscripts in Eqs (2.5) and (2.7).

The variables $S$, $D$ and $\Delta\sigma$ are approximated by linear functions, so:

$$dS(T, E, \Theta) = \left(\frac{\partial S}{\partial T}\right)_{E,\Theta} dT + \left(\frac{\partial S}{\partial E}\right)_{T,\Theta} dE + \left(\frac{\partial S}{\partial \Theta}\right)_{T,E} d\Theta$$

$$dD(T, E, \Theta) = \left(\frac{\partial D}{\partial T}\right)_{E,\Theta} dT + \left(\frac{\partial D}{\partial E}\right)_{T,\Theta} dE + \left(\frac{\partial D}{\partial \Theta}\right)_{T,E} d\Theta \qquad (2.8)$$

$$d\sigma(T, E, \Theta) = \left(\frac{\partial \sigma}{\partial T}\right)_{E,\Theta} dT + \left(\frac{\partial \sigma}{\partial E}\right)_{T,\Theta} dE + \left(\frac{\partial \sigma}{\partial \Theta}\right)_{T,E} d\Theta$$

Combining Eqs (2.6) and (2.7) results in:

$$dS = -\left(\frac{\partial^2 G}{\partial T^2}\right)_{\Theta,E} dT - \left(\frac{\partial^2 G}{\partial T\partial E}\right)_{\Theta} dE - \left(\frac{\partial^2 G}{\partial T\partial\Theta}\right)_{E} d\Theta$$

$$dD = -\left(\frac{\partial^2 G}{\partial E\partial T}\right)_{\Theta} dT - \left(\frac{\partial^2 G}{\partial E^2}\right)_{\Theta,T} dE - \left(\frac{\partial^2 G}{\partial E\partial\Theta}\right)_{T} d\Theta \qquad (2.9)$$

$$d\sigma = -\left(\frac{\partial^2 G}{\partial\Theta\partial T}\right)_{E} dT - \left(\frac{\partial^2 G}{\partial\Theta\partial E}\right)_{T} dE - \left(\frac{\partial^2 G}{\partial\Theta^2}\right)_{E,T} d\Theta$$

Now we have a set of equations connecting the (dependent) through-variables $S$, $D$ and $\sigma$ with the (independent) across-variables $T$, $E$ and $\Theta$. The system configuration (or the material) couples the conjugate variables of each pair. The second order derivatives in the diagonal represent properties in the respective domains: mechanical, electrical and thermal. For example, the top left second derivative in Eq. (2.9) represents the elasticity (or compliance) of the material (actually Hooke's law). All other derivatives represent cross effects. Note that these derivatives are pair-wise equal since (assuming linear equations) the order of differentiation is not relevant:

$$\frac{\partial}{\partial x}\left(\frac{\partial G}{\partial y}\right) = \frac{\partial}{\partial y}\left(\frac{\partial G}{\partial x}\right).$$

So the derivatives in Eq. (2.9) represent material properties; they have been given special symbols. The variables denoting constancy are put as superscripts, to make place for the subscripts denoting orientation.

$$S = s^{E,\Theta}T + d^{\Theta}E + \alpha^{E}\Delta\Theta$$
$$D = d^{\Theta}T + \varepsilon^{\Theta,T}E + p^{T}\Delta\Theta \qquad (2.10)$$
$$\Delta\sigma = \alpha^{E}T + p^{T}E + \frac{\rho}{T}c^{E,T}\Delta\Theta$$

For instance $s^{E,\Theta}$ is the compliance at constant electric field $E$ and constant temperature $\Theta$. The nine associated effects are displayed in Table 2.4.

Table 2.5 shows the associated material properties. The parameters for just a single domain ($\varepsilon$, $c_{\mathrm{p}}$ and $s$) correspond to those in Tables A.2, A.5 and A.8 of Appendix A. The other parameters denote 'cross effects' and describe the conversion from one domain to another. The piezoelectric parameters $p$ and $d$ will be discussed in detail in the chapter on piezoelectric sensors.

Note that direct piezoelectricity and converse piezoelectricity have the same symbol ($d$) because the dimensions are equal (m/V and C/N). The same holds for the pair pyroelectricity and converse pyroelectricity as well as for thermal expansion and piezocaloric effect.

Equations (2.7) and (2.10) can be extended just by adding other couples of conjugate quantities, for instance from the chemical or the magnetic domain. Obviously, this introduces many other material parameters. With three couples we have nine parameters, as listed in Table 2.3. With four couples of intensive and extensive quantities we have 16 parameters, so seven more (for instance the magnetocaloric effect, expressed as the partial derivative of entropy to magnetic field strength, see Appendix B). Further, Appendix B gives a visualization of these relations using Heckman diagrams.

**Table 2.4** Nine Physical Effects Corresponding to the Parameters in Eq. (2.10)

| | | |
|---|---|---|
| Elasticity | Converse piezoelectricity | Thermal expansion |
| Direct piezoelectricity | Permittivity | Pyroelectricity |
| Piezocaloric effect | Electro-caloric effect | Heat capacity |

**Table 2.5** Symbols, Parameter Names and Units of the Effects in Table 2.4

| Symbol | Property | Unit |
|---|---|---|
| $s$ | Compliance | m$^2$/N |
| $d$ | Piezoelectric constant | m/V = C/N |
| $\alpha$ | Thermal expansion coefficient | K$^{-1}$ |
| $p$ | Pyroelectric constant | C/m$^2$/K |
| $\varepsilon$ | Permittivity; dielectric constant | F/m |
| $c_{\mathrm{p}}$ | (Specific) Heat capacity | J/kg/K |

## 2.2  Sensor Classifications

A sensor (or input transducer) performs the conversion of information from the physical domain of the measurand to the electrical domain. Many authors have tried to build up a consistent classification scheme of sensors encompassing all sensor principles. Such a classification of the millions of available sensors would facilitate understanding of their operation and making proper choices, but a useful basis for a categorization is difficult to define. There are various possibilities:

- according to the measurand
- according to application fields
- according to a port model
- according to the conversion principle
- according to the energy domain of the measurand
- according to thermodynamic considerations.

These schemes will be briefly discussed in the next sections.

### 2.2.1  Classification Based on Measurand and Application Field

Many books on sensors follow a classification according to the measurand because the designer who is interested in a particular quantity to be measured can quickly find an overview of methods for that quantity. The more experienced designer may also consult books that deal with just one quantity (for instance temperature or liquid flow). Much information on sensors can also be found in books focusing on a specific application area, for instance (mobile) robots [1], industrial inspection [2], buildings [3], manufacturing [4], mechatronics [5], automotive, biomedical and many more. However, an application field provides no restricted set of sensors since in each field many types of sensors could be applied.

Figure 2.1 presents a list of physical quantities (measurands) [6]. The list is certainly not exhaustive, but it shows the many possible measurands. For each of these quantities one or more measurement principles are available.

### 2.2.2  Classification Based on Port Models

The distinguishing property in the classification based on port models is the need for auxiliary energy (Figure 2.2). Sensors that need no auxiliary energy for their operation are called direct sensors or self-generating sensors. Sensors that use an additional energy source for their operation are called modulating sensors or interrogating sensors.

Direct sensors do not require additional energy for conversion. Since information transport cannot exist without energy transport, a direct sensor withdraws the output energy directly from the measurement object. As a consequence, loss of information about the original state of the object may occur. There even might be energy loss too — for instance heat. An important advantage of a direct sensor is its

| Mechanical, solids | Mechanical, fluids | Nuclear radiation | Acoustic |
|---|---|---|---|
| Acceleration | Density | Ionization degree | Sound frequency |
| Angle | Flow direction | Mass absorption | Sound intensity |
| Angular velocity | Flow velocity | Radiation dose | Sound polarization |
| Area | Level | Radiation energy | Sound pressure |
| Diameter | Pressure | Radiation flux | Sound velocity |
| Distance | Rate of flow | Radiation type | Time of flight |
| Elasticity | Viscosity | | |
| Expansion | Volume | **Chemical** | **Magnetic, electrical** |
| Filling level | | Cloudiness | Capacity |
| Force | **Thermal** | Composition | Charge |
| Gradient | Enthalpy | Concentration | Current |
| Hardness | Entropy | Electrical conductivity | Dielectric constant |
| Height | Temperature | Humidity | Electric field strength |
| Length | Thermal capacity | Impurity | Electric power |
| Mass | Thermal conduction | Ionization degree | Electric resistance |
| Moment | Thermal expansion | Moisture | Frequency |
| Movement | Thermal radiation | Molar weight | Inductivity |
| Orientation | | Particle form | Magnetic field strength |
| Pitch | **Optical** | Particle size | Phase |
| Position | Colour | pH | Pulse duration |
| Pressure | Light polarization | Polymerization degree | Signal distortion |
| Proximity | Light wavelength | Reaction rate | |
| Rotation | Luminance | Redox potential | **Time** |
| Roughness | Luminous intensity | Thermal conductivity | Time |
| Shape | Reflection | Water content | Frequency |
| Tension | Refractive index | | Duty cycle |
| Torque | | | |
| Torsion | | | |
| Velocity | | | |
| Vibration | | | |
| Weight | | | |

**Figure 2.1** List of physical quantities.
*Source*: After Ref. [6].

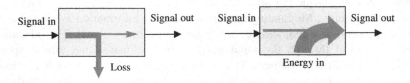

**Figure 2.2** Energy flow in direct and modulating sensors.

freedom from offset: at zero input the output is essentially zero. Examples of direct sensors are the piezoelectric acceleration sensor and the thermocouple.

Modulating or interrogating sensors use an additional energy source that is modulated by the measurand; the sensor output energy mainly comes from this auxiliary source, and just a fraction of energy is withdrawn from the measurement object. The terms modulating and interrogating refer to the fact that the measurand affects a specific material property which in turn is interrogated by an auxiliary quantity. Most sensors belong to this group: all resistive, capacitive and inductive

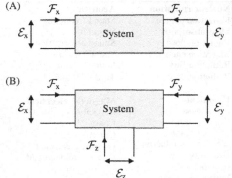

**Figure 2.3** Port models of a sensor: (A) two-port model; (B) three-port model.

sensors are based on a parameter change (e.g. resistance, capacitance and inductance) caused by the measurand. Likewise, most displacement sensors are of the modulating type: displacement of an object modulates optical or acoustic properties (e.g. transmission, reflection and interference), where light or sound is the interrogating quantity.

Energy (and thus information) enters or leaves the system through a pair of terminals making up a 'port'. We distinguish input ports and output ports. A direct sensor can be described by a two-port model or four-terminal model (Figure 2.3A).

The input port is connected to the measurand; the output port corresponds with the electrical connections of the sensor. Likewise, a modulating sensor can be conceived as a system with three ports: an input port, an output port and a port through which the auxiliary energy is supplied (Figure 2.3B). In these models the variables are indicated with across or effort variables $\mathcal{E}$ and through or flow variables $\mathcal{F}$, respectively. The subscripts $x$, $y$, and $z$ are chosen in accordance with the sensor cube, to be discussed in Section 2.2.4.

Direct sensors provide the information about the measurand as an output signal, an energetic quantity. Modulating sensors contain the information as the value of a material property, or a geometric quantity, not an energetic signal. The information enters the system through the input port, where the measurand affects specific material or geometric parameters. To extract the information from such a sensor, it has to be interrogated using an auxiliary signal. The information stored in the sensor is available latently, in the latent information parameters or LIP [7]. These parameters are *modulated* by the input signal and *interrogated* by the auxiliary or interrogating input.

At zero input the LIPs of a modulating sensor have initial values, set by the material and the construction. Generally, the input has only a small effect on these parameters, resulting in relatively small deviations from the initial values. Note that direct sensors too have LIPs, set by materials and construction. They determine the sensitivity and other transfer properties of the sensor. So the input port of all sensors can be denoted as the LIP input port. As a consequence, any sensor can be described with the three-port model of Figure 2.3B. Only the functions of the ports may differ, notably the LIP input port and the interrogating input port.

| | | Interrogating input | |
|---|---|---|---|
| | | Design controlled | Environment controlled |
| LIP input | Design controlled | Source | Direct sensor |
| | Environment controlled | Modulating sensor | Multiplying devices |

**Figure 2.4** Unified transducer classification.

According to the 'unified transducer model' as introduced in [7], an input port can be controlled either by design (it has a fixed value) or by the environment (the measurand or some unwanted input variable). So we have four different cases (Figure 2.4). The characteristics of these four transducer types are briefly reviewed:

- Design-controlled LIP input and design-controlled interrogating input.
  All inputs are fixed. This type represents a signal or information source, for instance a standard or a signal source with a constant or predetermined output. The output is totally determined by the construction and the materials that have been chosen. Any environmental effect on the output is (ideally) excluded.
- Design-controlled LIP input and environment-controlled interrogating input.
  Since the latent information parameters are fixed by design, the output depends only on what is connected to the interrogating input. When this is the measurand, the transducer behaves as a direct sensor. Examples:
  - *Thermocouple temperature sensor*: the Seebeck coefficient is fixed by the choice of the materials.
  - *Piezoelectric accelerometer*: the sensitivity is fixed by the seismic mass and the piezoelectric properties of the crystal.
- Environment-controlled LIP input and design-controlled interrogating input.
  The measurand affects particular material properties or geometric parameters. These changes are interrogated by a fixed or well-defined signal at the interrogating input. The transducer behaves as a modulating sensor. Examples:
  - *Strain gauge bridge*: strain alters the resistance of the strain gauges; a bridge voltage converts this resistance change into an output voltage;
  - *Linear variable differential transformer (LVDT)*: a displacement of an object connected to the moving core will change the transfer ratio of the differential transformer. An AC signal on the primary coil acts as interrogating quantity.
  - *Hall sensor*: the measurand is a magnetic induction field, which acts on moving charges imposed by a fixed (or known) current applied to the interrogating input.
- Environment-controlled LIP input and environment-controlled interrogating input.
  These are multiplying transducers: the output depends on the quantities at both inputs, often in a multiplicative relation. For instance a Hall sensor could act as such, when the interrogating input is not a fixed current (by design) but a current that is related to just another measurand.

It is important to note that any practical transducer shows all four types of responses. A strain gauge (a modulating transducer) produces, when interrogated, an output voltage related to the strain-induced change in resistance. But the circuit can also generate spurious voltages caused by capacitively or magnetically induced signals. A thermocouple (a direct transducer) produces an output voltage proportional to the measurand at the interrogating input. If, however, the material parameters change due to (for instance) strain or nuclear radiation (inputs at the LIP port), the measurement is corrupted.

Since just one response is desired, other responses should be minimized by a proper design. The universal approach helps to identify such interfering sensitivities.

## 2.2.3 Classification Based on Conversion Principles

The classification according to conversion principles is often used for the reason that the sensor performance is mainly determined by the physics of the underlying principle of operation. However, a particular type of sensor might be suitable for a variety of physical quantities and in many different applications. For instance a magnetic sensor of a particular type could be applied as displacement sensor, a velocity sensor, a tactile sensor and so on. For all these applications the performance is limited by the physics of this magnetic sensor, but the limitations manifest in completely different ways. A closer look at the various conversion effects may lead to the observation that the electrical output of a sensor depends either on a material property or the geometry or a movement. Figure 2.5 tabulates these three phenomena for various types of sensors.

| Type | Material property | Geometry (sensor examples) | Relative movement |
|---|---|---|---|
| Resistive | Resistivity (piezoresistor, LDR) | Relative length (potentiometer; metal strain gauge) | |
| Capacitive | Permittivity (fluid level sensor) | Relative electrode distance capacitive displacement (LVDC) | |
| Magnetic | Permeability (magnetoresistor) | Distance source-detector (magnetic displacement sensor) | Induction (magnetic velocity sensors) |
| Inductive | | Inductance self-inductance mutual inductance reluctance (inductive displacement sensors, LVDT and resolver) | Induction (inductive velocity sensors) |
| Optical | Index of refraction absorptivity (fibre optic sensors) | Distance transmitter–receiver (intensity modulation sensors, interferometer and TOF sensor) transmissivity and reflectivity (optical encoder and tachometer) | Doppler frequency (Doppler velocimeter) |
| Acoustic | Acoustic impedance | Distance transmitter–receiver (TOF displacement sensors) | Doppler frequency (Doppler velocimeter) |
| Piezoelectric | Polarization (piezoelectric sensors) | Deformation (piezoelectric sensors) | |

**Figure 2.5** Classification based on electrical conversion principles (and sensor examples).

## 2.2.4 Classification According to Energy Domain

A systematic representation of sensor effects based on energy domains involves a number of aspects. First, the energy domains have to be defined. Second, the energy domains should be allocated to both the sensor input and output. Finally, since many sensors are of the modulating type, the domain of the auxiliary quantity should also be considered. From a physical point of view, nine energy forms can be distinguished:

- Electromagnetic radiant energy
- Gravitational energy
- Mechanical energy
- Thermal energy
- Electrostatic and electromagnetic energy
- Molecular energy
- Atomic energy
- Nuclear energy
- Mass energy.

This classification is rather impractical for the description of sensors. Lion [8] has proposed only six domains and adopted the term signal domain. These six domains are: radiant, thermal, magnetic, mechanical, chemical and electrical. The number of domains is a rather arbitrary choice, so for practical reasons we will continue with the system of six domains and call them energy domains.

Information contained in each of the six domains can be converted to any other domain. These conversions can be represented in a $6 \times 6$ matrix. Figure 2.6 shows that matrix, including some of the conversion effects. An input transducer or sensor performs the conversion from a non-electrical to the electrical domain (the shaded column), and an output transducer or actuator performs the conversion from the

| | | OUTPUT DOMAIN | | | | | |
|---|---|---|---|---|---|---|---|
| | | Radiant | Thermal | Magnetic | Mechanical | Chemical | Electrical |
| **I N P U T  D O M A I N** | **Radiant** | Luminescense | Radiation heating | Photomagnetism | Radiation pressure | Photochemical process | Photoconductivity |
| | **Thermal** | Incandescense | Thermal conductivity | Curie–Weiss law | Thermal expansion | Endothermal reaction | Seebeck effect; pyroelectricity |
| | **Magnetic** | Faraday effect | Ettinghausen effect | Magnetic induction | Converse magnetostriction | | Hall effect |
| | **Mechanical** | Photo-elastic effect | Friction heat | Magnetostriction | Gear | Pressure-induced reaction | Piezoelectricity |
| | **Chemical** | Chemo-luminescense | Exothermal reaction | | Explosive reaction | Chemical reaction | Volta effect |
| | **Electrical** | Injection luminescense | Peltier effect | Ampere's law | Converse piezoelectricity | Electrolysis | Ohm's law |

**Figure 2.6** Physical domains and some cross effects.
*Source*: After Ref. [6].

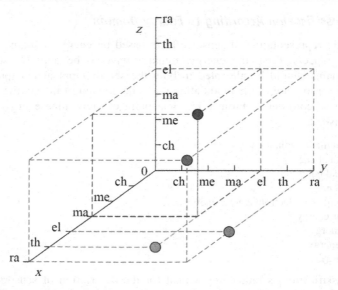

**Figure 2.7** Sensor cube; six domains: radiant (ra), thermal (th), electrical (el), magnetic (ma), mechanical (me) and chemical (ch).

electrical to another domain (the shaded row). The cells on the diagonal of the matrix indicate effects within a single domain.

This two-dimensional representation can be extended to three dimensions, when the interrogating energy domain is included. This gives 216 energy triplets. To get a clear overview of all these possible combinations, they can be represented in a 3D Cartesian space, the 'sensor cube' (Figure 2.7). The three axes refer to the input energy domain, the output energy domain and the interrogating input energy.

On each of the 216 elements of the $6 \times 6 \times 6$ matrix a conversion effect is located. When restricting to electrical transducers, there are 5 direct input transducers, 5 direct output transducers, 25 modulating input transducers and 25 modulating output transducers.

To facilitate notation, the transducers can be indicated by indices, like in crystallography, the so-called Miller indices: [x, y, z]. The x-index is the input domain, the y-index the output domain and the z-index the domain of the interrogating quantity. With these three indices a transducer can be typified according to the energy domains involved. Some examples are as follows:

*Direct input transducer*: thermocouple [th, el, 0]
*Modulating input transducer*: Hall sensor [ma, el, el]
*Direct output transducer*: LED [el, ra, 0]
*Modulating output transducer*: LCD [ra, ra, el].

These transducers are also visualized in Figure 2.7. The practical value of such a representation is rather limited. It may serve as the basis of a categorization for overviews or as a guide in the process of sensor selection.

# References to Cited Literature

[1] H.R. Everett: Sensors for mobile robots: theory and application; A.K. Peters, Wellesley Mass., 1995; ISBN 1-56881-048-2.

[2] C. Loughlin: Sensors for industrial inspection; Kluwer Academic Publishers, Dordrecht. 1993; ISBN 0-7923-2046-8.

[3] O. Gassmann, H. Meixner: Sensors in intelligent buildings; Wiley-VCH Verlag GmbH, Weinheim, FRG, 2001; ISBN 3-527-29557-7.

[4] H.K. Tönshoff, I. Inasaki: Sensors in manufacturing; Wiley-VCH Verlag GmbH, Weinheim, FRG, 2001; ISBN 3-527-29558-5.

[5] A.M. Pawlak: Sensors and actuators in Mechatronics[1]; CRC Press, Boca Raton FL, 2006; ISBN 0-8493-9013-3.

[6] S. Middelhoek, S.A. Audet: Silicon sensors; Academic Press, London, San Diego, New York, Berkeley, Boston, Sydney, Tokyo, Toronto, 1989; ISBN 0-12-495051-5.

[7] P.K. Stein: Classification system for transducers and measuring systems, Symposium on Environmental measurements: valid data and logical interpretation, 4—6 September 1963; US Department of Health Education and Welfare, Washington D.C. 1964, pp. 65—84.

[8] K. Lion: Transducers: problems and prospects, IEEE Trans. Ind. Electron. Control Instrum., IECI-16 (1969), 2—5.

# Literature for Further Reading

## Some books and articles on quantities and systems

[1] Q. Yang, C. Butler: An object-oriented model of measurement systems, IEEE Trans. Instr. Measurement, 47(1) (February 1998), 104—107.

[2] T. Kwaaitaal: The fundamentals of sensors, Sens. Actuators A, 39 (1993), 103—110.

[3] B.W. Petley: The fundamental physical constants and the frontier of measurement, Adam Hilger, Bristol (1985); ISBN 0-85274-427-7.

[4] B.S. Massey: Units, dimensional analysis and physical similarity; Van Nostrand Reinhold Comp., London, 1971; ISBN 0-442-05178-6.

---

[1] Mainly electromagnetic devices.

# 3 Uncertainty Aspects

No sensor is perfect. The mechatronic designer must be aware of the sensor's short-comings in order to be able to properly evaluate measurement results and to make a correct assessment of the system performance. Specifying sensor quality in terms of accuracy only is not sufficient: a larger number of precisely defined parameters is necessary to fully characterize the sensor's behaviour. Often a designer can reduce the effects of the intrinsic sensor limitations by the application of special configurations, procedures and methods. Similar measures can also be considered when environmental influences should be eliminated. This chapter reviews the most important terms to express sensor behaviour and presents some general design methods to reduce errors due to sensor deficiencies and environmental factors.

## 3.1  Sensor Specification

Imperfections of a sensor are usually listed in the data sheets provided by the manufacturer. These sensor specifications inform the user about deviations from the ideal behaviour. The user must accept technical imperfections, as long as they do not exceed the specified values.

Any measuring instrument, and hence any sensor, has to be fully specified with respect to its performance. Unfortunately, many data sheets show lack of clarity and completeness. Gradually, international agreements about formal error descriptions are being established. An exhaustive description of measurement errors and error terminology can be found in [1], along with an international standard on transducer nomenclature and terminology [2]. Various international committees are working towards a uniform framework to specify sensors [3]. Finally, a special document is in preparation, containing definitions of measurement-related terms: the International Vocabulary of Basic and General Terms in Metrology (short VIM) [4].

The characteristics that describe sensor performance can be classified into four groups:

- *Static characteristics*, describing the performance with respect to very slow changes.
- *Dynamic characteristics*, specifying the sensor response to variations in time and in the measurand (the quantity that has to be measured).
- *Environmental characteristics*, relating the sensor performance after or during exposure to specified external conditions (e.g. pressure, temperature, vibration and radiation).
- *Reliability characteristics*, describing the sensor's life expectancy.

**Sensors for Mechatronics. DOI: 10.1016/B978-0-12-391497-2.00003-0**
© 2012 Elsevier Inc. All rights reserved.

Errors that are specific for certain sensor types are discussed in the chapters concerned. In this section we first define some general specifications:

- Sensitivity
- Non-linearity and hysteresis
- Resolution
- Accuracy
- Offset and zero drift
- Noise
- Response time
- Frequency response.

### 3.1.1 Sensitivity

The sensitivity of a sensor is defined as the ratio between a change in the output value and the change in the input value that causes that output change. Mathematically, the sensitivity is expressed as $S = dy/dx$, where $x$ is the input signal (measurand) and $y$ is the output (an electrical signal). Usually a sensor is also sensitive to changes in quantities other than the intended input quantity, such as the ambient temperature or the supply voltage. These unwelcome sensitivities should be specified as well, for a proper interpretation of the measurement result. To have a better insight in the effect of such unwanted sensitivities, they are often related to the sensitivity of the measurement quantity itself.

> *Example 1*
> The sensitivity of a particular displacement sensor with voltage output is specified as 10 mV/mm. Its specified temperature sensitivity is 0.1 mV/K. Since 0.1 mV corresponds with a displacement of 10 μm, the temperature sensitivity can also be expressed as 10 μm/K. A temperature rise of 5°C results in an apparent displacement of 50 μm.
>
> *Example 2*
> The sensitivity of a particular type of temperature sensor is 100 mV/K, including the signal conditioning unit. The signal conditioning part itself is also sensitive to (ambient) temperature and appears to create an extra output voltage of 0.5 mV for each degree celcius rise in ambient temperature (not necessarily the sensor temperature). So, the unwanted temperature sensitivity is 0.5 mV/K or $0.5/100 = 5$ mK/K. A change in ambient temperature of $\pm 10°C$ gives an apparent change in sensor temperature equal to $\pm 50$ mK.

### 3.1.2 Non-linearity and Hysteresis

If the output $y$ is a linear function of the input $x$, the sensitivity $S$ does not depend on $x$. In the case of a non-linear transfer function $y = f(x)$, $S$ does depend on the input or output value. Often, a linear response is preferred to reduce computational burden in, for instance, multi-sensor control systems. In that case the sensitivity can be expressed with a single parameter. Furthermore, non-linearity introduces harmonics of the input signal frequencies, which may obstruct proper frequency analysis of the physical signals.

The transfer of a sensor with a slight non-linearity may be approximated by a straight line, to specify its sensitivity by just one number. The user should be informed about the deviation from the actual transfer; this is specified by the non-linearity error.

The linearity error of a system is the maximum deviation of the actual transfer characteristic from a *prescribed straight line*. Manufacturers specify linearity in various ways, for instance as the deviation in input or output units: $\Delta x_{max}$ or $\Delta y_{max}$, or as a fraction of FS (full scale): $\Delta x_{max}/x_{max}$. Non-linearity should *always* be given together with a specification of the straight line. The following definitions are in use:

- *Terminal non-linearity*: based on the terminal line: a straight line between 0% and 100% theoretical full-scale points.
- *End-point non-linearity*: based on the end-point line, the straight line between the calibrated end points of the range; coincides with the terminal (theoretical) line after calibration of zero and scale.
- *Independent non-linearity*: referring to the best-fit straight line, according to a specified error criterion, for instance the line midway between two parallel lines enclosing all calibration points; if the least square error criterion for the best-fit straight line is used, this linearity error is the:
- *Least square non-linearity*: based on the least square line, the line for which the summed squares of the residuals is minimized.

Hysteresis is the maximum difference in output signal when the measurand first increases over a *specified* range and then returns to the starting value. The travelled range should be specified because hysteresis strongly depends on it.

### 3.1.3  Resolution

The resolution indicates the smallest detectable increment of the input quantity. When the measurand varies continuously, the sensor output might show discontinuous steps. The value of the corresponding smallest detectable change in the input variable is the resolution: $\Delta x_{min}$. Sometimes this parameter is related to the maximum value $x_{max}$ that can be processed (full-scale value), resulting in the resolution expressed as $\Delta x_{min}/x_{max}$ or $x_{max}/\Delta x_{min}$. This mixed use of definitions seems confusing, although it is easy to see from the units or the value itself which definition is used.

*Example 1*

The resolution of a particular type of wire-wound linear potentiometer with a range of 10 cm is specified as $10^{-4}$; assuming this is the full-scale value, it means that the output changes discontinuously in steps equivalent to input displacements of 10 $\mu$m.

*Example 2*

A particular type of optical encoder has a resolution of 14 bit. The smallest change in angle that can be detected by this encoder is $2\pi/2^{14} \approx 3.8 \times 10^{-4}$ rad or 0.022°.

## 3.1.4 Accuracy

Formally, the accuracy reflects the closeness of the agreement between the actual measurement result and a true value of the measurand. The accuracy specification should include relevant conditions and other quantities. Many sensor manufacturers specify the sensor performance in terms of accuracy. This specification should be viewed with suspicion because it may or may not include particular imperfections of the sensor (e.g. non-linearity, hysteresis and drift) and may be only valid under strict conditions. *Precision* is not the same as accuracy. Actually, the term precision should not be used in any case, to avoid confusion.

## 3.1.5 Offset and Zero Drift

Most sensors are designed such that the output is zero at zero input. If the transfer characteristic does not intersect the origin $(x,y = 0,0)$, the system is said to have offset. The offset is expressed in terms of the input or the output quantity. Specifying the input offset is preferred, to facilitate a comparison with the value of the measurand.

*Example*
> The sensitivity of a particular type of force sensing system is 0.1 V/N. At zero force the output appears to be 3 mV. The (input) offset of this system is the output offset divided by the sensitivity, so 0.03 N.

A non-zero offset arises mainly from component tolerances. Offset compensation can be performed in the interface electronics or the signal processing unit. Once adjusted to zero, the offset may nevertheless change, due to temperature variations, changes in the supply voltage or aging effects. This relatively slow change in the offset is called *zero drift*. In particular, the temperature-induced drift (the *temperature coefficient*, or t.c., of the offset) is an important item in the specification list.

Sometimes a system is deliberately designed with offset. Many industrial transducers have a current output ranging from 4 to 20 mA. This facilitates the detection of cable fractures or a short circuit, producing a zero output clearly distinguishable from a zero input.

## 3.1.6 Noise

Electrical noise is a collection of spontaneous fluctuations in currents and voltages. They are present in any electronic system and arise from thermal motion of the electrons and from the quantized nature of electric charge. Electrical noise is also specified in terms of the input quantity, to show its effect relative to the value of the measurand. *White noise* (noise with constant power over a wide frequency range) is usually expressed in terms of spectral noise power (W/Hz), spectral noise voltage (V/$\sqrt{\text{Hz}}$) or spectral noise current (A/$\sqrt{\text{Hz}}$). Thermal noise is an example of 'white noise'.

Another important type of noise is $1/f$ *noise* (one-over-f noise), a collection of noise phenomena with a spectral noise power that is proportional to $f^{-n}$, with $n = 1-2$.

*Quantization noise* is the result of quantizing an analogue signal. The rounding off results in a (continuous) deviation from the original signal. This error can be considered as a 'signal' with zero mean and a standard deviation determined by the resolution of the AD converter.

### 3.1.7   Response Time

The response time is associated with the speed of change in the output on a step-wise change of the measurand. The specification of the response time needs *always* be accompanied with an indication of the input step (for instance FS — full scale) and the output range for which the response time is defined, for instance 10−90%. Creep and oscillations may make the specification of the response time meaning-less or at least misleading.

### 3.1.8   Frequency Response and Bandwidth

The sensitivity of a system depends on the frequency or rate of change of the measur-and. A measure for the useful frequency range is the frequency band. The upper and lower limits of the frequency band are defined as those frequencies for which the out-put signal has dropped to half the nominal value, at constant input power. For voltage or current quantities the criterion is $\frac{1}{2}\sqrt{2}$ of the nominal value. The lower limit of the frequency band may be zero; the upper limit has always a finite value. The extent of the frequency band is called the *bandwidth* of the system, expressed in Hz.

### 3.1.9   Operating Conditions

All specification items only apply within the operating range of the system, which should also be specified correctly. It is given by the measurement range, the required supply voltage, the environmental conditions and possibly other parameters.

> *Example 1*
> The frequency characteristics and noise behaviour of an accelerometer are important features. Table 3.1 is an excerpt from the data sheets of the QA-2000 accelerometer from Allied Signal Aerospace.
> *Example 2*
> Many humidity sensors have non-linear behaviour. Table 3.2 is an example of the spe-cifications of a humidity sensor EMD 2000 from Phys-Chem SCIENTIFIC Corp., NY; RH stands for relative humidity.

Despite the specified limitations of sensors, a sensing system can be configured in a way that the effect of some of these limitations are eliminated or at least reduced. We will consider various possibilities of error-reducing designs in the next section.

**Table 3.1** Selected Specifications of an Accelerometer

| Environmental: | Temperature | −55°C to +95°C |
|---|---|---|
| | Shock | 250 g, half-sine, 6 ms |
| | Vibration | MIL-E-5400 curve IV(A) |
| Frequency response: | 0−10 Hz | 0.01 dB |
| | 10−300 Hz | 0.45 dB |
| | Above 300 Hz | <5 dB peaking |
| | Natural frequency | >800 Hz |
| Noise: | 0−10 Hz | 10 nA rms |
| | 10−500 Hz | 100 nA rms |
| | 500 Hz to 10 kHz | 2 μA rms |

**Table 3.2** Selected Specifications of a Humidity Sensor

| | |
|---|---|
| Operating RH range | 10−98% RH |
| Operating temperature range | −10°C to 75°C |
| Response time | 10 s for a step change from 11% RH to 93% to reach 90% or better of equilibrium value |
| Hysteresis | ±0.3% RH at 25°C |

*Note*: Step response and hysteresis curves are included in the specification sheets.

## 3.2 Sensor Error Reduction Techniques

Any sensor system has imperfections, introducing measurement errors. These errors either originate from the system itself (for instance system noise, quantization and drift) or are due to environmental influences such as thermal, electromagnetic and mechanical interference. Sensor manufacturers try to minimize such intrinsic errors through proper design of the sensor layout and encapsulation; the remaining imperfections should be given in the data sheets of the sensor. The user (for instance the mechatronic designer) should minimize additional errors which could arise from improper mounting and faulty electronic interfacing. In this section we present some general concepts to minimize or to reduce the effect of the intrinsic errors when applying sensors.

Usually, a sensor is designed to be sensitive to just one specific quantity, thereby minimizing the sensitivity to all other quantities, despite the unavoidable presence of many physical effects. The result is a sensor that is sensitive not only to the quantity to be measured but also in a greater or lesser degree to other quantities; this is called the cross-sensitivity of the device. Temperature is feared most of all, illustrated by the saying that 'every sensor is a temperature sensor'.

Besides cross-sensitivities, sensors may suffer from many other imperfections. They influence the transfer of the measurement signal and give rise to unwanted output signals. Figure 3.1 shows a simplified model of a sensor system, with an

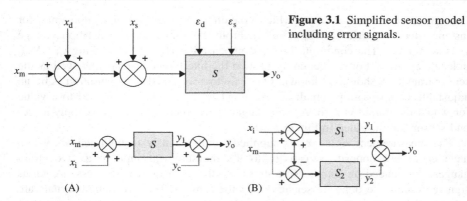

**Figure 3.1** Simplified sensor model including error signals.

**Figure 3.2** General layout of compensation: (A) compensation signal; (B) balanced configuration.

indication of several error sources. In this figure, $x_m$ is the measurement signal and $y_o$ the output signal. *Additive* error signals are modelled as additional input signals: $x_d$ and $x_s$ represent deterministic and stochastic error signals, respectively. They model all kind of interference from the environment and the equivalent error signals due to system offset and noise. The error inputs $\varepsilon_d$ and $\varepsilon_s$ represent *multiplicative* errors: these signals affect the sensitivity of the sensor. For this simplified model, the output signal of a sensor can be written according to Eq. (3.1):

$$y_o = S(1 + \varepsilon_d + \varepsilon_s)(x_m + x_d + x_s) \tag{3.1}$$

where $S$ is the nominal sensitivity. This model will be used to evaluate various error reduction methods. Some of these methods will reduce mainly additive sensors; others minimize multiplicative errors. Improvement of sensor performance can be obtained through use of a sophisticated design or simply through some additional signal processing. We will discuss five basic error reduction methods:

1. Compensation
2. Feedback
3. Filtering
4. Modulation
5. Correction

The methods not only apply to sensors but also to other signal handling systems as amplifiers and signal transmission systems.

### 3.2.1 Compensation

Compensation is a simple and effective method to minimize additive errors due to interference signals. The basic idea is as illustrated in Figure 3.2. In Figure 3.2A the output of the sensor is $y_1$, which contains unwanted signal components, for instance due to interference $x_i$ or offset. From this output a compensation voltage $y_c$ is subtracted. The condition for full compensation is $y_c = Sx_i$, making $y_o = Sx_m$, independent of $x_i$. For correct compensation the interference signal $x_i$ as well as the sensor transfer $S$ should be known. One way to accomplish compensation is by an adjustable compensation signal: at zero input $y_c$ is (manually) adjusted to a value for which the output is zero. A more elegant way to compensate is to apply a second sensor, as illustrated in Figure 3.2B.

The measurement signal $x_m$ is supplied simultaneously to both sensors which have equal but opposite sensitivity to the measurement input (e.g. two strain gauges: one loaded on compressive stress and the other on tensile stress). A minus sign represents the opposite sensitivity of the sensors. The two output signals are subtracted by a proper electronic circuit. Because of the anti-symmetric structure with respect to the measurement signal only, many interference signals appear as common output signals and thus are eliminated by taking the difference of the two outputs.

The effectiveness of the method depends on the degree of symmetry of the double sensor or differential sensor structure. From Figure 3.2 it follows:

$$y_1 = S_{m1}x_m + S_{i1}x_i$$

$$y_2 = -S_{m2}x_m + S_{i2}x_i \tag{3.2}$$

where $S_{mk}$ and $S_{ik}$ are the sensitivities for the measurement signal and the interference signal, respectively. So, the output signal of the sensor system equals

$$y_o = (S_{m1} + S_{m2})x_m + (S_{i1} - S_{i2})x_i \tag{3.3}$$

The sensitivities $S_{m1}$ and $S_{m2}$ are about equal, so the output signal becomes

$$y_o = 2S_m x_m + \Delta S_i x_i = 2S_m \left( x_m + \frac{\Delta S_i}{2S_m} x_i \right) \tag{3.4}$$

Analogous to the definition of the rejection ratio for differential amplifiers we can define a quality measure for the imbalance of the differential sensor:

$$H = \frac{2S_m}{\Delta S_i} \tag{3.5}$$

a parameter characterizing the system's ability to distinguish between measurand and interfering signals.

The method is illustrated with a two-active-element Wheatstone measurement bridge (Figure 3.3).

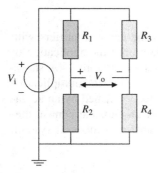

**Figure 3.3** Measurement bridge with two active elements $R_1$ and $R_2$.

In this half-bridge mode, $R_3$ and $R_4$ are fixed resistances and $R_1$ and $R_2$ are resistive sensors. Their resistance values change with a particular physical quantity as well as an interfering signal, according to:

$$R_1 = R(1 + S_{m1}x_{m1} + S_{i1}x_{i1})$$

$$R_2 = R(1 + S_{m2}x_{m2} + S_{i2}x_{i2}) \tag{3.6}$$

Here, $S_{mk}$ is the sensitivity of sensor $k$ ($k = 1,2$) to the measurand (for instance deformation), and $S_{ik}$ is the sensitivity to the interference signal (for instance temperature). Note that in this system the sensor resistance is just $R$ at a particular initial value of the measurand and at zero interference. Assuming both sensor parts experience the same measurement signal and have equal but opposite sensitivities to the measurand, the bridge output voltage satisfies, approximately, the equation:

$$\frac{V_o}{V_i} = \frac{1}{2}S_m x_m + \frac{1}{4}\Delta(S_i x_i) \tag{3.7}$$

If both sensor parts have equal sensitivity to interference (by a symmetric sensor design) and both sensor parts experience the interference equally, the error term in Eq. (3.7) is zero, and the interference is completely eliminated. Equation (3.7) is useful to make a quick assessment of the error due to asymmetry, relative to the measurement signal.

*Example*

The resistors $R_1$ and $R_2$ in Figure 3.3 are strain gauges with strain sensitivity $S_m = K$ (gauge factor) and t.c. $\alpha$ ($K^{-1}$). The measurement signal is the relative deformation or strain $\Delta l/l$; the interference signal is a change in temperature $\Theta$. So, the transfer of this bridge is:

$$\frac{V_o}{V_i} = -\frac{1}{2}K\frac{\Delta l}{l} + \frac{1}{4}(\alpha\Delta\Theta + \Theta\Delta\alpha) \tag{3.8}$$

where $\Delta\Theta$ is the temperature difference between the two sensor parts and $\Delta\alpha$ the difference in temperature sensitivity. In a proper design both sensor parts should have equal temperature and equal temperature sensitivity, over the whole operating range.

### 3.2.2  Feedback Methods

Feedback is an error reduction method originating from the early amplifiers with vacuum tubes. Their unstable operation was a real problem until the application of feedback [5], which reduces in particular multiplicative errors. Figure 3.4A shows the general idea. The sensor has a nominal transfer $S$, but due to multiplicative interference it has changed to $S(1 + \varepsilon_i)$. The feedback is accomplished by an actuator with an inverse transduction effect and a transfer $k$. From classical control theory the error reduction factor can easily be found. The transfer of the total system, $S_f$, is given by

$$S_f = \frac{S}{1 + S \cdot k} \tag{3.9}$$

A relative change $dS$ in the sensor transfer $S$ causes a relative change $dS_f$ in $S_f$ according to:

$$\frac{dS_f}{S_f} = \frac{1}{1 + kS} \cdot \frac{dS}{S} \tag{3.10}$$

So, the relative error in the forward part is reduced by a factor equal to the *loop gain* $S \cdot k$ of the system. The penalty for this improvement is a reduction of the overall sensitivity with the same factor.

Feedback also reduces additive interference signals, in a degree that depends on the point of injection in the system. Two cases are discussed in Figure 3.4B. The output due to two interfering signals $x_{i1}$ and $x_{i2}$ equals:

$$y_o = \frac{S}{1 + kS} \cdot x_{i1} + \frac{1}{1 + kS} \cdot x_{i2} \tag{3.11}$$

Obviously, signals entering at the input of the system are reduced by feedback as much as the measurement signal (so the SNR is not better). Interfering signals injected at the output of the sensing system are reduced by a factor $S$ more than the measurement signal, so the SNR is increased by the same amount.

Feedback reduces errors in the forward signal path: the transfer is mainly determined by the feedback path. Non-linearity in the forward path (for instance due to

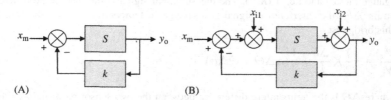

(A)                                    (B)

**Figure 3.4** Feedback system: (A) basic configuration; (B) with additive interference signals.

a non-linear sensor characteristic) is also reduced, provided a linear transfer of the feedback element. Prerequisites for an effective error reduction are as follows:

- High forward path transfer
- Stable feedback path transfer.

The application of this method to sensors requires a feedback element with a transfer that is the inverse of the sensor transfer. Imperfections of the sensor are reduced; however, the demands on the actuator are high. The method is illustrated with an example of a capacitive accelerometer system in which two error reduction methods are combined: compensation by a balanced sensor construction and feedback by an inverse transducer (Figure 3.5).

A displacement of the seismic mass $m$ results in a capacitance difference $\Delta C$; this value is converted to a voltage which is compared with a reference value (here this value is 0). The amplified voltage difference is supplied to an electromagnetic actuator that drives the mass back towards its initial position. When properly designed, the system reaches a state of equilibrium in which the applied inertial force is compensated by the electromagnetic force from the actuator. The current required to keep equilibrium is a measure for the applied force or acceleration.

A more detailed model of this system, for instance for stability analysis, is depicted in Figure 3.6. All transduction steps are visualized in separate blocks. Obviously, the feedback is performed in the mechanical domain by counteracting the inertial force $F_i$ with the electromagnetic force $F_a$ of the actuator.

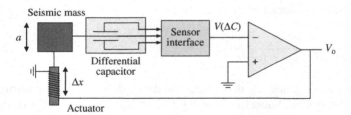

**Figure 3.5** Illustration of a feedback system to reduce sensor errors: capacitive accelerometer.

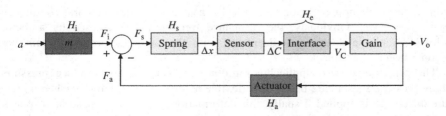

**Figure 3.6** Model of the differential capacitive accelerometer with feedback.

The transfer function of the whole system can easily be derived from this model and is given in Eq. (3.12):

$$V_o = \frac{H_s H_e}{1 + H_a H_s H_e} H_i a \tag{3.12}$$

In this equation, $H_i$ is the transfer from applied acceleration $a$ to inertial force $F_i$. $H_s$ is the transfer of the mechanical spring: from force to displacement $\Delta x$. The transfer of the electrical system (capacitive sensor, electronic interface and amplifier together) is $H_e$. Finally, $H_a$ is the transfer of the actuator. For $H_a H_s H_e \gg 1$, the transfer function of the total sensor system becomes:

$$V_o = \frac{H_i}{H_a} a \tag{3.13}$$

Indeed this is independent of the spring stiffness, the sensor transfer, the interface and the gain of the amplifier, according to the feedback principle. In equilibrium, the mass is at its centre position. Hence, no particular demands have to be made on the spring, the sensor and the interface circuit; the only requirement is a small zero error. Actually, the requirements with respect to the sensor quality have been transferred to those of the actuator. The system transfer depends only on the seismic mass $m$ and the actuator transfer $H_a$.

Feedback may also improve the dynamic performance of the sensor. Assume the sensor in Figure 3.6 is a mass-spring system described by the general second order (complex) transfer characteristic

$$S = \frac{S_0}{1 + 2z(j\omega/\omega_0) - (\omega^2/\omega_0^2)} \tag{3.14}$$

with $z$ the damping and $\omega_0$ the natural (undamped) frequency or resonance frequency of the system. Substitution in Eq. (3.9) results in the transfer of the system with feedback:

$$S_f = \frac{S_0}{1 + kS_0} \frac{1}{1 + 2z_f(j\omega/\omega_f) - (\omega^2/\omega_f^2)} \tag{3.15}$$

with a new damping factor $z_f = z/\sqrt{1 + kS_0}$ and a new natural frequency $\omega_f = \omega_0\sqrt{1 + kS_0}$. Obviously, the resonance frequency is increased by a factor $\sqrt{1 + kS_0}$ and the damping is decreased with the same factor: the system has become faster by feedback.

The principle is also applicable for other measurands, for instance pressure. Many pressure sensors are based on a flexible membrane that deforms when a pressure difference is applied. Usually, this deformation is measured by some displacement sensor, providing the output signal. The effect of the strong non-linear response of the membrane can be eliminated by feedback, using a (linear) actuator in the feedback path [6]. There are various sensors on the market that are based on the feedback principle. Some of them will be discussed in later chapters.

### 3.2.3 Filtering

Error signals can be reduced or avoided by filtering, to be performed either in the domain of the interfering signal (prior to transduction) or in the electrical domain (after transduction).

#### Filtering Prior to Transduction

The unwanted signals may be kept outside the sensor system by several filtering techniques, depending on the type of signal. Sometimes the technique is referred to as *shielding*. We briefly review some causes and the associated reduction methods:

- *Electric* injection of spurious signals, through a *capacitive* connection between the error voltage source and the system input; to be avoided by a *grounded* shield around the sensor structure and the input leads. The capacitively induced current flows directly to ground and does not reach the input of the sensor.
- *Magnetically* induced signals by a time-varying magnetic flux through the loop made up by the input circuit; can not only be reduced by a shield made up of a material with a high magnetic permeability but also by minimizing the area of the loop (leads close to each other or twisted).
- Changes in environmental *temperature*; thermal shielding (an encapsulation with high thermal resistance or a temperature-controlled housing) reduces thermal effects.
- Unwanted *optical* input (from lamps, the sun) can be stopped by optical filters; many filter types are available on the market. Measurand and interferences should have a substantial difference in wavelength.
- *Mechanical* disturbances (e.g. shocks and vibrations) are reduced by a proper mechanical construction with elastic mounting, performing suitable damping of the vibrations.

#### Filtering after Transduction

In case the interfering signals are of the same type (in the same signal domain) as the measurand itself, error reduction is accomplished with filters based on differences in particular properties of the signals. For electrical signals the distinguishing property is the frequency spectrum. Electrical interfering signals with a frequency spectrum different from that of the measurand can easily be filtered by electronic filters (Appendix C). If they have overlapping frequency spectra, this is not or only partially possible. A solution to this problem is to *modulate* (if possible) the measurement signal prior to transduction, to create a sufficiently large frequency difference enabling effective filtering in the electrical domain (see Section 3.2.4).

### 3.2.4  Modulation

Modulation of the measurement signal is a very effective way to reduce the effect of off-set errors and noise. We shortly review the basics of amplitude modulation and demodulation and show that this combination behaves as a narrow band-pass filter process.

Modulation is a particular type of signal conversion that makes use of an auxiliary signal, the *carrier*. One of the parameters of this carrier signal is varied analogously to the input (or measurement) signal. The result is a shift of the complete signal frequency band to a position around the carrier frequency. Due to this property, modulation is also referred to as frequency conversion.

A very important advantage of modulated signals is their better noise and interference immunity. In measurement systems modulation offers the possibility of bypassing offset and drift from amplifiers. Amplitude modulation is a powerful technique in instrumentation to suppress interference signals; therefore, we confine to amplitude modulation only.

A general expression for an AM signal with a sinusoidal carrier $v_c = \hat{v}_c\cos \omega_c t$ modulated by an input signal $v_i(t)$ is as follows:

$$v_m = \hat{v}_c\{1 + k \cdot v_i(t)\}\cos \omega_c t \tag{3.16}$$

where $\omega_c$ is the frequency of the carrier, and $k$ is a scale factor determined by the modulator. Suppose the input signal is a pure sine wave:

$$v_i = \hat{v}_i\cos \omega_i t \tag{3.17}$$

The modulated signal is as follows:

$$\begin{aligned}
v_m &= \hat{v}_c\{1 + k \cdot \hat{v}_i\cos \omega_i t\}\cos \omega_c t \\
&= \hat{v}_c\cos \omega_c t + \frac{1}{2}k \cdot \hat{v}_c\hat{v}_i\{\cos(\omega_c + \omega_i)t + \cos(\omega_c - \omega_i)t\}
\end{aligned} \tag{3.18}$$

which shows that this modulated signal has three frequency components: one with the carrier frequency ($\omega_c$), one with a frequency equal to the sum of the carrier frequency and the input frequency ($\omega_c + \omega_i$) and one with the difference between these two frequencies ($\omega_c - \omega_i$). Figure 3.7A shows an example of such an AM signal. The input signal is still recognized in the 'envelope' of the modulated signal, although its frequency component is not present.

When the carrier is modulated by an arbitrary input signal, each of the frequency components in its spectrum (Figure 3.7B) produces two new components, with the sum and the difference frequency. Hence, the whole frequency band is shifted to a region around the carrier frequency (Figure 3.7C). These bands at either side of the carrier are called the *side bands* of the modulated signal. Each side band carries the full information content of the input signal. The AM signal does not contain low-frequency components anymore. Therefore, it can be amplified without being disturbed by offset and drift. If such signals appear anyway, they can easily be removed from the amplified output by a high-pass filter.

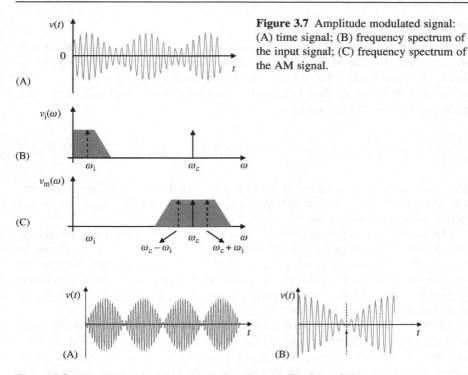

**Figure 3.7** Amplitude modulated signal: (A) time signal; (B) frequency spectrum of the input signal; (C) frequency spectrum of the AM signal.

**Figure 3.8** (A) AM signal with suppressed carrier and (B) phase shift.

There are many ways to modulate the amplitude of a carrier signal. We will discuss three methods: the multiplying modulator, the switching modulator and the bridge modulator.

## Multiplier as Modulator

Multiplication of two sinusoidal signals $v_c$ and $v_i$ (carrier and input) results in an output signal:

$$v_m = K \cdot \hat{v}_c \hat{v}_i \{\cos(\omega_c + \omega_i)t + \cos(\omega_c - \omega_i)t\} \tag{3.19}$$

with $K$ the scale factor of the multiplier. This signal contains only the two-side band components and no carrier; it is called an AM signal with *suppressed carrier*. For arbitrary input signals the spectrum of the AM signal consists of two (identical) side bands without carrier. Figure 3.8A shows an example of such an AM signal. Note that the 'envelope' is not identical to the original signal shape anymore. Further, the AM signal shows a phase shift in the zero crossings of the original input signal (Figure 3.8B).

## Switch Modulator

In the switch modulator the measurement signal is periodically switched on and off, a process that can be described by multiplying the input signal with a switch signal $s(t)$, being 1 when the switch is on and 0 when it is off (Figure 3.9).

To show that this product is indeed a modulated signal with side bands, we expand $s(t)$ into its Fourier series:

$$s(t) = \frac{1}{2} + \frac{2}{\pi} \left\{ \sin \omega t + \frac{1}{3}\sin 3\omega t + \frac{1}{5}\sin 5\omega t + \cdots \right\} \tag{3.20}$$

With $\omega = 2\pi/T$ and $T$ the period of the switching signal. For a sinusoidal input signal with frequency $\omega_i$, the output signal contains sums and differences of $\omega_i$ and each of the components of $s(t)$.

This modulation method produces a large number of side band pairs, positioned around odd multiples of the carrier frequency ($\omega_c$, $3\omega_c$, $5\omega_c$, ...), as shown in Figure 3.10. The low-frequency component originates from the multiplication by the mean of $s(t)$ (here ½). This low-frequency component and all components with frequencies $3\omega_c$ and higher can be removed by a filter. The resulting signal is just an AM signal with suppressed carrier.

Advantages of the switch modulator are its simplicity and accuracy: the side band amplitude is determined only by the quality of the switch. A similar modulator can be achieved by periodically changing the polarity of the input signal. This is equivalent to the multiplication by a switch signal with zero mean value; in that case there is no low-frequency band as in Figure 3.10.

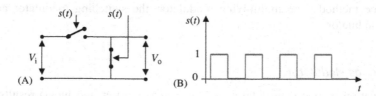

**Figure 3.9** (A) Series-shunt switch as modulator; (B) time representation of the switch signal.

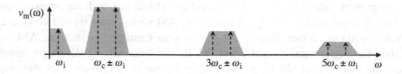

**Figure 3.10** Spectrum of an AM signal from a switching modulator.

The absence of DC and low-frequency components considerably facilitates the amplification of modulated signals: offset, drift and low-frequency noise can be kept far from the new signal frequency band. When very low voltages must be measured, it is recommended to modulate these prior to any other analogue signal processing that might introduce DC errors.

### Measurement Bridge as Modulator

The principle of the bridge modulator is illustrated with the resistance measurement bridge or Wheatstone bridge of Figure 3.11.

The bridge is connected to an AC signal source $V_i$. This AC signal (usually a sine or square wave) acts as the carrier. In this example we consider a bridge with only one resistance ($R_3$) that is sensitive to the measurand. Assuming equal values of the three other resistances, the signal $V_a$ is just half the carrier, whereas $V_b$ is an AM signal: half the carrier modulated by $R_3$. The bridge output $V_o$ is the difference between these two signals, so an AM signal with suppressed carrier.

This output can be amplified by a differential amplifier with high gain; its low-frequency properties are irrelevant; the only requirements are a sufficiently high bandwidth and a high CMRR for the carrier frequency to accurately amplify the difference $V_a - V_b$.

Modulation techniques also apply to many non-electric signals. An optical signal can be modulated using a LED or laser diode. If the source itself cannot be modulated, optical modulation can be performed by, for instance, a chopping wheel, as is applied in many pyroelectric measurement systems. Also, some magnetic sensors employ the modulation principle. Special cases are discussed in subsequent chapters.

### 3.2.5  Demodulation

The reverse process of modulation is demodulation (sometimes called detection). Looking at the AM signal with carrier (for instance in Figure 3.7), we observe the similarity between the envelope of the amplitude and the original signal shape. An obvious demodulation method would therefore be envelope detection or peak detection. Clearly, envelope detectors operate only for AM signals *with* carrier. In an AM signal *without* carrier, the envelope is not a copy of the input anymore. Apparently, additional information is required with respect to the phase of the input, for a full recovery of the original waveform.

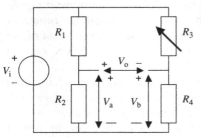

**Figure 3.11** Wheatstone bridge as modulator.

An excellent method to solve this problem, and which has a number of additional advantages, is *synchronous detection*. This method consists of multiplying the AM signal by a signal having the same frequency as the carrier. If the carrier signal is available (as is the case in most measurement systems), this synchronous signal can be the carrier itself.

Assume a modulated sinusoidal input signal with suppressed carrier:

$$v_m = \hat{v}_m \left\{ \cos(\omega_c + \omega_i)t + \cos(\omega_c - \omega_i)t \right\} \tag{3.21}$$

This signal is multiplied by a synchronous signal with a frequency equal to that of the original carrier, and a phase angle $\varphi$. The result is as follows:

$$\begin{aligned}
v_{dem} &= \hat{v}_m \hat{v}_s \left\{ \cos(\omega_c + \omega_i)t + \cos(\omega_c - \omega_i)t \right\} \cdot \cos(\omega_c t + \varphi) \\
&= \hat{v}_m \hat{v}_s \left[ \cos \omega_i t \cos \varphi + \frac{1}{2}\cos\left\{(2\omega_c + \omega_i)t + \varphi\right\} + \frac{1}{2}\cos\left\{(2\omega_c - \omega_i)t + \varphi\right\} \right]
\end{aligned} \tag{3.22}$$

With a low-pass filter, the components around $2\omega_c$ are removed, leaving the original component with frequency $\omega_i$. This component has a maximum value for $\varphi = 0$, i.e. when the synchronous signal has the same phase as the carrier. For $\varphi = \pi/2$ the demodulated signal is zero, and it has the opposite sign for $\varphi = \pi$. This phase sensitivity is an essential property of synchronous detection.

Figure 3.12 reviews the whole measurement process in terms of frequency spectra. The starting point is a low-frequency narrow band measurement signal (A). This signal is modulated and subsequently amplified. The spectrum of the resulting signal is depicted in (B), showing the AM spectrum of the measurement signal and some additional error signals introduced by the amplifier: offset (at DC), drift and $1/f$ noise (LF) and wide band thermal noise.

Figure 3.12C shows the spectrum of the demodulated signal. By multiplication with the synchronous signal, all frequency components are converted to a new position. The spectrum of the amplified measurement signal folds back to its original position, and the LF error signals are converted to a higher frequency range. A low-pass filter removes all components with frequencies higher than that of the original band.

An important advantage of this detection method is the elimination of all error components that are not in the (small) band of the modulated measurement signal. If the measurement signal has a narrow band (slowly fluctuating measurement quantities), a low cut-off frequency of the filter can be chosen. Hence, most of the error signals are removed, and a remarkable improvement of the S/N ratio is achieved, even with a simple first-order low-pass filter.

The low-pass filter with bandwidth $B$, acting on the demodulated signal, is equivalent to a band-pass filter acting on the modulated signal (around the carrier). This means that the effective bandwidth is $2B$. The selectivity of a band-pass filter is expressed with the quality factor $Q$, defined as the ratio between its central

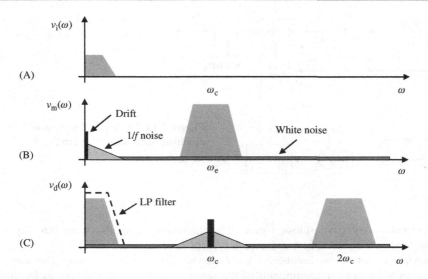

**Figure 3.12** Sequence of modulation and synchronous detection: (A) spectrum of original measurement signal; (B) spectrum of the modulated signal; (C) spectrum of the demodulated signal.

frequency $f_c$ and the ($-3$ dB) bandwidth $B$. So the synchronous detector can be conceived as a band-pass filter with quality factor equal to $f_c/B$. If, for example, the measurement signal has a bandwidth of 0.5 Hz, the cut-off frequency of the low-pass filter is also set to 0.5 Hz. Suppose the carrier frequency is 4 kHz, then the effective quality factor of the synchronous detector is 4000. Active band-pass filters can achieve $Q$-factors of about 100 at most, so synchronous detection offers a much higher selectivity compared to active filtering.

### 3.2.6   Correction Methods

An erroneous sensor signal can be corrected if knowledge about the causes of the errors or the value of the errors is available. Two different strategies can be distinguished:

- Static correction
- Dynamic correction.

In the first class of strategies correction is performed while leaving the sensor unaltered. Figure 3.13 shows the general configuration for two approaches. In the model-based approach, the sensor signal is corrected based on prior knowledge about the origin of the error, for instance non-linearity or a calibration curve, stored in a look-up table. If the errors are unknown (interference), the error signal could be measured separately by additional sensors. The output of these sensors is used

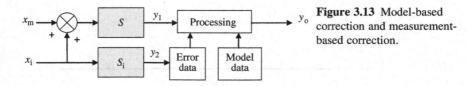

**Figure 3.13** Model-based correction and measurement-based correction.

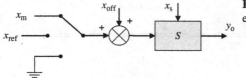

**Figure 3.14** Three-point measurement to eliminate scale and offset errors.

to correct the original sensor signal. The method is straightforward but requires additional sensors, at least one for each type of interference.

Dynamic correction involves a particular sensor design. Basically, the method involves a series of measurements of the same quantity, in such a way as to eliminate the errors by post-processing. This can be done in various ways, according to the type of quantity and error:

- multiple input signal measurements, eliminating scale and offset errors;
- cyclic interchanging of components (dynamic matching);
- cyclic changing the sensitivity direction of the sensor (flipping).

The first method is illustrated in Figure 3.14. Additive errors are represented by the signal $x_{off}$. Multiplicative errors due to the interfering signal $x_s$ are represented by the relative error $\varepsilon_s$. Alternately, the input of the system is connected to the measurand (yielding an output $y_{o1}$), to 'ground' (giving output $y_{o2}$) and to a reference (resulting in the output $y_{o3}$).

The three system outputs are as follows:

$$y_{o1} = S(1 + \varepsilon_s)(x_m + x_{off})$$
$$y_{o2} = S(1 + \varepsilon_s)(x_{off})$$
$$y_{o3} = S(1 + \varepsilon_s)(x_{ref} + x_{off})$$

(3.23)

From these equations the measurand can be calculated as follows:

$$x_m = \frac{y_{o1} - y_{o2}}{y_{o3} - y_{o2}} x_{ref}$$

(3.24)

Offset and scale errors are completely eliminated if the errors do not change during the sequence of the three measurements. The method requires a reference of the measurand type as well as a possibility to completely isolate the input from the measurand to find the offset error contribution. For most electrical quantities, like voltage, capacitance, and resistance, this is quite an easy task. This is not the

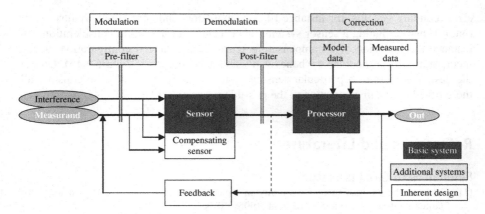

**Figure 3.15** Overview of error reduction methods.

case for many other physical quantities: repetitively short-circuiting the input of a magnetic field sensor or a force sensor, for instance, without removing the measurand itself requires complicated and therefore impractical shielding techniques.

All sensor systems suffer from offset and drift, which hampers the measurement of small, slowly varying signals. *Modulation*, as discussed before, is a very effective way to eliminate offset. However, modulation of the input signal is not always possible. An alternative way to reduce offset is *flipping* the directivity of the sensor, without changing the sign and value of the offset. Clearly, the method can only be applied for sensors that are sensitive to the direction of the input quantity (which should be a vector).

Flipping is performed either by changing the carrier signal in modulating sensors or by rotation of the sensor itself. An example of a modulating sensor (Chapter 2) is the Hall sensor for the measurement of magnetic field strength (Chapter 6). This is a modulating type sensor because the output is proportional to both the input signal and an auxiliary signal, in this case a DC bias current. Assuming the offset being independent of the bias, the offset reduction follows from Eq. (3.25), where $x_c$ is the carrier signal that is flipped between two successive measurements.

$$y_{o1} = S(x_c x_m + x_{off})$$
$$y_{o2} = S(-x_c x_m + x_{off})$$

(3.25)

Obviously, the difference between the two outputs is proportional to the input signal and offset independent. When the sensor is not of the modulating type, the same effect is obtained by just rotating the sensor [7]. Since this mechanical flipping requires some moving mechanism, the method is restricted to low-frequency applications.

All error reduction methods discussed in this section are illustrated schematically in Figure 3.15. In this scheme, the basic signal flow is shown in dark colour. The performance of a non-compensated sensor can be improved by adding error-reducing systems, for instance filters, modulators and correction circuitry (half-tone). When designing a sensor system, other techniques are worth consideration, for instance the inclusion of a compensating sensor in a balanced configuration, or the incorporation of a modulator whenever possible (white parts in Figure 3.15). During the design process of a particular sensing system, it is recommended to consider all these possibilities, in connection to the possible occurrence of errors.

# References and Literature

## References to Cited Literature

[1] Guide to the expression of uncertainty in measurement; International Organization for Standardization, Geneva, Switzerland, 1995; ISBN 92-67-10188-9.
[2] ISA S37.1: Electrical transducer nomenclature and terminology, 1975; ISBN 0-87664-113-3.
[3] Various documents of the ISA working groups, for instance:
ISA-S37.2, Specification and tests for piezoelectric acceleration transducers.
ISA-S37.3, Specification and tests for strain gage pressure transducers, 1982; ISBN 0-87664-378-0.
ISA-S37.5, Specification and tests for strain gage linear acceleration transducers.
ISA-S37.6, Specification and tests for potentiometric pressure transducers, 1982; ISBN 0-87664-380-2.
ISA-S37.12, Specification and tests for strain gage force transducers.
ISA-S37.10, Specification and tests for piezoelectric pressure transducers, 1982; ISBN 0-87664-382-9.
ISA-S37.12, Specification and tests for potentiometric displacement transducers, 1977; ISBN 0-87664-359-4.
[4] International Vocabulary of Metrology – Basic and general concepts and associated terms (VIM); Document produced by Working Group 2 of the Joint Committee for Guides in Metrology (JCGM/WG 2), JCGM 200: 2008.
[5] H.S. Black: Stabilized feedback amplifiers; Bell Syst. Tech. J. 13 (January 1934).
[6] B.M. Dutoit, P.-A. Besse, A.P. Friedrich, R.S. Popovic: Demonstration of a new principle for an active electromagnetic pressure sensor, Sens. Actuators, 81 (2000), 328–331.
[7] M.J.A.M. van Putten, M.H.P.M. van Putten, A.F.P. van Putten: Full additive drift elimination in vector sensors using the alternating direction method (ADM), Sens. Actuators A, 44 (1994), 13–17.

# Literature for Further Reading

## Many Books on Instrumentation and Measurement Comprise Chapters on Accuracy and Noise Reduction, for Instance

[1] A.F.P. van Putten: Electronic measurement systems; IOP Publishing, Bristol, Philadelphia, 1996; ISBN 0-7503-0340-9.
[2] P.P.L. Regtien, F. van der Heijden, M.J. Korsten, W. Olthuis: Engineering aspects of measurement science; Koogan Page Ltd., London, 2004; ISBN 1-9039-9658-9.

# 4 Resistive Sensors

In this chapter we discuss sensors based on resistive effects. These sensors behave as an electric resistor whose resistance is affected by a particular physical quantity. The resistance may change according to a change in material properties, a modification of the geometry or a combination of these. Quantities that can easily be measured using resistive effects are temperature (thermistors and metal thermometers), light (LDR or light dependent resistor), deformation (piezoresistors) and magnetic field strength (magnetoresistors). By a special construction or material choice, resistors can be made useful for sensing various quantities, for instance force, torque, pressure, distance, angle, velocity and acceleration.

After having defined resistance and resistivity, we discuss successively potentiometric sensors, strain gauges, piezoresistive and magnetoresistive sensors. Further, some thermal and optical sensors based on resistive effects are reviewed. Where appropriate, a discussion on interfacing is included.

## 4.1 Resistivity and Resistance

The electrical conductivity $\sigma$ (the inverse of electrical resistivity $\rho$) is defined as the ratio between current density $J$ (A m$^{-2}$) and electric field strength $E$ (V m$^{-1}$):

$$\sigma = \frac{1}{\rho} = \frac{J}{E} \tag{4.1}$$

Conductivity is a pure material property: it does not depend on shape or size of the device (skin effect and thin-layer effects are disregarded here). In this section all materials are assumed to be homogeneous. The resistance between the endpoints of a bar with length $l$ and constant cross section $A$ equals:

$$R = \rho \frac{l}{A} \tag{4.2}$$

The parameter $l$ is used to create distance sensors and angular displacement sensors: these sensors are called potentiometric sensors and are discussed in Section 4.2. Strain gauges are based on changes in $l/A$ (applied in force and pressure sensors, Section 4.3), while the parameter $\rho$ is the property of interest in piezoresistive, magnetoresistive, thermoresistive and optoresistive sensors (Sections 4.4–4.7).

**Sensors for Mechatronics.** DOI: 10.1016/B978-0-12-391497-2.00004-2
© 2012 Elsevier Inc. All rights reserved.

**Table 4.1** Overview of Resistive Sensor Types

| Domain | Measurand | Geometry | Resistivity |
|--------|-----------|----------|-------------|
| Mechanical | Linear displacement | Potentiometer | |
| | Angle | Potentiometer | |
| | Strain | | Metal strain gauge |
| | Force, torque, pressure | | Piezoresistor |
| | Acceleration | Potentiometric | |
| Magnetic | Magnetic field | | Magnetoresistor |
| Thermal | Temperature | | Thermistor, Pt100 |
| Optical | Light flux | | LDR |

**Figure 4.1** Various connections to the slider: (A) rotating shaft, (B) sliding rod and (C) flex cable.

Table 4.1 presents an overview of the various resistive sensors discussed in this chapter.

## 4.2  Potentiometric Sensors

### 4.2.1  Construction and General Properties

Potentiometric displacement sensors can be divided into linear and angular types, according to their purpose and associated construction. A potentiometric sensor consists of a (linear or toroidal) body which is either wire wound or covered with a conductive film. A slider (or wiper) can move along this conductive body, acting as a movable electrical contact. The connection between the slider and the object of which the displacement should be measured is performed by a rotating shaft (angular potentiometers), a moving rod, an externally accessible slider (sledge type) or a flexible cable that is kept stretched during operation. Figure 4.1 shows a schematic view of some of these constructions. In all cases the resistance wire or film and the wiper contacts should be properly sealed from the environment to min-imize mechanical damage and corrosion. This is an important issue when applied in mechatronic systems that operate in harsh environments. Robust potentiometers have a stainless steel shaft or rod, and a housing of, for instance anodized alumin-ium. The moving parts of the potentiometer are provided with bearings, to minimize the mechanical force needed to initiate movement of the slider and minimizing wear.

Potentiometric sensors are available in a wide variety of ranges. Linear types vary in length from several mm up to a few m, angular types have ranges from about $\pi/2$ rad up to multiples $(2-10)$ of $2\pi$ (multi-turn potentiometers), achieved by built-in gears or a spindle construction.

The specification of potentiometric sensors is standardized by the Instrument Society of America [1]. We list here the major items, in a short formulation. VR stands for voltage ratio, that is, the ratio between the wiper voltage and the full voltage across the resistor:

- range (linear distance or angle)
- linearity (in % VR over total range)
- hysteresis (in % VR over specified range)
- resolution (average and maximum)
- mechanical travel (movement from one stop to the other)
- electrical travel (portion of mechanical travel during which an output change occurs)
- operating temperature
- temperature error (in % VR per °C or in % VR over a quarter of the full range)
- frequency response (at given amplitude)
- cycling life (number of cycles at 1/4 of the maximum frequency)
- storage life (month, year)
- operating force or torque (break-out force or torque) to initialize movement
- dynamic force or torque (to continuously move the shaft after the first motion has occurred)
- shaft overload (at the extremities of the mechanical travel; without damage or degradation)
- shaft axial misalignment.

A manufacturer should mention all these specifications in the data sheets of the device. Table 4.2 lists the main specifications for various types of potentiometers. Besides the specifications listed in this table, many other parameters should be considered when using a potentiometer as displacement sensor, in particular in mechatronics applications. Some of them are listed below:

- maximum allowable force or torque on the wiper;
- minimum force or torque to move the wiper; typical starting torque is 0.1 Ncm, for a 'low-torque' potentiometer this can be less than 0.002 Ncm and for a robust type as high as 10 Ncm;
- maximum (rotational) wiper speed (usually about 1000 rev/min);
- maximum voltage across the resistance (typically 10 V);
- maximum current through the wiper contact (typically 10 mA).

## 4.2.2 Electrical Characteristics

Potentiometric displacement sensors can electrically be connected in two different ways: the potentiometric wiring and the rheostat wiring (Figure 4.2). In rheostat

**Table 4.2** Typical Specifications of Potentiometric Sensors

| Parameter | Linear | Rotational |
|---|---|---|
| Range | 2 mm to 8 m | 10–60° rev |
| Resistance | 1 kΩ to 1 MΩ ± 5% | As linear |
| Resolution | | |
|   Normal | ±0.1% FS | 0.2–2° |
|   Lowest, wire | 10 μm | |
|   Lowest, film | 0.1 μm | |
| Non-linearity | 0.01–1% FS | As linear |
| Temperature coefficient | $10^{-3}$ $K^{-1}$ | As linear |
| Temperature range | −20°C to 150°C | As linear |
| $v_{max}$ wiper | 1 m/s | 20 rev/s |
| Reliability, wire | $10^6$ movements | i.d., revolutions |
| Reliability, film | $10^7 - 10^8$ movements | i.d., revolutions |
| Maximum power | 0.1–50 W | As linear |

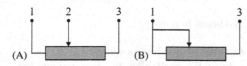

**Figure 4.2** (A) Potentiometric wiring and (B) rheostat wiring.

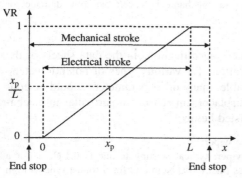

**Figure 4.3** Transfer characteristic of an ideal potentiometer.

mode the device acts just as a variable resistor (two terminal); the potentiometer mode is suitable for voltage division (two-port configuration; Figure 4.5).

The VR of a potentiometer equals $R_{12}/R_{13}$ which, for an ideal potentiometer, is equal to $x/L$, with $L$ the electrical stroke and $x$ the distance between the start position and the wiper position (Figure 4.3).

Non-linearity and resolution are the main causes of a deviation from this ideal transfer. A linear relationship between position and wiper voltage requires a wire or film with homogeneous resistivity over the whole range. The intrinsic non-linearity

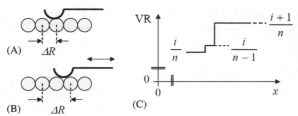

(A) $\Delta R$

(B) $\Delta R$

(C)

**Figure 4.4** (A) Slider on top a winding, (B) slider between two windings and (C) illustration of the resolution of a wire-wound potentiometer.

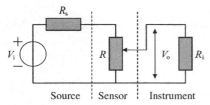

Source ⦙ Sensor ⦙ Instrument

**Figure 4.5** Interface circuit for a potentiometer.

can be as good as 0.01% (see Table 4.2). Improper interfacing may introduce additional non-linearity, as will be explained in Section 4.2.3.

The position resolution of a wire-wound potentiometer is set by the number of turns $n$. With $R$ the total resistance, the resistance of a single turn amounts $\Delta R = R/n$. As the wiper steps from one turn to the next, the VR changes leap-wise with an amount of $1/n$ when the wiper moves continuously (Figure 4.4A); hence the resolution equals $\Delta R/R = 1/n$. At wiper position on top of turn $i$, VR equals $i/n$; on top of the next turn it increases to $(i+1)/n$. Actually the wiper may short circuit one turn when positioned just between two windings (Figure 4.4B). In those particular positions the total resistance drops down to $(n-1)\Delta R$, hence VR $= i/(n-1)$, which is slightly more than $i/n$ as shown in Figure 4.4C.

The resolution can be increased (without change of outer dimension) by reducing the wire thickness. However this degrades the reliability because a thinner wire is less wear resistant. The resolution of a film potentiometer is limited by the size of the carbon or silver grains that are impregnated in the plastic layer to turn it into a conductor. The grain size is about 0.01 μm; the resolution is about 0.1 μm at best.

### 4.2.3  Interfacing

The interfacing of a potentiometric sensor is essentially simple (Figure 4.5).

To measure the position of the wiper, the sensor is connected to a voltage source $V_i$ with source resistance $R_s$; the output voltage on the wiper, $V_o$, is measured by an instrument with input resistance $R_i$. Ideally, the voltage transfer $V_o/V_i$ equals the VR. Due to the presence of a source resistance and load resistance, the transfer might differ from the VR. We will calculate the error introduced by both these effects.

First assume $R_s = 0$ and $R_i \to \infty$; under this condition, the output voltage of the sensor satisfies Eq. (4.3) for a linear potentiometer or Eq. (4.4) for an angular potentiometer:

$$V_o = \frac{x}{L} V_i \qquad (4.3)$$

$$V_o = \frac{\alpha}{\alpha_{max}} V_i \qquad (4.4)$$

where $L$ is the total electrical length and $\alpha_{max}$ the maximum electrical angle. The sensitivity of a linear sensor is:

$$S = \frac{dV_o}{dx} = \frac{V_i}{L} \quad (V/m) \qquad (4.5)$$

and apparently is independent of the resistance $R$ and proportional to the source voltage. The sensitivity can be increased by increasing the source voltage. Obviously, the maximally permissible power $P_{max} = V_{i,max}^2/R$ should be kept in mind. When ambient temperature increases, the maximum allowable dissipation drops; it is wise to carefully check the data sheets on this aspect.

Instability of $V_i$ results in an output change that is indiscernible to displacement. The effect is maximal for $x = L$, so a stability criterion for $V_i$ is:

$$\frac{\Delta V_i}{V_i} < \frac{\Delta x_m}{L} \qquad (4.6)$$

where $\Delta x_m$ is the smallest detectable displacement. Otherwise stated the stability of the source voltage should be better than the resolution of the potentiometer.

A non-zero value of the source resistance introduces a scale error. For $R_s \neq 0$ and $R_i \to \infty$, the sensor transfer is:

$$V_o = \frac{R}{R + R_s} \cdot \frac{x}{L} \cdot V_i \approx \left(1 - \frac{R_s}{R}\right) \cdot \frac{x}{L} \cdot V \qquad (4.7)$$

Hence, the sensitivity is reduced by an amount $R_s/R$ with respect to the situation with an ideal voltage source ($R_s = 0$).

A load resistance $R_i$ results in an additional non-linearity error. Now assuming $R_s = 0$, the voltage transfer is:

$$\frac{V_o}{V_i} = \frac{(x/L)}{1 + (x/L)(1 - (x/L))(R/R_i)} \approx \frac{x}{L} \left\{ 1 - \frac{x}{L} \left(1 - \frac{x}{L}\right) \frac{R}{R_i} \right\} \qquad (4.8)$$

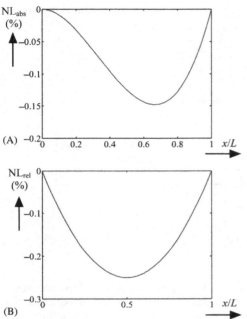

**Figure 4.6** Non-linearity error for $R/R_i = 0.01$: (A) absolute error and (B) relative error.

The approximation is valid when $R/R_i \ll 1$. The end-point linearity error (see Section 3.1), in this case the deviation from the ideal transfer $x/L$, amounts:

$$\text{NL (abs)} = \frac{V_o}{V_i} - \frac{x}{L} = -\frac{x^2}{L^2}\left(1 - \frac{x}{L}\right)\frac{R}{R_i} \tag{4.9}$$

The maximal error is $(-4/27)(R/R_i)$, occurring at $x/L = 2/3$. This error adds up to the intrinsic non-linearity error of the sensor, which is due to unevenly winding or inhomogeneity of the wire or film material. The *relative* non-linearity is:

$$\text{NL (rel)} = \frac{\text{NL (abs)}}{V_o/V_i} = -\frac{x}{L}\left(1 - \frac{x}{L}\right)\frac{R}{R_i} \tag{4.10}$$

from which follows a maximum relative deviation of $(-1/4)(R/R_i)$, occurring at the position $x/L = 0.5$ (midway between the end points). Both errors are proportional to the resistance ratio $R/R_i$. Figure 4.6 shows the absolute and relative non-linearity error for the case $R/R_i = 1\%$. In this case the (absolute) non-linearity is about 0.15%. If, for example, the absolute non-linearity due to loading should be less than 0.05% over the full range of the potentiometer, then the minimum value for the load resistance should be 300 times that of the potentiometer.

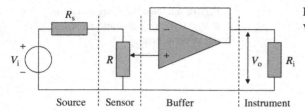

**Figure 4.7** Buffering the wiper voltage.

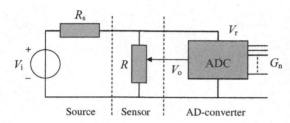

**Figure 4.8** Elimination of source voltage variations.

Additional non-linearity due to a possibly low-input resistance of the measurement instrument can be circumvented by inserting a voltage buffer between the wiper and the measurement instrument, as shown in Figure 4.7. Such a buffering alleviates the requirements on the input resistance of the read-out circuit.

The influence of source voltage changes can be reduced by applying a *ratio method*, as is illustrated in Figure 4.8. The transfer of the ADC is $V_a = G_n \times V_r$, where $G_n$ is the binary coded fraction between 0 and 1. Since the reference voltage of the ADC equals the voltage across the potentiometer, from Eq. (4.3) it follows $G_n = x/L$ which, in the ideal case, does not depend on $V_i$. The ADC output can be directly processed by a computer.

## 4.2.4 Contact-Free Potentiometers

Gradually film potentiometers grow in popularity over wire-wound types, because of higher resolution and a better resistance against moisture. However one important drawback still remains: wear due to the sliding contact of the wiper. A recent development is the contact-free potentiometer [2]. In this type the slider has been replaced by a floating electrode; the resistance track acts as counter electrode of the capacitor. Obviously, wear is minimized in this way.

The basic construction of potentiometers with capacitive readout is shown in Figure 4.9A. The floating wiper is capacitively coupled to the resistance track and a conductive return track. A voltage $V_i$ is connected across the resistance track and the return track is linked to the interface circuit.

Figure 4.9B shows the electric circuit model of the sensor and a simple read-out circuit: $C_1$ is the capacitance between the resistance track and the wiper; $C_2$ is the capacitance between the return track and the wiper and $C_3$ is the stray capacitance

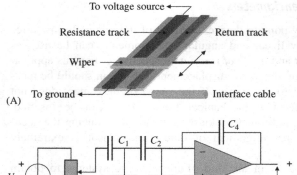

**Figure 4.9** Contact-free potentiometer: (A) basic layout and (B) electric circuit model and interface.

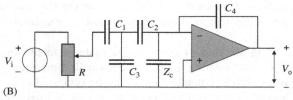

between wiper and ground. The inverting input of the operational amplifier is virtually at ground potential (Appendix C), so the voltage across the cable impedance $Z_c$ is zero. This makes the transfer of the device independent of the cable impedance.

When the stray capacitance $C_3$ is small compared to $C_1$ and $C_2$, the voltage transfer of this circuit is given by

$$\frac{V_o}{V_i} = -\frac{x}{L} \cdot \frac{C_w}{C_4} \cdot \frac{1}{1 + j\omega R_p C_w} \tag{4.11}$$

where $C_w = C_1 + C_2$ and $R_p$ is the parallel resistance of the two parts of the potentiometer. The latter varies with the wiper position; its maximum value occurs when the wiper is in the centre position of the potentiometer, and equals $R/4$. The transfer function shows a first-order low-pass behaviour, characterized by the time constant $R_p C_w$. Since the wiper capacitance $C_w$ is almost constant, the sensor's time constant varies with the wiper position; halfway through the electrical stroke the system has the lowest bandwidth. For low frequencies of the source voltage, the output voltage is directly proportional to the VR of the potentiometer, hence proportional to the displacement of the wiper. The low-frequency gain of the interface circuit amounts $C_w/C_4$ and can be given a specified value by a proper choice of $C_4$.

Note that the feedback capacitor $C_4$ should be shunted by a resistance to prevent the operational amplifier from running into saturation due to integration of its offset voltage and bias current (see also Appendix C.6). Together with $C_4$ this resistance introduces another time constant, so the associated cut-off frequency should be chosen well above the frequency of the voltage source.

Similar to the conventional types, the non-linearity of the contact-free potentiometer is determined by the inhomogeneity of the resistance track and is, therefore, not much better than the traditional types. However the transfer characteristic is somewhat smoothed due to the extension of the floating wiper: small local irregularities are averaged out.

## 4.2.5  Applications of Potentiometers

Potentiometric sensors are very popular devices because of the relatively low price and easy interfacing. Evidently linear and angular potentiometers can be used as sensors for *linear displacement* and *angle of rotation*, respectively. For this application, the construction element of which the displacement or rotation should be measured is mechanically coupled to the wiper. If the range of movement does not match that of the potentiometer, a mechanical transmission can be inserted. However, this might introduce additional backlash and friction, reducing the accuracy of the measurement. Proper alignment of object and sensor is extremely important here.

Potentiometers are widely used in all kind of mechatronic systems to obtain information about angular positions of rotating parts. When small dimensions should be combined with a high resolution, potentiometers may be preferred over the popular optical encoders (Chapter 7). For example, controlling a multi-fingered robotic hand requires several angular position sensors located at the joints of the fingers. Potentiometers have been considered as one of the possibilities to achieve this goal (see for instance Ref. [3]). Figure 4.10 shows how potentiometers, in combination with an elastic element, are used to measure joint angles $\alpha$ and $\beta$, from which follows the touch force exerted by the fingertip on an object.

A particular application of a potentiometric sensing device is an angular sensor for knee rotation, described in Ref. [4]. Two non-elastic wires are positioned along the leg, in parallel to the plane of rotation. At one end the wires are connected to each other and both other ends are connected to the sliders of two potentiometers. Springs keep the wires stretched. On bending, the free ends show a relative displacement related to the angle of rotation. The potentiometers form a Wheatstone bridge, so only the position difference of the wire ends is measured. The resolution of the measurement is better than $0.1°$ over a range of $100°$.

The use of potentiometers is not restricted to the measurement of linear and angular displacement. They can also be used for the measurement of acceleration, force, pressure and level. In these cases the measurand is transferred to a displacement by a proper mechanical construction. Combined with a spring, a potentiometer can act as a *force sensor*; with a seismic mass fixed to the slider the construction is sensitive to *acceleration*. Potentiometric *accelerometers* are available for ranges up to $10g$, inaccuracy around 2% FS, and a lateral sensitivity less than 1% (i.e. 1% of the sensitivity along the main axis).

Potentiometers find also application in *gyroscopes* with spinning rotor. When rotated, the axis of the rotor tends to maintain its original position hence making an

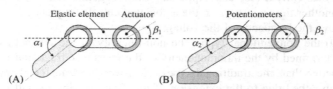

**Figure 4.10** Potentiometers in a robot finger for grip control (A) untouched (B) when touched.

angle with the sensor housing in proportion to the angular velocity. Gyroscopes are available in which two angles, roll and pitch, are measured by built-in potentiometers.

Another application is the measurement of *level*. For this purpose potentiometers of the type shown in Figure 4.1C are used. In case of a liquid the one end of the flexible cable is connected to a float and at the other end wound on a drum. When the level is raising or lowering, the drum winds or unwinds; a rotational potentiometer measures the rotation of the drum. In case of a granular material the measurement is usually performed intermittently: a weight on a flexible cable drops down until it reaches the top level of the material. The sudden change in the cable tension marks the distance over which the cable has been unrolled.

A potentiometer-like sensor for measuring *tilt* is presented in Ref. [5]. Although not a potentiometer in the usual sense, it functions in a similar way. The resistive material is an electrolytic solution residing in a micromachined cavity, and the three electrodes are fixed to the walls: two at both ends of the cavity and one in the middle. In horizontal position the resistances between the middle electrode and both end electrodes are equal, but when tilted the resistance ratio varies according to the angle of rotation. The potentiometric structure makes part of a Wheatstone bridge which is AC driven to prevent electrolysis. A resolution of better than $1°$ over an inclination range of $\pm 60°$ is achieved.

## 4.3  Strain Gauges

### 4.3.1  *Construction and Properties*

Strain gauges are wire or film resistors deposited on a thin, flexible carrier material. The wire or film is very thin, so it can easily be stressed (strain is limited to about $10^{-3}$). In 1843, in his first publication on the now well-known bridge circuit, C, Wheatstone mentioned that the resistance of a wire changes due to mechanical stress. Only 80 years later, the first strain gauges based on this effect were developed independently by E.E. Simmons (at CALTEC) and A.C. Ruge (at MIT). The latter fixed the measuring wire onto a carrier material, resulting in an independent measuring device for stress and strain [6]. Foil gauges, invented by P. Eisler, appeared in 1952.

Early strain gauges were made of thin wires that are folded to obtain high sensitivity in one direction, while keeping the total dimensions within practical limits. The transversal sensitivity is smaller but not unimportant. More popular are the foil gauges (Figure 4.11): a thin film of conducting material is deposited on an insulating backing material and etched to create a meanderlike structure (the grid). Foil gauges have many advantages over wire gauges:

- more accurate and cheaper production,
- better heat dissipation,
- special constructions possible.

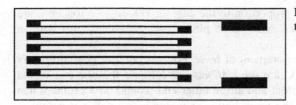

**Figure 4.11** General layout of a thin film strain gauge.

The sensitivity of a strain gauge is expressed in relative resistance change per unit of strain:

$$K = \frac{\mathrm{d}R/R}{\mathrm{d}l/l} \qquad\qquad (4.12)$$

$K$ is called the *gauge factor* of the strain gauge. Using the general equation for resistance Eq. (4.2) the relative resistance change due to positive strain equals:

$$\frac{\mathrm{d}R}{R} = \frac{\mathrm{d}\rho}{\rho} + \frac{\mathrm{d}l}{l} - \frac{\mathrm{d}A}{A} \qquad\qquad (4.13)$$

Hence the change of the resistance is caused by three parameters: the resistivity, the length and the cross-section area of the wire or film. In general all three parameters change simultaneously upon applying strain.

The gauge factor for metals can be calculated as follows. When an object is stressed in one direction, it experiences strain not only in this direction but also in perpendicular directions due to the Poisson effect. When a wire or thin film is stressed in the principal plane (longitudinal direction) it becomes longer but also thinner: its diameter shrinks. The ratio of change in length and change in diameter is the Poisson ratio:

$$\nu = -\frac{\mathrm{d}r/r}{\mathrm{d}l/l} \qquad\qquad (4.14)$$

where $l$ is the length of the wire and $r$ the radius of the circular cross section. Note that according to this definition $\nu$ is positive, because $\mathrm{d}r$ and $\mathrm{d}l$ have opposite signs. The area of a wire with cylindrical cross section is $A = \pi r^2$, so

$$\frac{\mathrm{d}A}{A} = 2\frac{\mathrm{d}r}{r} \qquad\qquad (4.15)$$

hence

$$\frac{\mathrm{d}R}{R} = \frac{\mathrm{d}\rho}{\rho} + \frac{\mathrm{d}l}{l} - 2\frac{\mathrm{d}r}{r} \qquad\qquad (4.16)$$

or, using Eq. (4.14):

$$\frac{dR}{R} = \frac{d\rho}{\rho} + (1 + 2\nu)\frac{dl}{l} \tag{4.17}$$

from which finally the gauge factor results:

$$K = 1 + 2\nu + \frac{d\rho/\rho}{dl/l} \tag{4.18}$$

To find a value for the Poisson ratio, we assume the volume $V = A \cdot l$ of the wire is unaffected by stress:

$$\frac{dV}{V} = \frac{dA}{A} + \frac{dl}{l} = 0 \tag{4.19}$$

resulting in

$$\frac{dl}{l} = -\frac{dA}{A} = -2\frac{dr}{r} \tag{4.20}$$

The Poisson ratio in this ideal case, therefore, equals $\nu = 0.5$. For metals of which the strain dependency of the resistivity can be neglected, the gauge factor equals $K = 1 + 2\nu$. So the gauge factor of a metal strain gauge is $K = 2$. In other words: *the relative resistance change equals twice the strain.* This rule of thumb is only a rough approximation. First of all the value of 2 for the Poisson ratio is a theoretical maximum: in practice the volume will increase somewhat when the wire is stressed, hence the Poisson ratio will be less than 0.5; actual values for the gauge factor range from 0.25 to 0.35. Further, common strain gauges, built from alloys, have a gauge factor larger than 2; typical values range from 2.1 to 3. This means that for such materials the stress dependency of the resistivity cannot be neglected, and the most right term in Eq. (4.18) can be as large as 1.5. The resistivity of a semiconductor material shows a much higher strain dependence; therefore, semiconductor strain gauges have a gauge factor much larger than 2; this will be discussed in Section 4.4 on piezoresistive sensors.

The maximum strain of a strain gauge is not large: about $10^{-3}$. For this reason strain is often expressed in terms of *microstrain* ($\mu$ strain): 1 $\mu$ strain corresponds to a relative change in length of $10^{-6}$. Consequently the resistance change is small too: a strain of 1 microstrain results in a resistance change of only $2 \times 10^{-6}$. The measurement of such small resistance changes will be discussed in the section on interfacing.

Table 4.3 lists some specifications of strain gauges. For comparison, a column for user mountable semiconductor strain gauges has been added. It should be noted that these gauges show a considerable non-linearity; even in a balanced bridge this effect cannot be neglected.

**Table 4.3** Typical Specifications of Strain Gauges

| Property | Metal | Semiconductor |
|---|---|---|
| Dimensions | $0.6 \times 1$ mm up to 150 mm | $0.8 \times 0.5$ mm (min) |
| Gauge factor | 2–2.5 | 200 (max) |
| t.c. gauge factor | Compensated | $4 \times 10^{-4}\ \mathrm{K}^{-1}$ |
| t.c. resistance | $20 \times 10^{-6}\ \mathrm{K}^{-1}$ (constantan) | |
| Linearity | 0.1% (up to 4000 μ strain) | 0.02% (min) |
| Breaking strain | 20.000 μ strain | |
| Fatigue life | up to $10^7$ strain reversals | |
| Temperature range | $-70°$C to $+200°$C (max 400) | $0-175°$C (max 400) |
| Power dissipation | 200 mW/cm$^2$ (max) | |

Typical (standardized) resistance values of strain gauges are 120, 350, 700 and 1000 Ω. Strain gauges have an intrinsic bandwidth of over 1 MHz. So the system bandwidth depends mainly on the (mechanical) interface and read-out electronics.

The resistance of a strain gauge changes not only with stress but also with temperature. Two parameters are important: the temperature coefficient of the resistivity and the thermal expansion coefficient. Both effects must be compensated for, because resistance variations due to stress are possibly much smaller than changes provoked by temperature variations. The effect on resistivity is minimized by a proper material choice for the strain gauge film or wire, for instance constantan, an alloy of copper and nickel with a low temperature coefficient. Residual temperature effects due to a non-zero temperature coefficient are further reduced by a proper interfacing (see Section 4.3.2).

To compensate for thermal expansion, manufacturers supply gauges that can be matched to the material on which the gauges are mounted (so-called *matched gauges*). When free gauges are employed, one should be aware of the difference in thermal expansion coefficient of the gauge and the test material. In this case the relative resistance change due to temperature effects is expressed by

$$\frac{\Delta R}{R} = \alpha_T \Delta T + (\alpha_s - \alpha_g)K\Delta T \tag{4.21}$$

where $\alpha_T$ is the temperature coefficient of the metal, $\alpha_s$ and $\alpha_g$ the thermal expansion coefficients of the specimen and the gauge, respectively. The effect of the different thermal expansion coefficients cannot be distinguished from an applied stress, and should therefore be minimized by using matched gauges when large temperature changes might occur.

Strain gauges are sensitive not only in the main or axial direction but also in the transverse direction. The transverse sensitivity factor $F_t$, defined as the ratio between the transverse sensitivity $K_t$ and the axial sensitivity $K_a$, is of the order of a few percent, and cannot always be neglected. The factor $F_t$ is determined by the manufacturer, using a specified calibration procedure. In general a force applied to an object in axial direction generates a biaxial strain field, due to the Poisson effect. The transverse sensitivity of the strain gauge is responsible for an additional resistance change as a result of this transverse strain. If only axial strain has to be measured, the output (resistance change) should be corrected for the transverse strain. This, however, requires knowledge of both the factor $F_t$ and the Poisson ratio $\nu_a$ of the object material. To simplify correction, strain gauge manufacturers specify the (overall) gauge factor $K$, based on a calibration with a test piece with a Poisson ratio of $\nu_o = 0.285$. So mounted on a material with the same Poisson ratio, no correction for transverse sensitivity is needed. If, on the other hand, the Poisson ratio differs substantially from the value during calibration, a correction factor

$$C_t = \frac{1 - \nu_o F_t}{1 - \nu_a F_t} \tag{4.22}$$

should be applied, for the most accurate measurement result.

When both axial and transverse strains have to be determined, a set of two strain gauges can be applied. Manufacturers provide multi-element strain gauges deposited on a single carrier (see Figure 4.12 for a few examples).

Other configurations are also available, for instance a strain gauge rosette (three gauges making angles of 120°). The three-element gauge is used when the principle strain axis is not known. From the multiple output of this set of gauges, all strain components of a biaxial strain field (including the shear component) can be calculated [7].

### 4.3.2 Interfacing

The resistance change of strain gauges is measured invariably in a bridge. The bridge may contain just one, but more often two or four active strain gauges, resulting in a 'half-' and 'full-bridge' configuration, respectively. The advantages of a half- and full-bridge are an effective temperature compensation and a better linearity of the sensitivity.

**Figure 4.12** Various combinations of strain gauges on a single carrier.

The general expression for the bridge circuit from Figure 3.3 is given in Eq. (4.23):

$$\frac{V_o}{V_i} = \frac{R_2}{R_1 + R_2} - \frac{R_4}{R_3 + R_4} \tag{4.23}$$

When in equilibrium, the bridge sensitivity is maximal if all four resistances are equal. We consider first the case of three fixed resistors and one strain gauge, for instance $R_2$. Assuming $R_1 = R_3 = R_4 = R$ and $R_2 = R + \Delta R$ (which means the strain gauge has resistance $R$ at zero strain), the bridge output voltage is:

$$V_o = \frac{\Delta R}{2(2R + \Delta R)} \cdot V_i = \frac{\Delta R}{4R} \cdot \frac{1}{1 + \Delta R/2R} \cdot V_i \tag{4.24}$$

The output is zero at zero strain. The transfer is non-linear; only for small relative resistance changes the bridge output can be approximated by

$$V_o \approx \frac{\Delta R}{4R} \cdot V_i \tag{4.25}$$

Better bridge behaviour is achieved when both $R_1$ and $R_2$ are replaced by strain gauges, in such a way that, upon loading, one gauge experiences tensile stress and the other compressive stress (compare the balancing technique as discussed in Chapter 3). In practice this can be realized, for instance in a test piece that bends upon loading: the gauges are fixed on either side of the bending beam, such that $R_1 = R - \Delta R$ and $R_2 = R + \Delta R$. The resulting bridge output becomes:

$$V_o = \frac{\Delta R}{2R} \cdot V_i \tag{4.26}$$

The transfer is linear, and twice as high compared to the bridge with only one gauge. It is easy to show that the transfer of a four-gauge or full bridge is doubled again:

$$V_o = \frac{\Delta R}{R} \cdot V_i = K \frac{\Delta l}{l} \cdot V_i \tag{4.27}$$

Although the temperature sensitivity of a strain gauge element is minimized by the manufacturer, the remaining temperature coefficient may cause substantial measurement errors at small values of the strain. In a bridge configuration, this temperature-induced interference can be partly reduced. When the gauges have equal gauge factors, equal temperature coefficients and operate at equal temperatures, the temperature sensitivity of the half- and full-bridge is substantially reduced compared to a bridge with just a single active element. Assume in a half-bridge the two active resistances vary according to $R_1 = R - \Delta R_S + \Delta R_T$ and $R_2 = R + \Delta R_S + \Delta R_T$, where $\Delta R_S$ is the change due to strain and $\Delta R_T$ the change due to temperature. The other two resistance values are $R_3 = R_4 = R$. Substitution of these values in Eq. (4.23) results in

$$V_o = \frac{1}{2} \frac{\Delta R_S}{R + \Delta R_T} \cdot V_i \tag{4.28}$$

In equilibrium ($\Delta R_S = 0$), the output voltage (the offset) is independent of $\Delta R_T$. A similar expression applies for the full bridge:

$$V_o = \frac{\Delta R_S}{R + 2\Delta R_T} \cdot V_i \tag{4.29}$$

Note that only the temperature coefficient of the offset is eliminated, not that of the bridge sensitivity, which is shown by rewriting Eq. (4.29) to be:

$$\frac{V_o}{V_i} \approx \frac{\Delta R_S}{R} \left( 1 - 2 \frac{\Delta R_T}{R} \right) \tag{4.30}$$

A condition for adequate temperature compensation is proper mounting of the two or four active bridge elements. The combined gauges in Figure 4.12 are very useful, since they guarantee optimal temperature matching.

Strain gauges in a differential bridge configuration allow the measurement of very small strain values, down to 0.1 μ strain. The problem of measuring such small strain is actually shifted to the bridge amplifier. Its offset, drift and low-frequency noise obscure the measurement signal. The way out is modulation: the bridge circuit supply voltage is not a DC but an AC voltage with fixed amplitude and frequency. The output is an amplitude-modulated signal with suppressed carrier, which can be amplified without difficulty: possible offset or low-frequency noise from the amplifier is removed by a simple high-pass filter (Chapter 3). Demodulation (by synchronous detection) yields the original, amplified signal. Using this method, strain down to 0.01 μ strain can be measured easily.

Intrinsic shortcomings of strain gauges like the temperature-dependent sensitivity and non-linearity can be reduced by dedicated digital signal processing. Manufacturers of strain gauge measurement transducers provide signal processing systems with these facilities. Current research is aiming at the development of integrated circuits in which both analogue and digital signal processing are combined, resulting in small-sized, low-cost and versatile interfaces for strain gauges. For example, in Ref. [8], four different algorithms are compared and evaluated, implemented with analogue, mixed and digital hardware, performing gain and offset compensation and correction.

### 4.3.3  Applications of Strain Gauges

Strain gauges are suitable for the measurement of all kind of force-related quantities, for example normal and shear force, pressure, torsion, bending and stress.

Strain gauges respond primarily on strain, $\Delta l/l$. Using Hooke's law the applied force is found from the value of the compliance or elasticity of the material on which the strain gauge is fixed.

There are two ways strain gauges are applied in practice:

1. mounted directly on the object whose strain and stress behaviour has to be measured; when cemented properly, the strain of the object is transferred ideally to the strain gauge (for instance to measure the bending of a robot arm);
2. mounted on a specially designed spring element (a bar, ring or yoke) to which the force to be measured can be applied (for instance to measure stress in driving cables and guys).

Strain gauges are excellent devices for the measurement of force and torque in a mechatronic construction. The unbound gauges are small and can be mounted on almost any part of the construction that experiences a mechanical force or moment.

Another approach is to include load cells in the construction. A *load cell* consists of a metal spring element on which strain gauges are cemented. The load is applied to this spring element. The position of the (preferably four) strain gauges on the spring element is chosen such that one pair of gauges is loaded with compressive stress and the other pair with tensile stress (differential mode). If the construction does not allow such a configuration, the two pairs of strain gauges are mounted in a way that one pair experiences the strain that has to be measured, while the other pair is (ideally) not affected by the strain. The second pair merely serves for offset stability and temperature compensation. Obviously, the construction of the spring element and the strain gauge arrangement determine the major properties of the device. Figure 4.13 shows several designs of spring elements for the measurement of force, for various ranges.

Figure 4.13A presents a typical construction for large loads. The central part of the transducer is a metal bar with reduced cross-sectional area where gauges are mounted. There are four strain gauges in a full-bridge configuration. Two strain gauges (front and back side) measure the axial force, with equal sensitivity; two other gauges measure the transverse force, also with equal sensitivity. The gauges

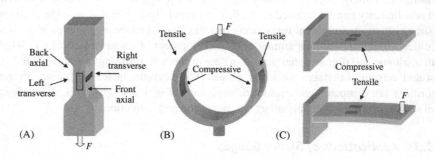

**Figure 4.13** Different spring element designs: (A) column type, (B) yoke type and (C) bending beam.

are arranged in the bridge of Figure 3.3, according to the sequence ($R_1$ front – $R_2$ left – $R_3$ back – $R_4$ right). The axially positioned gauges respond to a strain as

$$\frac{\Delta R_1}{R_1} = \frac{\Delta R_3}{R_3} = K\frac{\Delta l}{l} \qquad (4.31)$$

and the resistance of the transverse gauges change as

$$\frac{\Delta R_2}{R_2} = \frac{\Delta R_4}{R_4} = -\nu K\frac{\Delta l}{l} \qquad (4.32)$$

Substitution in Eq. (4.23) and assuming $\Delta R_i/R_i \ll 1$ for all four gauges, the bridge transfer is:

$$\frac{V_o}{V_i} \approx -K(1+\nu)\frac{\Delta l}{l} \qquad (4.33)$$

The applied force $F$ is found using Hooke's law or, rewritten in terms of force and deformation, $F = (\Delta l/l) \cdot A \cdot E$, with $A$ the cross-section area of the bar and $E$ Young's modulus of the material (or the elasticity $c$). Note that due to the Poisson ratio of the spring element, the compensating strain gauges (in the transverse direction) contribute significantly to the bridge output. With strain gauges mounted directly on a construction part, the same relations between the bridge output and the force apply.

Figure 4.13B shows a ring- or yoke-type spring element. Here the four strain gauges can be mounted in differential mode, compressive and tensile, with almost equal but opposite sensitivity to the applied force. For this arrangement Eq. (4.27) applies and full benefit from the advantages of a full bridge is achieved. The beam-type spring element of Figure 4.13C allows a differential mode measurement of the strain as well. Although two strain gauges are sufficient (one on the top and one at the bottom), four gauges are preferred to build up a full bridge.

*Torque* measurements are performed in a similar way as axial force measurements. The strain gauges are positioned on the beam or shaft under angles of 45° relative to the main axis; see Figure 4.14. Again four gauges are used to complete a full bridge: one pair experiences positive strain, the other negative strain. The transfer of a measurement bridge composed of these four gauges satisfies Eq. (4.27). Although the mutual position on the shaft is not critical, a close mounting is preferred to ensure equal temperatures of the gauges.

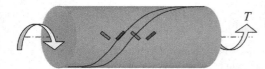

**Figure 4.14** Measurement of torque with strain gauges.

To determine the applied torque (or torsion moment) $T_M$ (N m) from the response of the strain gauges, we use the relation between the torque and the shear stress $T_{shear}$ (N m$^{-2}$) at the outer surface of the shaft. For a circular, solid shaft, this is:

$$T_{shear} = \frac{16 T_M}{\pi D^3} \tag{4.34}$$

with $D$ the shaft diameter. Further, with Hooke's law and assuming all normal stress components zero (pure torsion) the bridge output is found to be:

$$V_o = K \cdot T_{shear} \cdot \frac{1 + \nu}{c} \cdot V_i = K \cdot T_M \cdot \frac{1 + \nu}{c} \cdot \frac{16}{\pi D^3} \cdot V_i \tag{4.35}$$

where $K$ is the gauge factor of the strain gauges and $\nu$ the Poisson ratio of the shaft material.

When torque of a *rotating shaft* has to be measured, conventional wiring of the bridge circuit is not practical. Possible solutions are slip rings (wear sensitive) and transformer coupling (with coils around the rotating shaft). A contact-free torque measurement system is discussed in Chapter 7 on optical sensors.

Many motion control tasks require both torque and force information, up to all six degrees of freedom (d.o.f.). Special mechanical spring elements have been designed to simultaneously measure all these components. They are incorporated, for instance, between a link and the end effector of the machine. Such end-of-arm products are readily available from the market, in a variety of shapes, dimensions (down to 15 mm diameter), and measurement ranges.

A simple design of a *force—torque sensor* consists of a spring element with six or more strain gauges, mounted on a spring element of particular shape. A common spring type used for measuring all six components is the Maltese cross; see Figure 4.15A for a simplified sketch. The wheel-like device is inserted in the robot element of which torque and force components have to be determined. Both

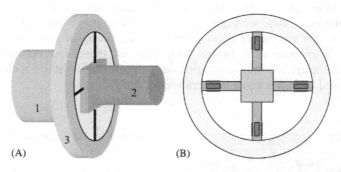

**Figure 4.15** Spring element for six-d.o.f. torque—force sensor: (A) mounting between mechanical interface, (B) positioning of strain gauges.

element ends 1 and 2 are rigidly connected to the sensor 3: part 1 to the outer side and part 2 on the hub. On each of the four 'spokes' two pairs of strain gauges are mounted (Figure 4.15B).

When a force or torque is applied to part 2 of the specimen, the eight differential pairs of strain gauges respond in proportion to these quantities, with a sensitivity that depends on the construction of the spring element. The gauges are arranged in a bridge configuration for obvious reasons. Each gauge pair responds to more than just one of the components $F_x$, $F_y$, $F_z$, $T_x$, $T_y$ and $T_z$. The sensitivity is described by the matrix equation:

$$[S] = [C] \cdot [F] \tag{4.36}$$

where $[F]$ is the $6 \times 1$ force–torque matrix, $[S]$ the $1 \times 8$ strain matrix (if there are eight strain gauge pairs) and $[C]$ is the $6 \times 8$ sensitivity matrix or coupling matrix of the spring element. The force and torque components are calculated using the inverse expression:

$$[F] = [C]^{-1} \cdot [S] \tag{4.37}$$

with $[C]^{-1}$ the decoupling matrix. Note that $[C]^{-1}$ strongly depends on the geometry of the spring element and on the position of the strain gauges.

Various other mechanical structures have been studied with the objective of combining minimal coupling between the force components with constructive simplicity. For instance Ref. [9] gives an extensive analysis of a six-d.o.f. force–torque sensor for a robot gripper. The construction consists of a set of parallel-plate beams provided with strain gauges and allows the simultaneous measurement of three force and three torque components. The force and torque ranges of this combined sensor are 50 N and 5 N m, respectively; interference (crosstalk) errors are around 1%.

Using finite element methods (FEM) techniques the sensitivity matrix can be calculated and the geometry can be optimized to a minimum coupling of the force– torque components (see for instance Refs [10,11]). However, tolerances in the positioning of the gauges and in the geometry of the spring element make calibration of the sensor necessary for accurate measurements of the force and torque vectors.

## 4.4 Piezoresistive Sensors

### 4.4.1 Piezoresistivity

Piezoresistive sensors are based on the change in electrical resistivity of a material when this is deformed. Many materials show piezoresistivity, but only those with a high sensitivity are suitable to be applied in sensors. Examples are semiconducting materials and some elastomers. The most popular piezoresistive semiconductor is silicon: this material can be used as the carrier of the sensor and, moreover, part of the interface electronics can be integrated with the sensor on the same carrier. Elastomers can be made piezoresistive by a special treatment, for instance by adding conductive particles to the non-conducting elastic material.

## Piezoresistive Silicon Sensors

The underlying physical principle of piezoresistivity in silicon goes back to the energy band structure of the silicon atom. An applied mechanical stress will change the band gap. Depending on the direction of the applied force with respect to the crystal orientation, the average mobility of electrons in n-type silicon is reduced, resulting in an increase of the resistivity. So the gauge factor of n-type silicon is negative and reaches values as high as −150. The absolute gauge factor increases with increasing doping concentration.

In p-silicon, holes are the majority carriers: their mobility is influenced by the position in the valance band. The gauge factor of p-silicon appears to be larger than n-silicon (at the same temperature and doping concentration), and has a positive value. In both cases the piezoresistive effect dominates the geometric effect (as used in metal strain gauges).

Table 4.4 shows numerical values for the gauge factor of p- and n-doped silicon, for three different crystal orientations, and a doping level corresponding to a resistivity of $1 \, \Omega \, cm$. Figure 4.16 displays these three main crystal orientations of silicon.

**Table 4.4** Some Gauge Factors for Silicon [12]

| Orientation | p-Type | n-Type |
| --- | --- | --- |
| [111] | 173 | −13 |
| [110] | 121 | −89 |
| [100] | 5 | −153 |

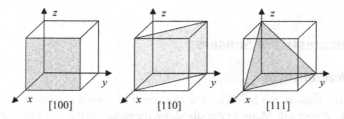

**Figure 4.16** Three orientation surfaces in silicon.

Although the resistance change is primarily caused by material deformation, it is common use to express the piezoresistivity of silicon in terms of pressure sensitivity:

$$\frac{\Delta R}{R} = K \cdot S = K \cdot s \cdot T = \pi \cdot T \tag{4.38}$$

with $K$ the gauge factor as defined in Eq. (4.12), $\pi$ the piezoresistivity ($m^2/N$) and $S$ and $T$ the mechanical strain and tension in the material, respectively. This is a simplified expression: the piezoresistive coefficient depends strongly on the direction of the applied force relative to the crystal orientation. The pressure sensitivity $\pi$ of piezoresistive sensors (in silicon) depends on three factors:

1. the conductivity (orientation dependent)
2. the direction of the applied force
3. the orientation of the resistors with respect to the crystal orientation.

The second factor in this list is related to the elastic behaviour of the material, as already described by the $6 \times 6$ compliance matrix (Appendix A):

$$S_k = s_{kl} T_l, \quad k, l = 1, \ldots, 6 \tag{4.39}$$

The combined orientation-dependent conductivity and compliance yields an expression for the relative resistance change $r$ as a function of the vector $T$:

$$r_i = \pi_{ij} \cdot T_j, \quad i, j = 1, \ldots, 6 \tag{4.40}$$

For silicon many of the matrix elements are zero due to crystal symmetry, and some are pair-wise equal. This results in the piezoresistivity matrix equation for silicon:

$$\begin{pmatrix} r_1 \\ r_2 \\ r_3 \\ r_4 \\ r_5 \\ r_6 \end{pmatrix} = \begin{pmatrix} \pi_{11} & \pi_{12} & \pi_{12} & 0 & 0 & 0 \\ \pi_{12} & \pi_{11} & \pi_{12} & 0 & 0 & 0 \\ \pi_{12} & \pi_{12} & \pi_{11} & 0 & 0 & 0 \\ 0 & 0 & 0 & \pi_{44} & 0 & 0 \\ 0 & 0 & 0 & 0 & \pi_{44} & 0 \\ 0 & 0 & 0 & 0 & 0 & \pi_{44} \end{pmatrix} \cdot \begin{pmatrix} T_1 \\ T_2 \\ T_3 \\ T_4 \\ T_5 \\ T_6 \end{pmatrix} \tag{4.41}$$

So there are only three independent components describing the piezoresistivity of silicon. Their numerical values are given in Table 4.5.

n-Type silicon appears to have a strong negative piezoresistivity in the $x$-direction, and about half as much positive in the $y$- and $z$-directions; p-silicon is less sensitive in these directions. However a shear force (with respect to an arbitrary direction) results in a large resistance change.

**Table 4.5** The Piezoresistivity Components of Silicon for p- and n-Type, in $10^{-12}\ Pa^{-1}$ [12]

| Material | p-Type | n-Type |
|----------|--------|--------|
| $\pi_{11}$ | 66 | −1022 |
| $\pi_{12}$ | −11 | 534 |
| $\pi_{44}$ | 1381 | −136 |

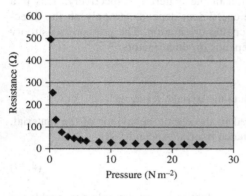

**Figure 4.17** Typical resistance−pressure characteristic of conductive elastomers: bulk resistance.

## Piezoresistive Elastomers

Piezoresistive elastomers are elastomers that are made conductive by impregnation with conductive particles (e.g. carbon and silver). The resistivity depends on the concentration of conductive particles, their mutual distance and the contact area between touching particles. When the material is pressed, more particles make contact, tending to decrease the resistivity of the material. The resistance between two adjacent or opposing points of the elastomer layer changes in a non-linear way with the applied pressure. Figure 4.17 presents a typical resistance−pressure characteristic, revealing that piezoresistive sensors are very sensitive (the resistance may change several orders of magnitude) but highly non-linear [13].

Unfortunately most piezoresistive materials show also hysteresis, poor reproducibility and creep, mainly because of permanent position changes of the conductive particles within the elastomer after being compressed, or lack of elasticity. The typical behaviour of conductive elastomers is still not fully understood; many attempts are being made to model the piezoresistive properties in relation to composition and manufacturing methods (see for instance Ref. [14]). Despite poor performance of the sensing properties, some useful designs have been constructed and applied for a variety of applications where accuracy is not an important issue.

Membranes and piezoresistors based on materials other than silicon have been studied as well. For instance Ref. [15] presents research on a MEMS pressure sensor (intended as touch sensor) using a Cr−Al membrane and indium tin oxide piezoresistors. The gauge factor of this material depends on various fabrication parameters and is of the order of 200.

## 4.4.2   Micromachined Piezoresistive Sensors

The unremitting progress in micromachining technology and the creation of micro-electromechanical devices (MEMS) have great impact on sensor development. Advantages of this technology are:

- all piezoresistors are deposited in one processing step, resulting in almost identical properties;
- the resistors on the membrane can be configured in a bridge;
- the resistors have nearly the same temperature, due to the high thermal conductivity of silicon;
- interface electronics, including further temperature compensation circuits, can be integrated with the sensor bridge on the same substrate;
- the dimensions of the device can be extremely small; they are mainly set by the package size and the mechanical interface.

Sensors based on this technology are now widely spread, but much research is still going on to further improve the (overall) performance. MEMS also have their limitations, and other technologies are also being investigated and further developed to create better or cheaper sensors. There is also a tendency to combine technologies to benefit from the advantages provided by each of them.

### Pressure and Force Sensors

Figure 4.18 shows the basic configurations of a silicon *pressure sensor* and a silicon *force sensor*, both based on piezoresistive silicon.

The sensor carrier or substrate is a silicon chip: a rectangular part cut from the wafer, with thickness about 0.6 mm. The wafer material is lightly doped with positive charge carriers, resulting in p-type silicon. On top of the wafer, a thin layer of n-type silicon is grown, called the epitaxial layer or *epilayer*. The substrate is locally etched away from the bottom up to the epitaxial layer, using selective etching technology. The result is a thin silicon membrane consisting of only the epitaxial layer, the thickness of which is some μm. This membrane acts as a deformable element.

Using standard silicon processing technology, piezoresistive sensors are deposited at positions on the membrane where the deformation is greatest. The gauge factor of silicon is much larger than that of metals; however the temperature

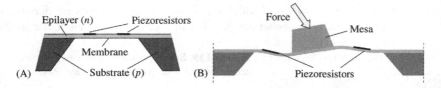

**Figure 4.18** Cross section of (A) a piezoresistive pressure sensor in silicon technology (simplified), (B) shear force sensor (not on scale).

coefficient of the resistivity is also higher. Therefore, silicon strain sensors are invariably configured in a full-bridge arrangement.

The silicon piezoresistors are deposited on the crystal surface, so they are technologically bound to a specific orientation (one of the orientations in Figure 4.16). Once the surface orientation being chosen, the resistors have to be positioned such that maximum sensitivity is obtained. Piezoresistors usually have a meanderlike elongated shape, to obtain directional sensitivity in a restricted area (as with metal strain gauges). To optimize sensitivity, the main axis of the resistor should coincide with the direction of maximal piezoresistivity on the membrane surface.

The membrane deforms under a pressure difference (as in Figure 4.18A) or on an applied force (Figure 4.18B). Special designs allow the measurement of both normal forces and shear forces. Independent measurement of the force components is accomplished by proper positioning of a set of piezoresistors on the membrane (see for instance Ref. [16]).

A common problem with all these types of sensors is the packaging. On one hand, the measurand (e.g. pressure and force) should have good access to the sensitive membrane in order to produce a deformation. On the other hand, the silicon chip must be protected against mechanical damage. Gas pressure sensors are encapsulated in a small box provided with holes that give the gas access to the membrane, at the same time shielding the chip from direct mechanical contact with the outside world. Force sensors require mechanical contact between the membrane and the (solid) object, which is accomplished by a force-transmitting structure, for instance a small steel ball, a small piece of material fixed in the centre of the membrane (a mesa) or a thin overlayer from an elastic material. In all cases this will affect the elastic properties of the membrane. In commercial force sensors the piezoresistors on the micromachined silicon are configured in a half- or full-bridge, to reduce temperature sensitivity.

## Piezoresistive Accelerometers

The same technology is used to construct silicon *accelerometers* and *gyroscopes*. Figure 4.19 shows the basic setup of an integrated accelerometer, consisting of a mass-and-spring element. All mechanical elements are created in silicon, by selective etching.

The seismic mass is an isolated part of the substrate, obtained by selective etching from the bottom side of the chip. The springs are silicon beams made from the epitaxial layer, by etching holes in the membrane. The mass suspends from the substrate by the thin beams, on top of which the piezoresistive elements are deposited. When the structure experiences an acceleration in a direction perpendicular to the

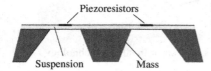

**Figure 4.19** Structure of an integrated silicon accelerometer with piezoresistive elements.

chip surface, the mass will move up or down due to its inertia. The beams bend, and the piezoresistive elements respond to this deformation.

Since the device is sensitive to inertial forces, it can be packed in a hermetically closed housing, making it resistant against mechanical and chemical influences from the environment.

Micromachined *gyroscopes* are constructed likewise. They basically consist of a vibrating mass connected by a thin beam to the sensor base. The micromass vibrates in resonance mode (for instance by an electrostatic drive) and the corresponding bending of the beams is measured in two directions (preferably with four piezoresistors in full-bridge configuration as discussed in Chapter 3). Upon rotation of the structure, the suspension undergoes torsion that is measured by piezoresistors integrated in the suspending beam. When the structure rotates the vibration modes change, resulting in a phase shift of the bridge signals. Examples of such micromachined angular rate sensors with piezoresistive sensors are presented in Refs [17] and [18].

### 4.4.3 Applications of Piezoresistive Sensors

Silicon accelerometers with micromachined mass, spring and integrated piezoresistors feature a high resonance frequency (up to 150 kHz), small dimensions and weight (down to 0.4 g) and low cost. They are available with one, two or three sensitive axes. Mounted at suitable spots on a mechatronic construction they furnish useful information on position, orientation and rotation of movable parts. The technology allows integration of the interface electronics with the sensor body, offering the prospect of very low-cost devices. One of the first silicon accelerometers with a complete interface circuit is described in Ref. [19]. The rectangular seismic mass is made using bulk micromachining and the circuit is realized in CMOS technology.

A first application example is a 3D *force sensor* for biomedical applications [20]. The sensor has a construction similar to the one shown in Figure 4.15 but is realized in silicon technology. Piezoresistors located at proper places on the square silicon frame (2.3 × 2.3 mm) measure three components of the force (0−2 N) applied to a short (one-half mm) protruding mesa. Evidently the sensor needs to be packaged in such a way that the externally applied force is properly transferred to the silicon mesa, at the same time providing an adequate mechanical and chemical protection.

Another application of a piezoresistive sensor is the displacement sensor described in Ref. [21], which is based on a conductive paste, deposited onto a piece of rubber. The high elasticity of the rubber and the paste allows large strain values, up to 40%. Its electrical resistance varies also roughly by this amount. As with all sensors based on piezoresistive elastomers, this sensor also exhibits a significant hysteresis. The sensor is designed for measuring the displacement of a soft actuator.

For the measurement of *body parameters* (e.g. posture, gesture and gait), wearable sensors are being developed. Yarn-based sensors are a good solution for this application since they can easily be integrated into fabrics for clothing.

Suitable sensor materials are piezoresistive polymers, rubbers and carbon as coating material. Fibres from these materials can be interwoven with the textile. Stretching and bending of the textile result in elongation of the fibres, and hence a change in resistance. An application example of such wearable sensing is given in Ref. [22], reporting about such sensors for the measurement of aspiration. The elongation of the fibre can be as large as 23%, resulting in a resistance change of about 300%.

Piezoresistive sensors are also found in *inclinometers*. Such sensors measure the tilt angle, that is the angle with respect to the earth's normal. Mounted on a robot, for instance, the sensor provides important data about its vertical orientation, which is of particular interest for walking (or legged) robots. In Ref. [23] a micromachined inclinometer was proposed, based on silicon piezoresistive sensors. The sensor consists of a micromass suspended on thin beams. Gravity forces the mass to move towards the earth's centre of gravity. The resulting bending of the beams is measured in two directions, by properly positioned integrated piezoresistors. The authors report an average sensitivity of about 0.1 mV per degree over a range of ±70° inclination.

## Tactile Sensors

Piezoresistive elastomers can be used for a variety of sensing tasks in robotics. The pressure sensitivity of the bulk resistance is useful to sense touch (recall the high sensitivity for small forces), to measure gripping force and for tactile sensing. The material is shaped in sheets, which is very convenient for the construction of flat sensors, and in particular for tactile sensors. Piezoresistive elastomers belong to the first *tactile sensors* in robot grippers [24,25]. They still receive much attention from designers of robots intended for human-like capabilities, in particular soft gripping (in horticultural applications, for example).

Most resistive tactile sensors are based on some kind of piezoresistive elastomer and are of the cutaneous type. At least one company offers such pressure-sensitive devices, which can also be used for the construction of tactile sensors [26]. Many researchers have reported on the usefulness of these so-called force-sensitive resistors (FSR) as tactile sensors, for instance in Refs [27–29].

Not only is the bulk resistivity a proper sensing parameter, but it is also possible to utilize the surface resistivity of such materials. The contact resistance between two conductive sheets or between one sheet and a conducting layer changes with pressure, mainly due to an increase of the contact area (Figure 4.20). The resistance–pressure characteristic is similar to that of Figure 4.17.

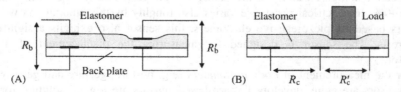

**Figure 4.20** Piezoresistive tactile sensors: (A) bulk mode and (B) contact mode.

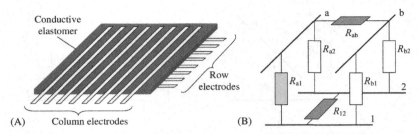

**Figure 4.21** Basic idea of row—column readout of a tactile matrix: (A) row and column electrodes on a piezoresistive elastomer and (B) model with four taxels, showing shunt resistances.

An important aspect of tactile sensors based on resistive sheets is the selection and readout of the individual pressure points (taxels). Most popular is the row—column readout, accomplished by a line grid of highly conductive electrodes on either side of the elastomer, making right angles and thus defining the pressure points of the sensor matrix (Figure 4.21A).

Individual taxels are addressed by selecting the corresponding row and column, for instance by applying a reference voltage on the column electrode and measuring the current through the selected row electrode. By using multiplexers for both the row and the column selection, the whole matrix can be scanned quickly. However due to the continuous nature of the resistivity of the sheet, the selected taxel resistance is shunted by resistances of all other taxels, as can be seen in the model of the tactile sensor given in Figure 4.21B. For instance when selecting taxel a-1, the taxel resistance $R_{a1}$ is shunted by resistances $R_{ab} + R_{b1}$ and $R_{a2} + R_{12}$, resulting in unwanted crosstalk between the taxels. Even when the inter-electrode resistances $R_{ab}$ and $R_{12}$ are large compared to the taxel resistances, the selected taxel resistance is shunted by $R_{a2} + R_{2b} + R_{b1}$. As a consequence when one or more of the taxels a-2, b-2 and b-1 are loaded, the unloaded taxel a-1 is virtually loaded. This phenomenon is denoted by 'phantom images'. In an $n \times m$ matrix many of such phantom images are seen by the selected taxel, an effect that is more pronounced when loading multiple taxels of the tactile device.

Crosstalk and phantom images are reduced by actively guarding the non-selected rows and columns, thus zeroing the potential over all non-selected taxels. The principle is illustrated with the simple $2 \times 2$ matrix of Figure 4.22.

In Figure 4.22A selection of taxel a-1 is performed by connecting a voltage $V_a$ to column electrode **a**, and measuring the resulting current through row electrode **1**, while all other rows and columns remain floating. Obviously since an additional current component $I_2$ flows through the other three taxel resistances, the apparent taxel resistance amounts $V_a/I_1 = R_{a1}//(R_{a2} + R_{2b} + R_{b1})$. In Figure 4.22B the non-selected row electrodes are all connected to ground; the additional current, which is now $I_2'$, flows directly to ground, so the measured resistance is $V_a/I_1 = R_{a1}$, which is just the resistance of the selected taxel. Note that the current through the selected column electrode **a** can be quite large, in particular when many taxels on this

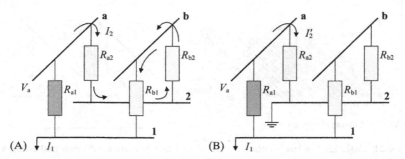

**Figure 4.22** Reduction of crosstalk: (A) original selection mode and (B) improved selection mode.

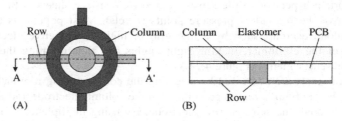

**Figure 4.23** Taxel layout (A) top view, (B) cross section through A–A′.

column are loaded and thus have low resistance values to ground. Although not strictly necessary, grounding of the non-selected column electrodes is preferred too, to prevent possible interference due to the large resistance of non-loaded taxels. Various alternative schemes have been investigated aiming at increased effectiveness of this solution and reduction of electronic complexity, for instance in Refs [30–32].

Glued onto the elastic layer, the conductive electrodes affect significantly the elastic properties of the pressure sensitive material. Therefore the use of anisotropic elastomers has been proposed: this material shows a high conductivity in one direction and a low conductivity in perpendicular direction (see for instance Ref. [33]). Another solution is given in Ref. [32], where the functions of pressure sensitivity and read-out electrodes are combined: the sensor consists of two orthogonally positioned arrays of conductive fibres. The $8 \times 8$ taxels are defined by the crossing points of the fibres. The contact resistance changes with pressure, mainly due to a reduction in contact area.

A completely different approach is to avoid having electrodes on top of the elastic layer, leaving the front side of the tactile sensor free to be accessed by objects. One of the solutions, applied in, for instance, the tactile sensor described in Ref. [34], is based on a double-sided printed circuit board. The $16 \times 16$ taxel sensor responds to the contact resistance between a highly conductive elastomeric sheet and an electrode pattern deposited on the printed circuit board. The layout of one taxel on the PCB is given in Figure 4.23, showing how the crossings between column electrodes and row electrodes can be realized. The pressure sensitive part of each taxel consists of the two crossing areas between row and column.

An often overlooked aspect is mechanical crosstalk between adjacent taxels, due to the stiffness of the elastic layer or an additional protective layer. The sheet material acts as a spatial low-pass filter in the transduction from applied force at the top to the measurement side at the bottom of the sheet [35]. Hence the spatial resolution is not equal to the pitch of the electrodes but might be substantially lower.

Many attempts have been undertaken to increase the spatial resolution by applying silicon technology. One of the first results of this approach is described in Ref. [36], where the conductive elastomer is mounted on top of a silicon wafer, provided with electrodes and electronic circuitry to measure the (pressure sensitive) contact resistance. An all-silicon tactile sensor lacks the piezoresistive elastomer: it consists of an array or matrix of pressure sensors, similar to the devices shown in Section 4.4. A few examples of such completely integrated resistive tactile sensors are given here. The first one [37] consists of $32 \times 32$ piezoresistive bridges integrated on a single chip measuring $10 \times 10$ mm. The chip contains signal processing circuitry as well. To prevent damage when pressed, the fragile sensor surface is covered with an elastic protective layer. A further example [38] concerns a $4 \times 8$ tactile sensor specifically designed for a large force range (up to 50 N). Finally in Ref. [39] a $4 \times 4$ tactile sensor with a different architecture is presented. The diaphragm layer is deposited over a matrix of small cavities at the top of the wafer, building up a matrix of membranes (without etching from the back side). Each element contains one piezoresistor. These resistors are connected sequentially to a Wheatstone bridge by controlling a set of on-chip CMOS switches.

A last aspect to be discussed here is the measurement of shear forces using a tactile sensor. Most of the devices discussed so far are only sensitive to normal forces, that is, a force pointing perpendicular to the sensor surface. A shear sensor should also be sensitive to tangential forces. Few resistive tactile sensors thus far offer this ability [40,41].

# 4.5  Magnetoresistive Sensors

## 4.5.1  Magnetoresistivity

Some ferromagnetic alloys show anisotropic resistivity, which means that the resistivity depends on the direction of the current through the material. This property arises from the interaction between the charge carriers and the magnetic moments in the material. As a consequence the resistivity depends on the magnetization of the material. This effect, called anisotropic magnetoresistivity (AMR), was discovered by W. Thomson in 1856, and is manifest in materials as iron, nickel and alloys of those metals (Permalloy). The effect is rather small and requires very thin layers to be useful for sensing applications. In 1970 the first magnetoresistive head for magnetic recording appeared, but it was only in 1985 that a commercial tape drive with the magnetoresistive head was introduced [42].

To describe anisotropic resistivity, Eq. (4.1) is rewritten into

$$\begin{pmatrix} E_x \\ E_y \end{pmatrix} = \begin{pmatrix} \rho_x & 0 \\ 0 & \rho_y \end{pmatrix} \begin{pmatrix} J_x \\ J_y \end{pmatrix} \tag{4.42}$$

Here $\rho_x$ and $\rho_y$ are the components of the resistivity in $x$- and $y$-directions. In general the resistivity is large in the direction of the magnetization and small in the perpendicular direction. Assume the magnetization is in the $x$-direction. If $\phi$ is the angle between the current density vector $J$ and the magnetization vector $M$, the resistivity can be described by

$$\rho(\phi) = \rho_x + \rho_y \cos 2\phi \tag{4.43}$$

or in a different way:

$$\rho(\phi) = \rho_0(1 + \beta \cos^2 \phi) \tag{4.44}$$

The material parameters $\rho_0$ and $\beta$ determine the sensitivity of the device. The angle $\phi$ runs from 0° (where the resistivity is maximal) to 90° (minimum resistivity). As an example, for a 50 nm thick layer of Permalloy, $\rho_x$ is about $2 \times 10^{-7}$ $\Omega$ m and $\rho_y$ about $3 \times 10^{-9}$ $\Omega$ m [43]. The sensitivity of an AMR material is usually expressed as

$$\frac{\Delta\rho}{\rho} = \frac{\rho_x - \rho_y}{\rho_x + \rho_y} \tag{4.45}$$

so the maximum relative change in resistivity (at room temperature) amounts to only a few percent. An external magnetic field in the $y$-direction will change the direction of the magnetization. So when the current remains flowing along the $z$-axis, the angle $\phi$ changes, and hence the resistivity of the layer (Figure 4.24).

The magnetization vector $M$ has a preferred direction, set by the shape of the ferromagnetic body. For a film-shaped body (length $l$, width $w$ and thickness $t$), the preferred magnetization is along the $z$-direction, with a strength that can be expressed by an effective magnetic field $- H_0 = (t/w)M_s$. The relative change in

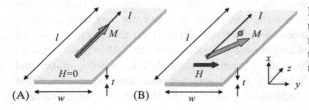

Figure 4.24 Magnetoresistivity: the angle $\phi$ is affected by a magnetic field $H_y$ perpendicular to $I$; this changes the resistivity of the device in the direction of $z$.

resistance can now be expressed in terms of $H_0$ and the external field $H_y$, according to the equations:

$$\frac{\Delta R}{(\Delta R)_{max}} = 1 - \left(\frac{H_y}{H_0}\right)^2 \quad (H_y < H_0) \tag{4.46}$$

$$\frac{\Delta R}{(\Delta R)_{max}} = 0 \quad (H_y > H_0) \tag{4.47}$$

So the resistance decreases with the external magnetic field in both positive and negative $y$-direction. For sensor purposes, this symmetric sensitivity is very unlucky: it is strongly non-linear and multi-valued. To improve sensor behaviour, an angle offset is required to shift the zero to, for instance, $-45°$. One way to achieve this is the introduction of an extra magnetic field. A more elegant way is to offset the direction of the measurement current $I$. This is accomplished by a special arrangement of the electrodes, called the 'Barber pole' construction [44], after the similarity to the red-white spinning poles at barber shops (Figure 4.25). Due to the low resistance of the electrodes (deposited as thin layers on the ferromagnetic sheet) the current is forced to flow through the device under an angle of approximately $45°$. At zero field the angle between the current $I$ and the magnetization $M$ is $45°$, corresponding to a resistivity of about half the minimum value. When a magnetic field is applied in positive $y$-direction, the angle decreases, so the resistivity increases and vice versa.

Commercial barber pole magnetic sensors contain four meanderlike structures as in Figure 4.25, arranged in a full bridge to compensate for common mode errors. Typical data: measurement range $0-10^4$ A/m (when specified in gauss, the range refers to magnetic induction $B$ rather than magnetic field strength $H$ and a conversion factor according to $B = \mu_0\mu_r H$ must be applied; see Chapter 6 on inductive and magnetic sensors); temperature range $-40°$C up to $150°$C, non-linearity error 1% at FS sweep. Resolution and offset are mainly determined by the interface electronics; the resolution is good enough to measure the earth's magnetic field, allowing compass heading (mobile robots) and attitude sensing (walking robots). Three-axis devices are available as well: they actually consist of three orthogonally positioned magnetosensors in one package.

Around 1980 magnetoresistive devices were realized that exhibit a much larger sensitivity, according to the so-called *giant magnetic resistance* effect (GMR) and

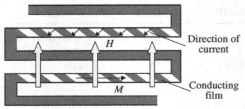

**Figure 4.25** Introducing angle offset by forcing the current to follow a 45° path (barber pole).

Direction of current

Conducting film

the *giant magnetic impedance* effect (GMI). A GMR device consists of a multi-layer of ferromagnetic thin films (typically 1−10 nm thick), sandwiched between conductive but non-ferromagnetic interlayers. The physical principle differs fundamentally from the AMR sensors. In the absence of an external field, the magnetization in the successive layers of a GMR structure is anti-parallel. In this state the resistance is maximum. At a sufficiently strong external magnetic field the magnetizations are in parallel, corresponding to a minimum resistance. The GMI sensor employs the skin effect, the phenomena in which conduction at high frequencies is restricted to a thin layer at the surface of the material. The axial impedance of an amorphous wire may change as much as 50% due to an applied magnetic field [45]. Present research is aiming at higher sensitivity, better thermal stability and wider frequency range (see for instance Refs [46,47]).

Commercial GMR sensors became available around 1995. Current research focuses on even higher sensitivity: in layered structures of amorphous ribbons made up of particular alloys, GMR ratios of up to 2400% have been reported [48]. Since GMR reading heads for magnetic discs are gradually replacing the older thin film AMR heads, there is some confusion about the terminology: GMR is now often called just MR or AMR.

In 1993 another group of materials, manganese-based perovskites, was discovered; these exhibit an even stronger magnetoresistive effect, at least at low temperatures and high magnetic fields. For this reason the effect is called colossal magnetoresistivity (CMR) [49]. Unfortunately at room temperature the effect is much smaller, but nevertheless can be employed for sensor purposes [50].

Table 4.6 presents some specifications of these three types of magnetoresistive sensors. The resistors in the AMR sensor are configured as two Wheatstone bridges, allowing the measurement of the magnetic field in two orthogonal directions. The selected GMR sensor uses a Wheatstone bridge as well. Its output is a differential voltage, and a differential amplifier is required to convert this to a single-sided output. The GMI sensor in Table 4.6 consists of an amorphous metal wire with a pickup coil wrapped around the wire. This wire is driven by an AC current of 200 kHz. The frequency range of this sensor is, however, much lower compared to GMR and AMR: DC to 1 kHz. The sensor package contains three GMI sensing elements; the output signals are buffered in a sample-hold circuit, amplified and multiplexed.

**Table 4.6** Comparison of Magnetoresistive Sensors

| Type[a]          | AMR  | GMR  | GMI  | Units    |
|------------------|------|------|------|----------|
| Sensitivity      | 5    | 75   | 240  | mV/Gauss |
| Linear range     | 6    | 2    | >2   | Gauss    |
| Frequency range  | 5000 | 1000 | 1    | kHz      |
| Number of axes   | 2    | 1    | 3    | −        |

[a]AMR: HMC1052 (Honeywell); GMR: AAH00202 (NVE Corp.); GMI: AGMI302 (Aichi Corp.).

### 4.5.2   Applications of Magnetoresistive Sensors

A magnetoresistive sensor measures essentially magnetic field strength. Therefore many applications are feasible, for instance as a compass (for mobile robots). Combined with a magnetic source (e.g. permanent magnet, active coil and earth magnetic field), magnetic field sensors are useful devices for the measurement of a variety of quantities in the mechanical domain.

Various possibilities for linear and angular position detection using AMR sensors are presented in Ref. [51]. In Ref. [52] AMR sensors are applied for vehicle detection: the sensors measure the change in earth magnetic field when a vehicle is passing the sensor. The system is used to monitor the traffic on highways.

GMR (and GMI) sensors too have been extensively tested for measuring a variety of mechanical quantities, for instance contact-free position [53], angle and speed [54], stress (in steel bars, [55]) and torque [56]. They appear also useful for non-destructive testing [57] and, because of their small size, for biomedical applications; see for instance Refs [58] and [59].

Properties of commercial magnetic sensors and applications based on the combination of a magnetic source and magnetic sensors will be discussed further in the chapter on inductive and magnetic sensors.

## 4.6   Thermoresistive Sensors

### 4.6.1   Thermoresistivity

The resistivity of a conductive material depends on the concentration of free-charge carriers and their mobility. The mobility is a parameter that accounts for the ability of charge carriers to move more or less freely throughout the atom lattice; their movement is constantly hampered by collisions. Both concentration and mobility vary with temperature, at a rate that depends strongly on the material.

In intrinsic (or pure) *semiconductors*, the electrons are bound quite strongly to their atoms; only a very few have enough energy (at room temperature) to move freely. At increasing temperature more electrons will gain sufficient energy to be freed from their atom, so the concentration of free-charge carriers increases with increasing temperature. As the temperature has much less effect on the mobility of the charge carriers, the resistivity of a semiconductor decreases with increasing temperature: its resistance has a negative temperature coefficient.

In *metals* all available charge carriers can move freely throughout the lattice, even at room temperature. Increasing the temperature will not affect the concentration. However at elevated temperatures the lattice vibrations become stronger, increasing the possibility that the electrons will collide and hamper free movement throughout the material. Hence the resistivity of a metal increases at higher temperature: their resistivity has a positive temperature coefficient (PTC).

The temperature coefficient of the resistivity is used to construct temperature sensors. Both metals and semiconductors are applied. They are called (metal) resistance thermometers and thermistors, respectively.

## 4.6.2 Resistance Thermometer

The construction of a high-quality resistance thermometer requires a material (metal) with a resistivity temperature coefficient that is stable and reproducible over a wide temperature range. By far the best material is platinum, due to a number of favourable properties. Platinum has a high melting point (1769°C), is chemically very stable, resistant against oxidation, and available with high purity. Platinum resistance thermometers are used as international temperature standard for temperatures between the triple point of the hydrogen equilibrium (13.8033 K) and the freezing point of silver (+961.78°C) but can be used up to 1000°C.

A platinum thermometer has a high linearity. Its temperature characteristic is given by

$$R(T) = R(0)\{1 + aT + bT^2 + cT^3 + dT^4 + \cdots\} \qquad (4.48)$$

with $R(0)$ the resistance at 0°C. The values of the coefficients $R(0)$, $a$, $b$, ... are specified according to various standards and temperature ranges. As an example the resistance of a Pt100 is characterized, according to the DIN-IEC 751 standard, as

$R(0) = 100.00 \ \Omega$
$a = 3.90802 \times 10^{-3} \ \mathrm{K}^{-1}$
$b = 5.8020 \times 10^{-7} \ \mathrm{K}^{-2}$
$c = 4.2735 \times 10^{-10} \ \mathrm{K}^{-3}$

for a temperature range $-200°C < T < 0°C$; for temperatures from 0°C to 850°C the parameters $a$ and $b$ are the same, and $c = 0$. The tolerances are also specified in the standard definition. For instance the resistance value of a 'class A' Pt100 temperature sensor at 0°C is 0.06%, which corresponds to a temperature tolerance of 0.15 K. At the upper end of the range (850°C) the tolerance is 0.14% or 1.85 K. Class B sensors have wider specified tolerances, for instance 0.3 K at 0°C.

There also exist resistance temperature sensors with other values than 100 $\Omega$ at 0°C, for instance the Pt1000 with a resistance value of 1000 $\Omega$ at 0°C. However, the Pt100 is most popular for industrial applications.

The sensitivity of a Pt100 is approximately 0.4%/K or 0.39 $\Omega$/K. The parameters $b$ and $c$ account for the non-linear relationship between temperature and resistance. Within a limited temperature range the non-linearity might be neglected, when a high accuracy is not required.

The tolerances and non-linearity errors discussed so far relate to the sensor only. Additional errors may arise due to:

- self-heating
- errors of the interface circuit
- connecting cable resistance.

Self-heating should be minimized by reducing the current through the sensor and a low thermal resistance to the environment. A measurement current $I$

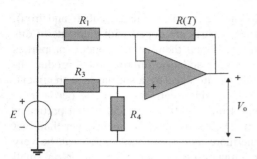

**Figure 4.26** Interface circuit using a single operational amplifier.

introduces heat dissipation of $I^2 R(T)$ in the sensor. For example at 0°C and 1 mA current the dissipation is 0.1 mW. In order to limit the error due to self-heating to 0.1°C, the sensor should be mounted in such a way that the thermal resistance is less than $10^3$ K/W.

Special integrated circuits are available that facilitate the interface between resistive sensors and a microprocessor or microcontroller. They are based on either single-, two-, or three-point calibration techniques (see Section 3.2.6). Present interface modules have high resolution and accuracy and are suitable for multi-sensor applications as well. For instance the interface circuit for a Pt1000 resistive temperature sensor presented in Ref. [60] has a resolution of 0.1 $\Omega$, a relative systematic error less than 0.01%, and a temperature coefficient of about $1.5 \times 10^{-7}$ K$^{-1}$ over the range $10-50°$.

A simple analogue interface circuit for a resistance thermometer is shown in Figure 4.26.

The output voltage of this circuit (assuming ideal properties of the operational amplifier and using the equations from Appendix C) is:

$$V_{o} = E \frac{R_3}{R_3 + R_4} \left( \frac{R_4}{R_3} - \frac{R(T)}{R_1} \right) \tag{4.49}$$

With the resistors $R_3$ and $R_4$ the output can be adjusted to zero at an arbitrary temperature (for instance 0°C). The resistor $R_1$ sets the sensitivity of the circuit. Obviously the accuracy of this interface is set by the tolerances of the voltage source $E$ and the resistances $R_1$, $R_3$ and $R_4$, which all should be chosen in accordance to the required performance of the measurement system.

Resistive sensors can also be connected in a bridge configuration, in order to reduce non-linearity and unwanted common mode interferences. The effect of cable resistance might be substantial, in particular when the sensor is located far from the measuring circuit. Its influence can be compensated for (see Chapter 3).

### 4.6.3 Thermistors

A thermistor (contraction of the words thermally sensitive resistor) is a resistive temperature sensor built up from ceramics. Commonly used materials are sintered oxides from the iron group (e.g. chromium, manganese, nickel, cobalt and iron); the most popular material is $Mn_3O_4$. These oxides are doped with elements of different valence to obtain a lower resistivity, giving them semiconductor properties (mainly of the p-type). Several other oxides are added to improve the reproducibility. To obtain a stable sensitivity, thermistors are aged by a special heat treatment. A typical value of the drift in resistance after aging treatment is +0.1% per year.

Thermistors cover a temperature range from $-100°C$ to $+350°C$, but particular types go down to 2 K (ruthenium oxide). Their sensitivity is much larger than that of metal resistance thermometers. Furthermore the size of thermistors can be very small, so that they are applicable for temperature measurements in or on small objects. Compared to metal resistance thermometers, a thermistor is less stable in time and shows a much larger non-linearity.

The resistance of most semiconductors has a negative temperature coefficient. This also applies to thermistors. That is why a thermistor is also called an NTC-thermistor or just an NTC, although also thermistors with a PTC do exist.

The temperature dependence of an NTC is given by

$$R(T) = R(T_0)e^{B((1/T) - (1/T_0))} \qquad (4.50)$$

where $R(T_0)$ is the resistance at a reference temperature $T_0$ (usually 25°C), $B$ is a constant that depends on the type of NTC.

From this equation it follows for the temperature coefficient (or sensitivity) of an NTC:

$$S = \frac{1}{R}\frac{dR}{R} = -\frac{B}{T^2} \quad (K^{-1}) \qquad (4.51)$$

Further it should be realized that due to the non-linearity of the temperature characteristic, the sensitivity varies with temperature (as also appears from the expression given above). The parameter $B$ is in the order of $2000-5000$ K. For instance at $B = 3600$ K and room temperature ($T = 300$ K) the sensitivity amounts to $-4\%$ per K. At 350 K the sensitivity has dropped to $-3\%$ per K.

The resistance value of a thermistor at the reference temperature depends on the material, the doping type and concentration, the device dimension and geometry. Values of $R_{25}$ (the resistance at 25°C) range from a few ohms to some 100 k$\Omega$. The accuracy of thermistors is in the order of $\pm1°C$ for standard types and $\pm0.2°C$ for high precision types.

Like metal resistive sensors, NTCs too suffer from self-heating. The current through the sensor should be kept low in order to avoid large errors due to self-heating. Generally a current less than 0.1 mA is acceptable in most cases (this is called zero-power mode). The effect of self-heating also depends on the encapsulation material and on the dimensions and shape of the sensor. In particular the condition of the environment (gas or liquid, still or flowing) determines the self-heating. A measure for this effect is the thermal dissipation constant, a figure that is provided by the manufacturer for various sensor types and various environmental conditions. It ranges from about 0.5 mW/K to some 10 mW/K.

The basic shape of a thermistor is a chip: a slice of ceramic material with metallized surfaces for the electrical contacts. The device is encapsulated in a coating of thermally conductive epoxy to ensure mechanical protection and a low thermal resistance to the measurement object. There are various shapes available: disc, glass bead, probe, surface mount (SMD). The smallest devices have a size down to 1 mm.

To measure the resistance value of an NTC, the device can be connected in a single-element bridge (one NTC and three fixed resistances), or in a half- or full-bridge configuration (with two respectively four NTCs). This type of interfacing is used when small temperature changes or temperature differences have to be measured.

Thermistors are small-sized and low-cost temperature sensors, and useful in applications where accuracy is not a critical design parameter. They can be found in all kind of systems, for the purpose of temperature monitoring and control, for temperature compensation in electrical circuits (e.g. amplifiers and oscillators) and many other uses.

Besides NTC thermistors there are also PTC thermistors. The base material is barium or strontium titanate, made semiconductive by adding particular impurities. The temperature effect differs essentially from that of a thermistor. PTCs have a PTC over a rather restricted temperature range. Altogether they cover a temperature range from 60°C to 180°C. Within the range of a PTC, the characteristic is approximated by

$$R(T) = R(T_0)e^{BT}, \quad T_1 < T < T_2 \tag{4.52}$$

The sensitivity in that range is $B$ ($K^{-1}$) and can be as high as 60% per K. PTCs are rarely used for temperature measurements, because of lack of reproducibility and the restricted temperature range of the individual device. They are mainly applied as safety components to prevent overheating at short circuits or overload.

## 4.7  Optoresistive Sensors

The resistivity of some materials, for instance cadmium sulphide (CdS) and cadmium selenide (CdSe), depends on the intensity of incident light. This is called the photo-resistive effect. A resistor made up of such a material is the LDR (photoresistor). In the absence of light the concentration of free-charge carriers is low, hence the resistance of the LDR is high. When light falls on the material, free-charge carriers are generated; the concentration increases and thus the resistance decreases with increasing intensity. The resistance of the LDR varies as

$$R(E_d) = a \cdot E_d^{-b} \tag{4.53}$$

where $E_d$ is the incident power per unit area, $a$ and $b$ being constants that depend on the material and the shape of the resistor.

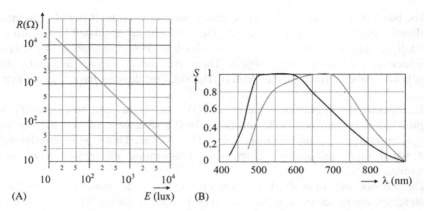

**Figure 4.27** Characteristics of an LDR: (A) sensitivity characteristic and (B) spectral sensitivity.

The sensitivity of an LDR depends on the wavelength of the light, so it is expressed in terms of spectral sensitivity ($\Delta R/E_d$ per unit of wavelength). Figure 4.27B shows the relative spectral sensitivity for two types of CdS resistors, one typical and one with high sensitivity for red light. The sensitivity is normalized to the maximum sensitivity within the bandwidth. Apparently below 400 nm and above 850 nm the LDR is not sensitive. The spectral sensitivity of this material matches rather well that of the human eye. Therefore the sensitivity is also expressed in terms of lux. For other wavelengths other materials are used: in the near-IR region (1−3 μm) PbS and PbSe, and in the medium and far IR (up to 1000 μm), InSb, InAs and many other alloys. Figure 4.27A shows the resistance−illuminance characteristic of a typical CdS LDR. Note the strongly non-linear response of the device.

Even in complete darkness, the resistance appears to be finite; this is the dark resistance of the LDR, which can be more than 10 MΩ. The light resistance is usually defined as the resistance at an intensity of 1000 lux; it may vary from 30 to 300 Ω for different types. Photoresistors change their resistance value rather slowly: the response time from dark to light is about 10 ms; from light to dark, the resistance varies only about 200 kΩ/s, resulting in a response time of about 1.5 s.

LDRs are used in situations where accuracy is not an important issue (dark-light detection only), for example in alarm systems, in the exposure-meter circuit of automatic cameras, and in detectors for switching streetlights on and off.

# References and Literature

## References to Cited Literature

[1] ISA-S37.12: Specifications and tests for potentiometric displacement transducers; Instrument Society of America; Pittsburgh, USA; 1977; ISBN 87664-359-4.

[2]  X. Li, G.C.M. Meijer: A novel smart resistive-capacitive position sensor, IEEE Trans. I&M, 44(1) (1995), 768−770.

[3]  I. Yamano, K. Takemura, T. Maeno: Development of a robot finger for five-fingered hand using ultrasonic motors, Proceedings of the 2003 International Conference on Intelligent Robots and Systems, Las Vegas, Nevada, October 2003, pp. 2648−2653.

[4]  R. Roduit, P.-A. Besse, J.-P. Micallef: Flexible angular sensor, IEEE Trans. Instrum. Meas., 47(4) (1998), 1020−1022.

[5]  H. Jung, C.J. Kim, S.H. Kong: An optimized MEMS-based electrolytic tilt sensor, Sens. Actuators A, 139 (2007), 23−30.

[6]  J. Vaughan: Application of B&K equipment to strain measurements; Brüel & Kjaer; Nærum, Denmark; 1975; ISBN 87-87355-08-6.

[7]  H J.W. Dally, W.F. Riley, K.G. McConnell: Instrumentation for engineering measurements; Wiley & Sons, New York, NY, 1984; ISBN 0-471-04548-9.

[8]  H. Rabah, S. Poussier, S. Weber: Toward a generic on chip conditioning system for strain gage sensors, Measurement, 39 (2006), 320−327.

[9]  J.J. Park, G.S. Kim: Development of the 6-axis force/moment sensor for an intelligent robot's gripper, Sens. Actuators A, 118 (2005), 127−134.

[10]  Lu-Ping Chao, Kuen-Tzong Chen: Shape optimal design and force sensitivity evaluation of six-axis force sensors, Sens. Actuators A, 63 (1997), 105−112.

[11]  Gab-Soon Kim, Dea-Im Kang, Se-Hun Ree: Design and fabrication of a six-component force/moment sensor, Sens. Actuators A, 77 (1999), 209−220.

[12]  S. Middelhoek, S.A. Audet: Silicon sensors; Academic Press, London, 1989; ISBN 0-12-495051-5.

[13]  Measurement data from student projects, unpublished, University of Twente, The Netherlands, 1997−1998.

[14]  M. Hussain, Y.-H. Choa, K. Niihara: Conductive rubber materials for pressure sensors, J. Mater. Sci. Lett., 20 (2001), 525−527.

[15]  A. Wisitsoraat, V. Patthanasetakul, T. Lomas, A. Tuantranont: Low cost thin film based piezoresistive MEMS tactile sensor, Sens. Actuators A, 139 (2007), 17−22.

[16]  L. Wang, D.J Beebe: A silicon-based shear force sensor: development and characterization, Sens. Actuators A, 84 (2000), 33−44.

[17]  S. Sassen, R. Voss, J. Schalk, E. Stenzel, T. Gleissner, R. Gruenberger, F. Neubauer, W. Ficker, W. Kupke, K. Bauer, M. Rose: Tuning fork silicon angular rate sensor with enhanced performance for automotive applications, Sens. Actuators A, 83 (2000), 80−84.

[18]  H. Yang, M. Bao, H. Yin, S. Shen: A novel bulk micromachined gyroscope based on a rectangular beam-mass structure, Sens. Actuators A, 96 (2002), 145−151.

[19]  H. Seidel, U. Fritsch, R. Gottinger, J. Schalk: A piezoresistive silicon accelerometer with monolithically integrated CMOS circuitry, Proceedings of the Eurosensors IX Conference on Solid State Sensors and Actuators Stockholm, Sweden; June 1995, pp. 597−600.

[20]  L. Beccai, S. Roccella, A. Arena, F. Valvo, P. Valdastri, A. Menciassi, M. Chiara Carrozza, P. Dario: Design and fabrication of a hybrid silicon three-axial force sensor for biomechanical applications, Sens. Actuators A, 120 (2005), 370−382.

[21]  K. Kure, T. Kanda, K. Suzumori, S. Wakimoto: Flexible displacement sensor using injected conductive paste, Sens. Actuators A, 143 (2008), 272−278.

[22]  C.T. Huang, C.L. Shen, C.F. Tang, S.H. Changa: A wearable yarn-based piezo-resistive sensor, Sens. Actuators A, 141 (2008), 396−403.

[23] U. Mescheder, S. Majer: Micromechanical inclinometer, Proceedings of the Eurosensors X Conference on Solid State Sensors and Actuators; Leuven, Belgium; September 1996, pp. 1133−1136.

[24] W.E. Snyder, J. St.Clair: Conductive elastomers as sensor for industrial parts handling equipment, IEEE Trans. Instrum. Meas., IM-27(1) (1978), 94−99.

[25] M.H.E. Larcombe: Tactile perception for robot devices, First Conference on Industrial Robot Technology, Nottingham, UK, 27-29 March 1973, pp. R16-191 to R16-196.

[26] M. Witte, H. Gu: Force and position sensing resistors: an emerging technology, Company information Interlink Electronics Europe, Echternach, Luxembourg.

[27] E.M. Petriu, W.S. McMath, S.S.K. Yeung, N. Trif: Active tactile perception of object surface geometric profiles, IEEE Trans. Instrum. Meas., 41(1) (1992), 87−92.

[28] T.H. Speeter: A tactile sensing system for robotic manipulation, Int. J. Rob. Res., 9(2) (1990), 25−36.

[29] H. Liu, P. Meusel, G. Hirzinger: A tactile sensing system for the DLR three-finger robot hand, ISMCR '95, Proceedings, 1995, pp. 91−96.

[30] H. van Brussel, H. Belien: A high resolution tactile sensor for part recognition, Proceedings 6th International Conference on Robot Vision and Sensory Controls, 1986, pp. 49−59.

[31] W.D. Hillis: A high-resolution imaging touch sensor, Int. J. Rob. Res., 1(2) (1982), 33−44.

[32] B.E. Robertson, A.J. Walkden: Tactile sensor system for robotics, Proceedings 3rd International Conference on Robot Vision and Sensory Controls, November 1983, pp. 572−577.

[33] L.H. Chen, S. Jin, T.H. Tiefel: Tactile shear sensing using anisotropically conductive polymer, Appl. Phys. Lett., 62(19) (1993), 2440−2442.

[34] E.G.M. Holweg: Autonomous control in dexterous gripping, PhD thesis, Delft University of Technology, 1996, ISBN 90.9009962-X.

[35] M. Shimojo: Mechanical filtering effect of elastic cover for tactile sensor, IEEE Trans. Rob. Autom., 13(1) (1997), 128−132.

[36] M.H. Raibert, J.E. Tanner: Design and implementation of a VLSI tactile sensing computer, Int. J. Rob. Res., 1(3) (1982), 3−18.

[37] S. Sugiyama, K. Kawahata, M. Yoneda, I. Igarashi: Tactile image detection using a 1k-element silicon pressure sensor array, Sens. Actuators A, 23 (1990), 397−400.

[38] T. Mei, W.J. Li, Y. Ge, Y. Chen, L. Ni, M.H. Chan: An integrated MEMS three-dimensional tactile sensor with large force range, Sens. Actuators A, 80 (2000), 155−162.

[39] L. Liu, X. Zheng, Z. Li: An array tactile sensor with piezoresistive single-crystal silicon diaphragm, Sens. Actuators A, 32 (1993), 193−196.

[40] F. Zee, E.G.M. Holweg, P.P.L. Regtien, W. Jongkind: Tactile sensor and method for determining a shear force and slip with such a tactile sensor, Patent WO97/40339, 1997.

[41] C.-C. Wen, W. Fang: Tuning the sensing range and sensitivity of three axes tactile sensors using the polymer composite membrane, Sens. Actuators A, 145−146 (2008), 14−22.

[42] F.B. Shelledy, J.L. Nix: Magnetoresistive heads for magnetic tape and disk recording, IEEE Trans. Magn., 28(5) (1992), 2283−2288.

[43] C.J.M. Eijkel, J.H.J. Fluitman, H. Leeuwis, D.J.M. van Mierlo: Contactless angle detector for control applications, Journal A, 32 (1991), 43−52.

[44] K.E. Kuijk, W.J. van Gestel, F.W Gorter: The barber pole, a linear magnetoresistive head, IEEE Trans. Magn., 11 (1975), 1215.

[45] D. Atkinson, P.T. Squire, M.G. Maylin, J. Gore: An integrating magnetic sensor based on the giant magneto-impedance effect, Sens. Actuators A, 81 (2000), 82−85.

[46] R.M. Kuźmiński, K. Nesteruk, H.K. Lachowicz: Magnetic field meter based on giant magnetoimpedance effect, Sens. Actuators A, 141 (2008), 68−75.

[47] P. Jantaratana, C. Sirisathitkul: Giant magnetoimpedance in silicon steels, J. Magn. Magn. Mater., 281 (2004), 399−404.

[48] F. Amalou, M.A.M. Gijs: Giant magnetoimpedance in trilayer structures of patterned magnetic amorphous ribbons, Appl. Phys. Lett., 81(8) (2002), 1654−1656.

[49] C.L. Chien: Standing out from the giants and colossi, Phys. World, (2000), 24−26.

[50] O.J. González, E. Castaño, J.C. Castellano, F.J. Gracia: Magnetic position sensor based on nanocrystalline colossal magnetoresistances, Sens. Actuators A, 91 (2001), 137−143.

[51] D.J. Adelerhof, W. Geven: New position detectors based on AMR sensors, Sens. Actuators A, 85 (2000), 48−53.

[52] M.H. Kang, B.W. Choi, K.C. Koh, J.H. Lee, G.T. Park: Experimental study of a vehicle detector with an AMR sensor, Sens. Actuators A, 118 (2005), 278−284.

[53] G. Rieger, K. Ludwig, J. Hauch, W. Clemens: GMR sensors for contactless position detection, Sens. Actuators A, 91 (2001), 7−11.

[54] C. Giebeler, D.J. Adelerhof, A.E.T. Kuiper, J.B.A. van Zon, D. Oelgeschläger, G. Schulz: Robust GMR sensors for angle detection and rotation speed sensing, Sens. Actuators A, 91 (2001), 16−20.

[55] W. Ricken, J. Liu, W.-J. Becker: GMR and eddy current sensor in use of stress measurement, Sens. Actuators A, 91 (2001), 42−45.

[56] H. Wakiwaka, M. Mitamura: New magnetostrictive type torque sensor for steering shaft, Sens. Actuators A, 91 (2001), 103−106.

[57] W. Sharatchandra Singh, B.P.C. Rao, S. Vaidyanathan, T. Jayakumar, B. Raj: Detection of leakage magnetic flux from near-side and far-side defects in carbon steel plates using a giant magneto-resistive sensor, Meas. Sci. Technol., 19 (2008) doi:10.1088/0957-0233/19/1/01559.

[58] K. Totsu, Y. Haga, M. Esashi: Three-axis magneto-impedance effect sensor system for detecting position and orientation of catheter tip, Sens. Actuators A, 111 (2004), 304−309.

[59] D.K. Wood, K.K. Ni, D.R. Schmidt, A.N. Cleland: Submicron giant magnetoresistive sensors for biological applications, Sens. Actuators A, 120 (2005), 4−6.

[60] F. Reverter, J. Jordana, M. Gasulla, R. Pallàs-Areny: Accuracy and resolution of direct resistive sensor-to-microcontroller interfaces, Sens. Actuators A, 121 (2005), 78−87.

# Literature for Further Reading

## Books and Reviews on Magnetoresistive Sensors and Applications

[1] P. Ripka: Magnetic sensors and magnetometers, Artech House Publ. 2000; ISBN 1−580533-057-5.

[2] K.-M.H. Lenssen, D.J. Adelerhof, H.J. Gassen, A.E.T. Kuiper, G.H.J. Somers, J.B.A.D. van Zon: Robust giant magnetoresistance sensors, Sens. Actuators A, 85 (2000), 1−8.

[3]  M.J. Caruso, L.S. Withanawasam: Vehicle detection and compass applications using AMR magnetic sensors; Company information Honeywell, Plymouth, MN, 1999.

[4]  J.E. Lenz: A review of magnetic sensors, Proc. IEEE, 78(6) (1990), 973–989.

## Books and Review Articles on Tactile Sensors

[5]  H.R. Nicholls (ed.): Advanced tactile sensing for robotics, World Scientific, 1992; ISBN 981-02-0870-7.

[6]  R.A. Russell: Robot tactile sensing, 1990; ISBN 0–13-781592-1.

[7]  A. Pugh (ed.): Robot sensors, Vol. 2: Tactile and non-vision, 1986; ISBN 0–948507-02-0.

[8]  L.D Harmon: Automated tactile sensing, Int. J. Rob. Res., 1(2) (1982), 3–32.

# 5 Capacitive Sensors

Capacitive sensors for displacement and force measurements have a number of advantages. A capacitor consists of a pair of conductors; since no other materials are involved, capacitive sensors are very robust and stable and applicable at high temperatures and in harsh environments. The dimensions of capacitive sensors may vary from extremely small (in MEMS) up to very large (several metres). The theoretical relation between displacement and capacitance is governed by a simple expression, which in practice can be approximated with high accuracy, resulting in a very high linearity. Using special constructions, the measurement range of capacitive sensors can be extended almost without limit while maintaining the intrinsic accuracy. Moreover, because of the analogue nature of the capacitive principle, the sensors have excellent resolution.

This chapter starts with resuming the notions capacitance and permittivity. Next, we discuss basic configurations for capacitive sensors. The capacitance value changes with variation in the geometry. We will focus in particular on linear and angular displacement sensors and on force sensors. Next, integrated silicon capacitive sensors are briefly reviewed and some interface circuits are given to measure small capacitance changes.

## 5.1 Capacitance and Permittivity

The capacitance (or capacity) of an isolated conducting body is defined as $Q = C \cdot V$, where $Q$ is the charge on the conductor and $V$ the potential (relative to 'infinity', where the potential is zero by definition). To put it differently, when a charge $Q$ from infinite distance is transferred to the conductor, its potential becomes $V = Q/C$.

In practice we have a set of conductors instead of just one conductor. When a charge $Q$ is transferred from one conductor to another, the result is a voltage difference $V$ equal to $Q/C$; the conductors are oppositely charged with $Q$ and $-Q$, respectively (Figure 5.1). Again, for this pair of conductors, $Q = C \cdot V$, where $V$ is the voltage difference between these conductors.

Usually, the capacitance is defined 'between' two conductors: this capacitance is determined exclusively by the geometry of the complete set of conductors and the dielectric properties of the (non-conducting) matter in between the conductors and not by the potential of the conductors. The set of conductors is called a capacitor.

Sensors for Mechatronics. DOI: 10.1016/B978-0-12-391497-2.00005-4
© 2012 Elsevier Inc. All rights reserved.

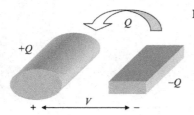

**Figure 5.1** Relation between voltage and charge.

**Table 5.1** Dielectric Constants of Various Environmental and Construction Materials [1,2]

| Material | $\varepsilon_r$ | Material | $\varepsilon_r$ |
|---|---|---|---|
| Vacuum | 1 | $Al_2O_3$ | 10 |
| Dry air (0°C, 1 atm) | 1.000576 | $SiO_2$ | 3.8 |
| Water (0°C) | 87.74 | Mica | 5–8 |
| Water (20°C, $\omega \to 0$) | 80.10 | Teflon | 2.1 |
| Water (20°C, $\omega \to \infty$) | 4.5–6 | PVC | 3–5 |
| Ice (0°C, $\omega \to 0$) | 260 | Silicon (20°C) | 11.7 |
| Ice (0°C, $\omega \to \infty$) | 3.2 | Glass, Pyrex 7740 | 5.00 |

A well-known configuration is a set of two parallel plates, the flat-plate capacitor, with capacitance

$$C = \varepsilon \cdot \frac{A}{d} \tag{5.1}$$

where $A$ is the surface area of the plates and $d$ the distance between the plates. This is only an approximation: at the plate's edges the electric field extends outside the space between the plates, resulting in an inhomogeneous field near the edge (*stray field* or *fringe field*).

The parameter $\varepsilon$ is called the *permittivity* and describes the dielectric properties of the matter in between the conductors. It is written as the product of $\varepsilon_0$, the permittivity of free space or vacuum (about $8.8 \cdot 10^{-12}$ F/m) and $\varepsilon_r$, the relative permittivity. The latter is also called the *dielectric constant* of the medium between the conductors. For vacuum $\varepsilon_r = 1$, for air and other gases the dielectric constant is fractionally higher, liquids and solids have larger values (Table 5.1).

For all kind of configurations the capacitance can be calculated [3]. Two straight parallel conductors (Figure 5.2A) with radius $a$ and centres at spacing $d$ have a capacitance per unit length of:

$$C' = \frac{\pi \varepsilon_0 \varepsilon_r}{\ln \{(d + \sqrt{d^2 - 4a^2})/2a\}} \quad \text{or, for } d \gg a, \quad C' = \frac{\pi \varepsilon_0 \varepsilon_r}{\ln (d/a)} \quad \text{(F/m)} \tag{5.2}$$

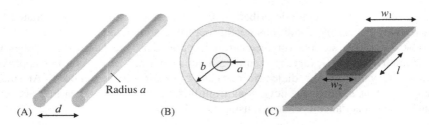

**Figure 5.2** Various configurations: (A) parallel conductors; (B) coaxial conductors; (C) parallel plates.

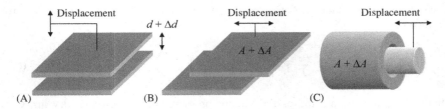

**Figure 5.3** Capacitive displacement sensors: (A) and (B) flat plate; (C) cylindrical.

A coaxial cable (two concentric cylinders, Figure 5.2B) with outer diameter of the central conductor $a$ and inner diameter of the shield $b$ has a capacitance per unit length equal to

$$C' = \frac{2\pi\varepsilon_0\varepsilon_r}{\ln\,(b/a)}\quad(\text{F/m}) \tag{5.3}$$

The capacitance of other, less symmetrical configurations can be calculated analytically, but results in rather complex expressions [3]. For instance consider a flat-plate capacitor (Figure 5.3C) with rectangular top electrode (width $w_2$ and length $l$) opposed to a long strip electrode (width $w_1$) at distance $d$. The capacitance of this structure (as often encountered in capacitive displacement sensors) is expressed as follows [4]:

$$C = \frac{\varepsilon_0\varepsilon_r l}{\pi}\ln\left[\frac{\cosh\,(\pi(w_1+w_2)/4d)\cosh\,(-\pi(w_1+w_2)/4d)}{\cosh\,(-\pi(w_1-w_2)/4d)\cosh\,(\pi(w_1-w_2)/4d)}\right] \tag{5.4}$$

Anyhow, for a homogeneous dielectric, the capacitance can always be written in the form

$$C = \varepsilon \cdot G \tag{5.5}$$

where $G$ is a geometric factor. For the parallel plate capacitor with plate distance $d$ and infinite plate dimensions, $G = A/d$. When the plates have finite dimensions, as for instance for the structure described by Eq. (5.4), $G$ can still be approximated by $A/d$; we will discuss the conditions for this approximation later.

Most capacitive sensors for mechatronic applications are based on changes in the geometric factor $G$, thereby assuming a constant value of $\varepsilon$. As a fair approximation, the value for the dielectric constant $\varepsilon_r$ of air can be set to 1. At small changes of $A$ or $d$, however, deviations from this value must be taken into account because the capacitance $C$ changes, using Eq. (5.1) as:

$$\frac{dC}{C} = \frac{\partial \varepsilon}{\varepsilon} + \frac{\partial A}{A} - \frac{\partial d}{d} \tag{5.6}$$

As a matter of fact, $\varepsilon$ depends somewhat on the air composition (see for instance [5]):

$$\varepsilon_r - 1 = A_d \cdot \frac{p_d}{T} + A_c \cdot \frac{p_c}{T} + \frac{p_w}{T} \cdot \left( A_w + \frac{B_w}{T} \right) \tag{5.7}$$

where $p_d$, $p_c$ and $p_w$ are the partial vapour pressure of, respectively, dry air, $CO_2$ and water vapour in mbar. $A_d$, $A_c$, $A_w$ and $B_w$ are empirical parameters, and $T$ the absolute temperature.

Further, due to relaxation effects, $\varepsilon$ appears to be frequency dependent. The dielectric constant is expressed in complex notation: $\varepsilon = \varepsilon' - j\varepsilon''$, where both the real and imaginary part are frequency dependent. For water the frequency dependency is expressed as [2]:

$$\varepsilon_w(\omega) = \varepsilon_w(\infty) + \frac{\varepsilon_w(0) - \varepsilon_w(\infty)}{1 + (\omega\tau_r)^2} \tag{5.8}$$

where $\tau_r$ is the relaxation time. For liquid water at 0°C $\tau_r$ amounts $17.8 \times 10^{-12}$ s, $\varepsilon_w(\infty) \approx 5$ and $\varepsilon_w(0) \approx 88$. In mechatronics we will consider only the static dielectric constant, that is, the value at frequencies far below the relaxation frequency.

Most capacitive displacement sensors operate in air. The influence of atmospheric changes on the dielectric constant can be summarized as follows [5]: starting with standard air ($T = 293.15$ K, $p_d = 100$ kPa, $p_w = 117.5$ Pa and $p_c = 30$ Pa), the dielectric constant changes by:

$-2.3 \times 10^{-6}$ per K in temperature,
$5.3 \times 10^{-6}$ per kPa absolute pressure,
$90 \times 10^{-6}$ per kPa water vapour pressure,
$9 \times 10^{-6}$ per kPa $CO_2$ pressure.

Temperature effects can be minimized using balancing techniques (Section 3.2) and proper electrode materials. Humidity cannot be ignored in high precision capacitive displacement sensors; either the temperature should be kept sufficiently far above the dew-point temperature or humidity effects must be avoided through the use of special designs [6].

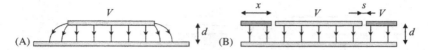

**Figure 5.4** (A) Fringe fields and (B) active guarding reduces fringe fields.

Many capacitive displacement sensors are based on flat-plate capacitors (Figure 5.3). Either the plate distance $d$ or the effective plate area $A$ is used as a position-dependent parameter. Other shapes are also used as displacement sensor, for instance the cylindrical one in Figure 5.3C. When using the approximated expression (5.1), an error is made due to stray fields (fringes) at the edges of the plates. The practical implication of this error is a non-linear relationship between displacement and capacitance.

The effect of stray fields can be reduced by the application of *guarding*. One electrode is grounded; the other, active electrode of the capacitor, is completely surrounded by an additional conducting electrode in the same plane and isolated from the active electrode (Figure 5.4). The potential of the guard electrode is made equal to that of the active electrode (*active guarding*), using a buffer amplifier. The result is that the electric field is homogeneous over the total area of the active electrode, assuming infinite guard electrodes and a zero gap width between the two electrodes.

However, since the guard electrode has finite dimensions and the gap width is not zero, a residual error occurs. This error depends on the dimensions of the guard electrode and the gap width [3]:

$$\frac{\Delta C}{C} \le e^{-\pi x/d}$$

$$\frac{\Delta C}{C} \le e^{-\pi d/s}$$

(5.9)

with $d$ the plate distance, $x$ the width of the guard electrode and $s$ the gap width between guard and active electrodes. As a rule of thumb, for $x/d$ and $d/s$ equal to 5, these errors are less than 1 ppm.

Finally, when the plates are not exactly in parallel, Eq. (5.1) does not hold anymore. The relative error at a skew angle $\alpha$ is of the order $\alpha^2/3$ [7]. For 1° (or 0.017 rad) this corresponds to an error of about $10^{-4}$ only.

## 5.2 Basic Configurations of Capacitive Sensors

### 5.2.1 Flat-Plate Capacitive Sensors

Capacitive sensors are very useful for the measurement of displacement utilizing variation in the geometric factor $G$ in Eq. (5.5). If one of the plates is part of a

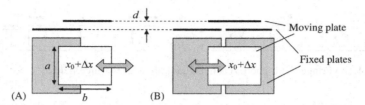

**Figure 5.5** Basic capacitive displacement sensors with variable surface area: (A) single mode and (B) differential mode.

deformable membrane, pressure can also be measured using the capacitive principle. Generally, electric fields are better manageable than magnetic fields: by (active) guarding, it is easy to create electric fields that are homogeneous over a wide area. This is the major reason that displacement sensors based on capacitive principles have excellent linearity.

Figure 5.5 shows, schematically, two basic configurations for linear displacement. The moving object of which the displacement has to be measured is connected to the upper plate. In these examples the parameter $A$ (effective surface area) varies with displacement.

In Figure 5.5A a linear displacement of the upper plate in the indicated direction introduces a capacitance change which is, ideally, $\Delta C = \varepsilon \Delta x a / d$, or a relative change $\Delta C / C = \Delta x / x$. All other parameters need to be constant during movement of the plate. Capacitive sensors for rotation can be configured in a similar way, with segmented plates rotating relatively to each other, where the distance $d$ remains constant. To assure a linear relation between the displacement and the capacitance change, active guarding has to be applied in all cases.

Figure 5.5B shows a differential transducer (see Chapter 3): a displacement of the upper plate causes two capacitances to change simultaneously but with opposed sign: $\Delta C_1 = -\Delta C_2 = \varepsilon \Delta x a / d$. The initial position is defined as the position for which $C_1 = C_2$. For this position ($\Delta x = 0$) a change in $a$ or $d$ (due to for instance play, backlash or temperature change) does not introduce a zero error, as long as both capacitance changes are equal.

An additional advantage of the differential configuration is the extended dynamic range: in the reference or initial position the capacitances of the two capacitors are equal. Since only the difference is processed, the initial output signal is zero. A small displacement results in a small output signal that can be electronically amplified without overload problems. A single capacitance in the initial position may produce a considerable output signal, which cannot be amplified much more, unless it is first compensated by an equal but opposite offset signal. However, compensation by an equally shaped sensor is more stable than compensation in the electrical domain.

A disadvantage of the configurations shown in Figure 5.5 is the electrical connection to the moving plate, required for the supply and transmission of the measurement signals. Figure 5.6 shows a configuration without the need for such a connection.

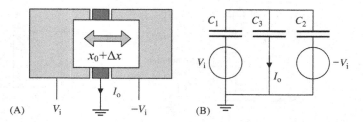

**Figure 5.6** Differential capacitive displacement sensor with current read-out: (A) top view of a flat-plate capacitor and (B) electronic model.

The moving plate acts here as a coupling electrode between two equal, flat electrodes and a read-out electrode (in the middle). All fixed electrodes are in the same plane. In the initial position the coupling electrode is just midway the two active plates, supplied by sinusoidal signals with opposite phase. In this position, both (balanced) signals are coupled to the read-out electrode where they cancel, resulting in a zero output signal. When the top electrode shifts to an off-centre position, the coupling becomes asymmetrical, resulting in an output signal proportional to the displacement, and with a phase indicating the direction of the movement. Using the simplified model of Figure 5.6B, it is easy to verify that the current through the read-out electrode satisfies:

$$I_{out} = \frac{j\omega C_1 V_1 + j\omega C_2 V_2}{1 + C_1/C_3 + C_2/C_3} \tag{5.10}$$

Note that $C_3$ does not change with displacement: its surface area remains unaltered. When $V_1 = -V_2 = V_i$, the output current $I_{out}$ is directly proportional to the capacitance difference $C_1 - C_2$. In a differential configuration, $C_1 = C_0 + \Delta C$ and $C_2 = C_0 - \Delta C$, hence

$$I_{out} = \frac{2j\omega\Delta C}{1 + 2C_0/C_3} \cdot V_i \tag{5.11}$$

The output current appears to be proportional to $\Delta C$ and consequently to the displacement $\Delta x$. Its phase is either $+\pi/2$ or $-\pi/2$ according to the direction of the displacement. It should be pointed out that proportionality only holds when fringe field effects are small or when these fields are eliminated by guarding.

An alternative read-out scheme is shown in Figure 5.7A. Two adjacent plates are supplied with two sinusoidal voltages with phase difference $\pi/2$. The voltages are capacitively coupled to the common read-out electrode, via the moving plate of the capacitor, similar to the design in Figure 5.6A. When the moving electrode is just opposite the left plate, the phase of the output current is just $\pi/2$, according to Eq. (5.10). When the electrode arrives just above the right plate, the phase is $\pi$. When moving from one electrode to the next, the output phase changes gradually over $\pi/2$ rad.

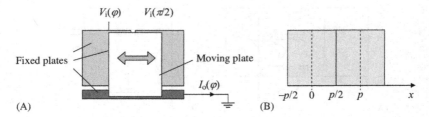

**Figure 5.7** Phase read-out of a capacitive displacement sensor: (A) structure and (B) co-ordinates.

To calculate the phase of the output current in dependence of the displacement we define the co-ordinates as in Figure 5.7B. The centre of the moving plate runs from $x = 0$ to $x = p$. For simplicity all plates have equal width $p$. Suppose the input voltages are $V_1 = \hat{V}\sin\omega t$ and $V_2 = \hat{V}\cos\omega t$. According to Eq. (5.10) the output current equals

$$I_o(t) = \frac{C_1\omega\hat{V}\cos\omega t + C_2\omega\hat{V}\sin\omega t}{1 + C_1/C_3 + C_2/C_3} \tag{5.12}$$

The capacitances $C_1$ and $C_2$ have the value $\varepsilon aw/d = c'w$, with $a$ the (constant) plate length and $w$ the effective width, running from $p$ to 0 for $C_1$ and from 0 to $p$ for $C_2$. With Figure 5.7B the capacitances are found:

$$\begin{aligned} C_1(x) &= c'(p - x) \\ C_2(x) &= c'x \end{aligned} \tag{5.13}$$

Substitution in Eq. (5.12) results in

$$I_o(t) = K\frac{(p - x)\cos\omega t + x\sin\omega t}{1 + c'p/C_3} = \hat{I}\cos(\omega t + \varphi) \tag{5.14}$$

where $K$ is a constant. The phase angle satisfies the relation

$$\varphi = \arctan\frac{x}{p - x} \tag{5.15}$$

The phase varies with displacement in a slightly non-linear way (Figure 5.8). Stray fields cause further deviations from linearity, but even with guarding the relation is essentially nonlinear. However, by modifying the electrode shapes it is possible to get an almost linear relationship.

Capacitive displacement sensors can also be configured in a cylindrical configuration, as in Figure 5.3C. The principle is the same, but a cylindrical set-up is more compact, has less stray capacitances and therefore a better linearity. This type of capacitive sensor is called a linear variable differential capacitor (LVDC). The sensor exhibits extremely good linearity (better than 0.01%) and a low temperature sensitivity (down to 10 ppm/K).

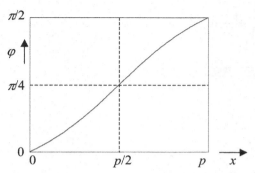

**Figure 5.8** Phase of the output current versus displacement.

The rotational version of the LVDC is called RVDC (rotational variable differential capacitor). It is based on the same principle but configured for angular displacements. Angular sensitivity is obtained using triangularly shaped electrodes. Further specifications on these sensors are given in the overview table (Table 5.2) in Section 5.4.

The lower limits of linearity, stability and accuracy of these precision sensors are set by the air humidity: water vapour in the air may condensate in edges and small gaps of the construction (capillary condensation), thereby locally replacing air (with dielectric constant of about 1) by liquid water (with $\varepsilon_r \approx 80$, see Table 5.1). Even at temperatures above dew point, water vapour may condensate due to contamination: hygroscopic particles attract water molecules and act as condensation nuclei. This means that for the highest performance the temperature of the sensor should be kept well above the dew-point temperature.

### 5.2.2 Multiplate Capacitive Sensors

The measurement range of the linear capacitive sensors discussed so far is limited to approximately the width of the moving plate (Figures 5.5–5.7). However, the range can simply be extended by repeating the basic structure of the previously discussed configuration. Figure 5.9A shows this concept [8]. It consists of an array of sections similar to the one in Figure 5.7. When the moving plate approaches the end of one section, it simultaneously enters the next section.

The fixed electrodes in the array are connected to sine wave voltages with phase differences of $\pi/2$. When the moving plate travels along four successive sections, the phase of the voltage on this plate changes from 0 to $2\pi$ (Figure 5.9B). This is repeated for each group of four electrodes. Hence the output phase changes periodically with displacement, with a period of four times the plate pitch $p$. An unambiguous output is obtained by keeping track of the number of passed cycles, using an incremental counter.

The straight lines in Figure 5.9B are actually four s-shaped curves from Figure 5.8 in series, but nonlinearity can be compensated for by appropriate shaping of the fixed electrodes. Phase differences can be measured with high resolution, down to 0.1°. By periodically repeating the structure a large range is obtained

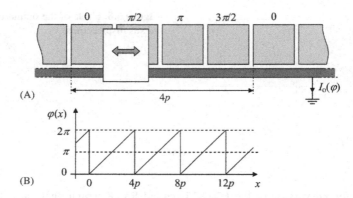

(A)

(B)

**Figure 5.9** Basic set-up of a multiplate capacitive displacement sensor with phase read-out: (A) range extension by multiple electrodes; (B) linearized transfer characteristic.

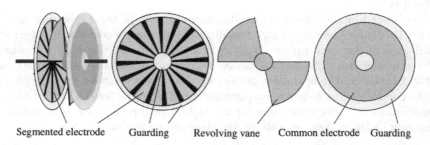

Segmented electrode    Guarding    Revolving vane    Common electrode    Guarding

**Figure 5.10** Basic structure of a capacitive multiplate angular sensor; the vane rotates between the fixed segmented and common electrodes.

combined with high resolution. A well-known application is found in some electronic calliper rules: these have a range of about 20 cm and a resolution of 0.01 mm.

Capacitive angular sensors operate in a similar way (Figure 5.10). The fixed segmented plates are arranged in a circular array and the moving plate is a vane connected to the rotating shaft.

The challenge of the designer is to obtain the highest possible resolution over a fixed range of $2\pi$ rad. Various solutions have been proposed, for instance multiple electrodes on the moving part. Evidently, guard electrodes are indispensable here to reduce stray fields. The resolution of capacitive multiplate angular sensors can be as high as 18 bit or 5 s of an arc [9].

## 5.2.3 Silicon Capacitive Sensors

Silicon micromachining technology offers the possibility to construct extremely small capacitive sensors. The technology allows the fabrication of very thin membranes, similar to piezoresistive sensors as described in Chapter 4. Instead of adding piezoresistors to the spring element, the flexible membrane is opposed to a fixed flat electrode, building a miniaturized flat-plate capacitor. An applied force or pressure difference causes the membrane to deflect, resulting in a capacitance change that can easily be measured. Advances in the technology of MEMS have resulted in the realization of microsensors for other quantities as well, for instance acceleration. These systems now find widespread applications in automotive and other large volume market segments.

These technologies are highly attractive for accelerometers, rate gyroscopes and (gas) pressure sensors but less for force sensors because of the susceptibility to mechanical damage of the tiny movable parts. Accelerometers get much attention because the force is introduced simply by the inertia of a seismic mass fabricated on the same chip. Capacitive sensors for other measurands are also considered to be implemented in MEMS technology, e.g. the angular sensor discussed in Ref. [10].

From a technological point of view, however, a capacitive sensor is more difficult to realize compared to its resistive counterpart. The surfaces of the various parts making up a capacitance should be provided with a conducting layer acting as one of the capacitor's flat plates. Further, the spacing between the plates should be made small, to obtain a reasonable capacitance and sensitivity, and the alignment of the two electrodes is also a critical processing step.

## Silicon Pressure Sensors

The first silicon capacitive pressure sensors consist of a stack of three layers: a support layer, a membrane (etched silicon) and a top layer with the fixed counter electrode. The support layer and top layer can be made from silicon or glass. The electronic circuitry resides on the membrane chip or the (silicon) top layer. The layers are connected using special techniques. Figure 5.11 shows some examples [11–13] of integrated capacitive pressure sensors (vertical scale much larger than horizontal scale). In the configuration in Figure 5.11B the electronics and the membrane are made from separate wafers, to enhance technological freedom: the processes for etching membranes and deposition of electronic circuitry are not fully compatible. The figures show only a general layout; details can be found in the cited papers.

A large variety of alternative structures have been proposed over the last three decennia. In [14] the membrane chip is positioned flipped at the top, whereas a reference capacitance is included in the structure. A structure without separate top layer is described in Ref. [15]. Both capacitor plates are constructed on the membrane chip using surface micromachining. The gap is realized by selective etching of a so-called sacrificial layer, positioned between the two electrode layers. Here, too, a reference capacitor (insensitive to pressure) has been integrated.

Due to the small dimensions of the structures (in the order of millimetres) the capacitance value is small too, and the change caused by the measurand is even smaller. A larger (initial) capacitance may be realized by reducing the gap distance between the plates. However, electrostatic (attractive) forces between the plates can easily result in 'buckling' and short-circuiting of the capacitor. The alternative is to

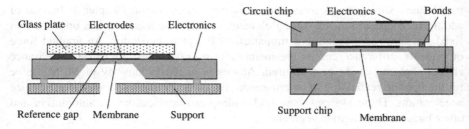

**Figure 5.11** Basic structures of integrated capacitive pressure sensors: (A) with glass top plate and (B) with three silicon layers.
*Source*: (A) After Refs [11] and [12]; (B) After Ref. [13].

increase the effective surface area. This can be realized by a supporting dielectric structure, as is applied in [16]. The sensor is made up of two processed wafers, put together face-to-face and resulting in a narrow gap between them. After bonding the two wafers, the substrate of the top layer is removed by selective etching, leaving a thin membrane (2.5 µm) as the remainder of the top wafer. The membrane is supported by a hexagonal grid of a dielectric material that prevents it from collapsing. In this way a relatively large area combined with a narrow gap is achieved, and hence a larger capacitance and sensitivity can be obtained than with other techniques.

## Silicon Accelerometers

Many *accelerometers* are based on capacitive sensing techniques, for instance the accelerometer with feedback as introduced in Chapter 3. A few decennia ago it was shown that the complete system can be integrated on silicon, resulting in a small, easy-to-mount device [17].

Capacitive accelerometers made in MEMS technology are basically structured as given in Figure 4.19 (a piezoresistive accelerometer): they consist of a seismic mass suspended on flexible beams. Instead of integrated piezoresistors, electrodes are deposited on proper places in the structure forming flat-plate capacitors. The mass moves due to an inertial force, resulting in a change in distance between the plates, and this causes the capacitance to change. By a special configuration with three or four masses and associated electrodes it is possible to realize a 3D accelerometer in a single piece of silicon [18]. Together with the MEMS sensor, additional electronics and signal processing, for instance voltage regulators or an ASIC (application-specific integrated circuit) and a temperature sensor for offset compensation can be housed in a small package [19].

Micromachined acceleration sensors suffer from small capacitances too and hence have low sensitivity. A larger (initial) capacitance might be realized by a dielectric support structures (see the preceding section on pressure sensors). Another way is to increase the effective surface area. This is realized in MEMS

**Figure 5.12** Principle of interdigitated fingers (comb structure) in MEMS technology.

technology, allowing the construction of so-called comb structures. Figure 5.12 shows the basic configuration.

When the two combs move relative to each other in the direction of the arrow, the effective surface area of the capacitor changes. The more fingers, the larger the surface area and the higher the capacitance. Such structures, and even more complicated configurations, can be readily made using advanced MEMS technology. The same structure is also used for capacitive actuators, so-called comb drives.

## 5.3   Interfacing

In high-resolution sensors and microsensors the capacitance changes that have to be measured can be extremely small (down to fractions of a femtofarad or $10^{-15}$ F). Adequate interfacing is required to preserve the intrinsic performance of the sensor system. We discuss four major interface methods for capacitive sensors (Figure 5.13):

1. Impedance measurement (preferably in a bridge)
2. Current−voltage measurement (with operational amplifier)
3. Frequency measurement ($C$ determines the frequency in an $LC$-oscillator)
4. Time measurement (charging and discharging $C$ with constant current).

For each of these methods we give a brief analysis (see Appendix C for the transfer of operational amplifier configurations). The bridge transfer (Figure 5.13A) is:

$$\frac{V_o}{V_i} = \frac{1}{2} - \frac{C_1}{C_1 + C_2} = \frac{1}{2} \cdot \frac{C_2 - C_1}{C_1 + C_2} \tag{5.16}$$

In a single sensor configuration, $C_1 = C + \Delta C$ and $C_2 = C$, hence

$$\frac{V_o}{V_i} \approx -\left(\frac{\Delta C}{4C}\right) \cdot \left(1 - \frac{\Delta C}{2C}\right) \tag{5.17}$$

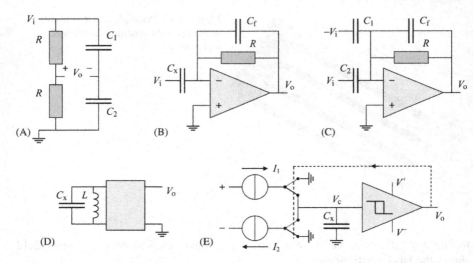

**Figure 5.13** Methods for the measurement of capacitance changes: (A) bridge; (B) and (C) current–voltage measurement – single and differential; (D) frequency measurement; (E) time measurement.

The transfer is frequency independent but nonlinear. In a differential or balanced configuration $C_1 = C + \Delta C$ and $C_2 = C - \Delta C$, thus

$$\frac{V_o}{V_i} = -\frac{\Delta C}{2C} \tag{5.18}$$

Clearly, the differential mode yields better linearity and a wider dynamic range (see also Chapter 4 on resistance bridges). A proper measurement of the bridge output voltage requires a differential amplifier with high input impedance and high common mode rejection ratio.

Interface circuits based on current–voltage measurements are shown in Figure 5.13B and C: the first for single mode operation and the second for differential mode operation. If resistance $R$ is disregarded, the transfer function of the single mode circuit is

$$\frac{V_o}{V_i} = \frac{C + \Delta C}{C_f} \tag{5.19}$$

and for the differential mode:

$$\frac{V_o}{V_i} = 2\frac{\Delta C}{C_f} \tag{5.20}$$

Again the advantage of a balanced configuration appears immediately from expression (5.20), which is linear, and the output is zero in the initial state (equal capacitances). The transfer is frequency independent if resistance $R$ in Figure 5.13C is left out. However, the circuit does not work properly in that case (see Appendix C) due to integration of the offset and bias currents of the operational amplifier. The resistance $R$ prevents the circuit from running into overload but causes the transfer to fall off with decreasing frequency from about $f = 1/2\pi RC_f$. This means that the frequency of the interrogating signal $V_i$ should be chosen well above this cut-off frequency.

When the maximum capacitance change is small, the feedback capacitance $C_f$ should be chosen small too, for a reasonable gain of the circuit (as follows from Eq. (5.20)). However, stray capacitances (wiring, input capacitance of the amplifier) set a limit to the minimum value for $C_f$ (to some pF). In integrated capacitive sensors these stray capacitances are very small, allowing a small feedback capacitance, hence high gain.

The sensitivity is limited by the stability of the initial state (where the output should be zero). Neither the amplitude drift nor the frequency instability of the oscillator affects the zero stability since they are common to both differential sensors. The performance of the measurement is mainly limited by the phase difference between the two AC voltage sources. The output appears to be very sensitive for deviations from the required 180° phase difference, a fact that can be used to create a sensitive capacitance-to-phase converter [20].

A simple though rather inaccurate method for measuring capacitance changes is the conversion to frequency, shown in Figure 5.13D. The capacitance is part of an LC-oscillator circuit that generates a periodic signal with frequency

$$f = \frac{1}{2\pi\sqrt{LC_x}} \approx f_0\left(1 - \frac{\Delta C}{2C}\right) \tag{5.21}$$

where $f_0$ is the frequency in the initial state. An advantage of the oscillator method is easy signal processing with a microprocessor: the frequency can be determined by simply counting the number of clock pulses during one cycle of the oscillator signal. Obviously, the other circuit elements must be stable and the oscillation process should be guaranteed over the entire capacitance range. A detailed analysis of a circuit based on this principle can be found in [21]. The interface contains a microprocessor controlled feedback system with a VCO (voltage-controlled oscillator), which keeps the actuation frequency equal to the resonance frequency of the LC-circuit. This interface has a sensitivity of 100 Hz/fF and a resolution of about 0.1 fF.

An interface circuit that combines the advantages of a time signal output and high linearity is given in Figure 5.13E, an example of a relaxation oscillator. The unknown capacitance is periodically charged and discharged with currents $I_1$ and $I_2$. During charging, $V_c$ increases linearly with time until the upper hysteresis level of the Schmitt trigger (Appendix C) is reached. At that moment the switches turn over and the capacitance is discharged until the lower hysteresis level is reached. The result is a periodic triangular output signal with fixed amplitude between the levels of the Schmitt trigger and a frequency related to the capacitance according to:

$$f = \frac{I}{2C_x V_s} \tag{5.22}$$

where $V_s$ is the hysteresis range of the Schmitt trigger. The method yields accurate results because the frequency depends only on these fixed levels and the charge current.

Based on one or more of the previously described principles, many other, more advanced interface circuits have been developed, aiming at a higher stability, sensitivity or processing speed and the reduction of parasitic capacitances. We mention some of these works.

In Ref. [22] an interface circuit based on a three-phase time sequence is described, applied to a differential capacitive sensor pair. A very high linearity is claimed (24 ppm) and a high processing speed (10 kHz sampling rate). Hu et al. [23] give an extended version of the circuit in Figure 5.13C, with switchable feedback capacitances and a rectifier. By selecting a suitable value for the feedback capacitor (like $C_f$ in Figure 5.13C) the transfer (gain) can be adopted to the needs, maximizing signal-to-noise ratio. The paper gives an extensive analysis of the noise behaviour, and experimental verification is performed using a capacitive pressure sensor. A resolution of 0.005% is reported.

In Ref. [24] an integrated interface chip for capacitive sensors with a programmable configuration of the circuit is introduced. The feedback capacitance, a reference capacitance and the voltage gain can be programmed to optimize circuit performance. This facility makes the chip applicable for a variety of capacitive sensors.

Finally, we mention an advanced version of the circuit in Figure 5.13E [25]. In this circuit, two capacitors instead of just one are periodically charged and discharged, resulting in a periodic signal of which the duty cycle is related to the capacitance to be measured. A careful analysis of possible errors resulted in a circuit design having an inaccuracy less than 0.04%.

## 5.4 Applications

### 5.4.1 Capacitive Sensors for Position- and Force-Related Quantities

Capacitive sensors are suitable for the measurement of all sorts of position-related quantities: displacement, rotation, speed, acceleration and tilt. Combined with an elastic element, force, pressure, torque and mass sensors can also be realized using capacitive principles. Most of these sensors are based on Eq. (5.1), indicating the relation between the capacitance value and a geometric parameter (e.g. effective surface area and plate distance). Capacitive sensors are particularly suitable for

**Table 5.2** Typical Specifications of Various Capacitive Sensors

| Type | Range | Resolution | t.c. Gain $(K^{-1})$ | $T_{max}$ $(°C)$ | Bandwidth $(Hz)$ |
|------|-------|-----------|----------|--------|-----------|
| LVDC | 2.5−250 mm | 1(0.01) μm | $10^{-4}$ | 80 | 250 |
| RVDC | 70° | 0.1 arcsec | | 150 | |
| Acceleration (feedback) | ± 0.5... ±150 g | $10^{-6}$ g | $10^{-4}$ | 125 | 300 |
| Tilt | ±80° | 0.01° | | 80 | |

applications with high demands (for instance extreme temperatures). The mechanical construction is simple and robust. Furthermore, extremely small capacitance changes (down to 1 fF) can be measured with simple interface circuits. Finally, stray capacitances and other capacitances due to wiring and amplifiers can easily be eliminated using guarding, virtual grounding and proper interfacing.

As with all displacement sensors, capacitive displacement sensors can be modified to make them suitable for the measurement of angular rate, acceleration, force, torque, mass and pressure. Differential configurations are recommended and, where appropriate, a feedback principle as well, to reduce intrinsic sensor errors (refer to Chapter 3 for a capacitive accelerometer based on feedback). Table 5.2 resumes some specifications of commercial capacitive sensors for various measurement quantities.

Using the circular multiplate capacitor discussed in Section 5.2.2, *angular position* and *angular speed* can be determined with high accuracy, as has been demonstrated in many papers, for instance in Refs [26−28]. With a careful design, angular position sensors can have a resolution of 18 bit, a reproducibility of 17 bit and an absolute accuracy better than 14 bit. A micromachined contact-free angle sensor was already mentioned in Section 5.2.3 [10]: this angle sensor combines an inductive principle for the transduction from angle (with respect to a fixed magnetic field) to the deflection of a polysilicon micromass and the capacitive measurement of this deflection. A planar micromachined *gyroscope* is described in Ref. [29] with a balanced capacitive read-out circuit.

With capacitive sensors, the *torque* on (rotating) axes can be measured [30−32]. In some of these applications the sensor on the axis is purely passive, which means that no power supply on the axis is needed. The information transfer to and from the axis is accomplished either inductively or with pure capacitive techniques.

The capacitive *mass* sensor in Ref. [33] uses a metal spring as the elastic element; this sensor is designed for a low temperature sensitivity, for which thermally stable materials have been selected. The capacitive *force* sensor in Ref. [34] contains an elastic dielectric; its basic configuration is depicted in Figure 5.14.

It is sensitive in two shear directions and the normal direction. At zero force, all four capacitances $C_1 - C_4$ are equal. A pure normal force will increase each capacitance equally, and a pure shear force results in an increased difference between adjacent capacitances (differential method), according to the direction of the

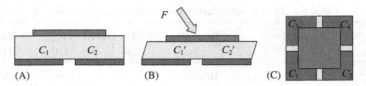

**Figure 5.14** 3D force sensor layout with elastic dielectric: (A) and (B) cross section; (C) top view.

applied shear force. The interface should be able to distinguish between these force components.

## 5.4.2    Particular Applications

In this section examples of capacitive measurements for a variety of applications are discussed, illustrating the versatility of the capacitive sensing method. Selected measurands are distance, displacement, thickness, rotation, strain, force, level and flow. Some special applications concern the detection of persons and rain on a car's windscreen.

Capacitive sensors offer the possibility of highly accurate *distance* or *displacement* measurements, as in controlled $x-y$ tables. A capacitive sensor designed for product inspection is given in Ref. [35]. The goal is to measure the inner diameter of cylinders using a co-ordinate measuring machine with touch probe. In this work the bending of the touch probe (a thin stylus) is determined by capacitive means. The method allows measuring of internal diameters less than 0.3 mm, with an accuracy of 1 μm.

Coating *thickness* can be measured in several ways. For a non-conductive coating on a conductive plate, the capacitive method is suitable to measure the thickness of such a coating. An example of this method is given in Ref. [36], proposing a capacitive probe and giving an analysis of the various sources of errors.

In Section 5.2.2 multiple plate capacitive sensors were discussed, to enlarge the range of a (linear) displacement sensor. An alternative solution is given in Ref. [37], where a configuration is presented similar to the inductosyn given in Chapter 6 on inductive sensors. The sensor consists of two printed circuit boards (PCBs) each provided with an array of electrodes (transmitter and receiver), able to slide linearly over each other as in an optical encoder. The signal transfer from transmitter to receiver is determined by the configuration of the electrodes and shows a triangular course with displacement. Although realized with simple technology (e.g. PCBs, copper electrodes and coating), the resolution of the sensor appears to be about 126 nm, over a range of 20 mm.

Not only linear displacement but also *rotation* can be measured accurately by capacitive means. An example was given in Figure 5.10. A modified version is presented in Ref. [38], having a completely isolated rotor. The main advantage of this design is that no electrical connections to the rotating part are needed.

A totally different application of a capacitive sensor is found in Ref. [39]: the measurement of tyre *strain*. Car tyres are usually reinforced with layers of steel wires. When the tyre is loaded, the resulting strain causes the distance between these wires to increase. Hence, the capacitance between the wires changes accordingly. The paper describes how to measure this capacitance change. The interface is based on an oscillator (like in Figure 5.13D), but instead of an *LC*-oscillator, an *RC*-oscillator is chosen. Wireless read-out allows strain measurement during rotation of the wheel with tyre. Another approach to measure tyre strain using a capacitive method is described in Ref. [40]. The authors have fabricated a particular strain gauge, consisting of a pair of interdigitated flat electrodes on a flexible polyimide carrier. The gauge, glued onto the inside tyre, responds to strain by a change in capacitance [41].

As mentioned before, capacitive sensors can also be applied for *force* and *pressure* measurements. A particular application is on-line weighing of (for instance) cars [42]. The sensor consists of two rubber layers sandwiched between (three) conductive sheets. When pressed, the distance between the conductive sheets becomes smaller, producing an increase in capacitance. The sensor is flexible, light-weight and easy to carry. It can simply be positioned on the road and is capable of dynamically measuring the weight of a (slowly) passing car.

Thus far we have used the geometric factor $G$ in Eq. (5.5) as the basis for a capacitive measurement. The dielectric constant $\varepsilon_r$ is another parameter determining the capacitance value of a capacitive structure and hence a workable method for sensing. Measuring liquid *level* in a tank is an obvious example. A pair of flat-plate electrodes hanging vertically in a tank represents a capacitance that depends on the dielectric constant of the fluid between these plates (Figure 5.15A).

In an empty tank this is air (or another gas with $\varepsilon_r \approx 1$), in a full tank the liquid (for instance water with $\varepsilon_r \approx 80$). A partly filled capacitor can be conceived as two capacitors in parallel: one filled with the liquid up to the upper level ($C_L$), and one with air from the level to the top of the plates ($C_A$). The total capacitance is the sum of these two capacitances, and hence a measure for the level. Obviously, this method holds only for non-conducting liquids. Since water is both dielectric and conductive, at least one of the electrodes should be isolated from the liquid, and

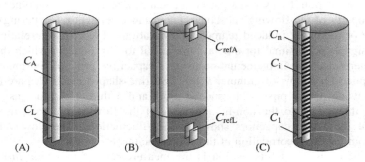

**Figure 5.15** Capacitive level measurements.

special measures have to be taken when the container is grounded, which is explained in Ref. [43]. By inserting two reference capacitors, one at the top (in air) and another at the bottom (in the liquid), as shown in Figure 5.15B, the influence of several parameter variations can be eliminated [44]. Instead of one continuous capacitor, an array of smaller capacitors $C_i$ can be conceived, resulting in a discrete output (Figure 5.15C). In Ref. [45] such a solution is presented, resulting in a resolution of 0.1 mm in level. Further, Ref. [46] shows how a capacitive sensor structure can be used to detect unbalanced load in a washing machine.

The capacitive method makes it possible to detect water drops on a windscreen: a *rain sensor*. It consists basically of a pair of flat coplanar electrodes; the dielectric is composed of three parts: at the back side glass, at the front side air and/or water. To be less sensitive to temperature variations and other possible common mode interferences, a differential structure is preferred: one capacitor is exposed to water, the other is not and serves as a reference capacitor [47].

The dielectric constant of a human being differs substantially from that of air, which makes it possible to capacitively detect the presence or passage of *persons*. Pavlov et al. [48] describe a configuration with small electrodes located in a doorway, giving a capacitance change of a few tenths pF when a person passes. The interface is based on the oscillation method and with special signal processing it is also possible to extract particular features of the passing person, for instance high or low speed and direction. George et al. [49] discuss a capacitive method to detect the presence of a person on a car seat. The seat is provided with a set of electrodes in the seat as well as the back rest. The sensor configuration enables the discrimination between an adult and a child, and between persons and objects (for instance bottles with water). Another capacitive technique for occupancy detection is presented in Ref. [50]. The method is based on an impedance model of biological tissue, being part of the dielectric between two electrodes in the seat. The measured impedance parameters are compared with the model parameters, from which information about the occupancy is derived. An extension to *position* of humans is presented in Ref. [51]; large flat electrodes are integrated in the floor and walls of a $2 \times 2$ m room. A detailed analysis of the changes in the various capacitors building up this configuration shows that it is possible to locate persons with reasonable resolution.

One of the many ways to measure *flow velocity* of a substance moving through a pipe is correlation: two sensors placed a distance $d$ apart from each other measure some property of the flowing substance, for instance reflection (scattering), acoustic noise or thermally induced temperature variations. The cross correlation of the (noisy) signals is maximal for a delay time equal to $\tau = d/v$ from which the velocity is determined. An implementation using a capacitive principle is found in Ref. [52], applied for a flow of granular solid. Two ring-shaped electrodes are mounted at the outside of the pipe, a distance $d$ apart and a third electrode just midway between them. Due to the non-homogeneity of the flowing material, the capacitances of these two capacitors show random fluctuations. The flow velocity is found from the cross correlation of the two capacitance signals.

The last example of this section is the measurement of *magnetic flux* using a capacitive principle [53]. A varying magnetic flux in a (ferromagnetic) material produces eddy currents (see Chapter 6), which in turn create a voltage difference across the material. This voltage can be measured using, for instance, the needle method: two needle-like probes are pressed on the surface of the material under test, and the potential difference is a measure for the eddy currents and hence the magnetic flux. To be able to measure in this way, the needles must make electrical contact, so any possible isolating coating should be removed at the contact places. In Ref. [53] the needles are replaced by conductive pads acting as pick-up electrodes: the voltage is coupled capacitively to the measurement instrument. Calculations show that the voltages are small (mV range) but measurable, if precautions are taken as outlined in the previous sections: differential configuration, guarding and proper filtering.

This section clearly shows that the capacitive sensing principle can be applied in numerous situations. The configuration is simple, a balanced mode is preferred, and proper interfacing may reduce the influence of stray and cable capacitances on the sensitivity and stability of the transfer.

# References and Literature

### References to Cited Literature
[1] J.B. Hasted: Aqueous dielectrics (Series: Studies in chemical physics). VII, Chapman and Hall, London, 1973.
[2] D. Eisenberg, W. Kauzmann: The structure and properties of water; Oxford University Press, London, 1969.
[3] W. Chr: Heerens: application of capacitance techniques in sensor design, J. Phys. E Sci. Instrum., 19 (1986), 897−906.
[4] G.W. de Jong: Smart capacitive sensors, PhD thesis, Delft University Press, Delft, the Netherlands, 1994.
[5] M.H.W. Bonse: Capacitive position transducers; theoretical aspects and practical applications, PhD thesis, Delft University of Technology, Delft, the Netherlands, 1995; ISBN 90.370.0135.1.
[6] G.A. Bertone, Z.H. Meiksin, N.L. Carroll: Elimination of the anomalous humidity effect in precision capacitance based transducers, IEEE Trans. Instrum. Meas., 40(6) (1991), 897−901.
[7] F.N. Toth, G.C.M. Meijer: A low-cost, smart capacitive position sensor, IEEE Trans. Instrum. Meas., 41(6) (1992), 1041−1044.
[8] K.B. Klaassen, J.C.L. van Peppen, R.E. Sandbergen: Thin-film microdisplacement transducer; in: W.A. Kaiser, W.E. Proebster (eds.), From electronics to microelectronics, North-Holland Publishing Company, 1980, pp. 648−650.
[9] X.J. Li, G.C.M. Meijer, G.W. de Jong: An accurate low cost capacitive absolute angular-position sensor with full-circle range, IEEE Trans. Instrum. Meas., IM-45 (1996), 516−520.
[10] J.R. Kaienburg, R. Schellin: A novel silicon surface micromachining angle sensor, Sens. Actuators A, 73 (1999), 68−73.
[11] C.S. Sander, J.W. Knutti, J.D. Meindl: A monolithic capacitive pressure sensor with pulse-period output, IEEE Trans. Electron Devices, ED-27(5) (1980), 927−930.

[12] Y.S. Lee, K.D. Wise: A batch-fabricated silicon capacitive pressure transducer with low temperature sensitivity, IEEE Trans. Electron Devices, ED-29(1) (1982), 42−56.

[13] A. Hanneborg, T.-E. Hansen, P.A. Ohlckers, E. Carlson, B. Dahl, O. Holwech: An integrated capacitive pressure sensor with frequency-modulated output, Sens. Actuators A, 9 (1986), 345−351.

[14] S.T. Moe, K. Schjølberg-Henriksen, D.T. Wang, E. Lund, J. Nysaether, L. Furuberg, M. Visser, T. Fallet, R.W. Bernstein: Capacitive differential pressure sensor for harsh environments, Sens. Actuators A, 83 (2000), 30−33.

[15] C.H. Mastrangelo, X. Zhang, W.C. Tang: Surface micromachined capacitive differential pressure sensor with lithographically-defined silicon diaphragm; Proceedings of the Eurosensors IX, Stockholm, Sweden, 25−29 June 1995, Vol. 1, pp. 612−615.

[16] T. Pedersen, G. Fragiacomo, O. Hansen, E.V. Thomsen: Highly sensitive micromachined capacitive pressure sensor with reduced hysteresis and low parasitic capacitance, Sens. Actuators A, 154 (2009), 35−41.

[17] L.B. Wilner: A high performance, variable capacitance accelerometer, IEEE Trans. Instrum. Meas., IM-37(4) (1988), 569−571.

[18] H. Rödjegård, C. Johansson, P. Enoksson, G. Andersson: A monolithic three-axis SOI-accelerometer with uniform sensitivity, Sens. Actuators A, 123−124 (2005), 50−53.

[19] H. Rödjegård, C. Rusu, K. Malmström, G.I. Andersson: Frequency and temperature characterization of a three axis accelerometer; Proceedings Eurosensors XIX, Barcelona, Spain, 11−14 September 2005, MP27.

[20] R.F. Wolffenbuttel, P.P.L. Regtien: Capacitance-to-phase angle conversion for the detection of extremely small capacities, IEEE Trans. Instrum. Meas., IM-36(2) (1987), 868−872.

[21] C. Ghidini, D. Marioli, E. Sardini, A. Taroni: A 15 ppm resolution measurement system for capacitance transducers, Meas. Sci. Technol., 7 (1996), 1787−1792.

[22] A.S. Hou, S.X.P. Su: Design of a capacitive-sensor signal processing system with high accuracy and short conversion time, Sens. Actuators A, 119 (2005), 113−119.

[23] H.L. Hu, T.M. Xu, S.E. Hui: A high-accuracy, high-speed interface circuit for differential-capacitance transducer, Sens. Actuators A, 125 (2006), 329−334.

[24] W. Bracke, P. Merken, R. Puers, C. Van Hoof: Design methods and algorithms for configurable capacitive sensor interfaces, Sens. Actuators A, 125 (2005), 25−33.

[25] N.M. Mohan, V.J. Kumar: Novel signal conditioning circuit for push-pull type capacitive transducers, IEEE Trans. Instrum. Meas., IM56(1) (2007), 153−157.

[26] T. Fabian, G. Brasseur: A robust capacitive angular speed sensor, IEEE Trans. Instrum. Meas., IM-47(1) (1998), 280−285.

[27] G. Brasseur: A capacitive 4-turn angular-position sensor, IEEE Trans. Instrum. Meas., IM-47(1) (1998), 275−279.

[28] P.L. Fulmek, F. Wandling, W. Zdiarsky, G. Brasseur: Capacitive sensor for relative angle measurement, IEEE Trans. Instrum. Meas., 51(6) (2002), 1145−1149.

[29] S.E. Alper, T. Akin: A planar gyroscope using a standard surface micromachining process; Proceedings Eurosensors XIV, 27−30 August, Copenhagen, Denmark, pp. 387−390.

[30] J.D. Turner: The development of a thick-film non-contact shaft torque sensor for automotive applications, J. Phys. E Sci. Instrum., 22 (1989), 82−88.

[31] A. Falkner: A capacitor-based device for the measurement of shaft torque, IEEE Trans. Instrum. Meas., IM-45(4) (1996), 835−838.

[32] R.F. Wolffenbuttel, J.A. Foerster: Noncontact capacitive torque sensor for use on a rotating axle, IEEE Trans. Instrum. Meas., IM-39(6) (1990), 1008−1013.

[33]  A.A. Al Aish, M. Rehman: Development of a capacitive mass measuring system, Sens. Actuators A, 151 (2009), 113S–117S.

[34]  J.G. Vieira da Rocha, P.F. Antunes da Rocha, S. Lanceros-Mendez: Capacitive sensor for three-axis force measurements and its readout electronics, IEEE Trans. Instrum. Meas., 58(8) (2009), 2830–2836.

[35]  S. Yang, S. Li, M.J. Kaiser, F.H.K. Eric: A probe for the measurement of diameters and form errors of small holes, Meas. Sci. Technol., 9 (1998), 1365–1368.

[36]  R.J. Zhang, S.G. Dai, P.A. Mu: A spherical capacitive probe for measuring the thickness of coatings on metals, Meas. Sci. Technol., 9 (1997), 1028–1033.

[37]  M. Kim, W. Moon: A new linear encoder-like capacitive displacement sensor, Measurement, 39 (2006), 481–489.

[38]  V. Ferrari, A. Ghisla, D. Marioli, A. Taroni: Capacitive angular-position sensor with electrically floating conductive rotor and measurement redundancy, IEEE Trans. Instrum. Meas., 55(2) (2006), 514–520.

[39]  R. Matsuzakia, A. Todorokib: Wireless strain monitoring of tires using electrical capacitance changes with an oscillating circuit, Sens. Actuators A, 119 (2005), 323–331.

[40]  R. Matsuzaki, A. Todoroki: Wireless flexible capacitive sensor based on ultra-flexible epoxy resin for strain measurement of automobile tires, Sens. Actuators A, 140 (2007), 32L–42L.

[41]  R. Matsuzakia, T. Keating, A. Todoroki, N. Hiraoka: Rubber-based strain sensor fabricated using photolithography for intelligent tires, Sens. Actuators A, 148 (2008), 1–9.

[42]  L. Cheng, H. Zhang, Q. Li: Design of a capacitive flexible weighing sensor for vehicle WIM system, Sensors, 7 (2007), 1530–1544.

[43]  F. Reverter, X.J. Li, G.C.M. Meijer: Liquid-level measurement system based on a remote grounded capacitive sensor, Sens. Actuators A, 138 (2008), 1F–8F.

[44]  H. Canbolat: A novel level measurement technique using three capacitive sensors for liquids, IEEE Trans. Instrum. Meas., 58(10) (2009), 3762–3768.

[45]  F.N. Toth, G.C.M. Meijer, M. van der Lee: A planar capacitive precision gauge for liquid-level and leakage detection, IEEE Trans. Instrum. Meas., 46(2) (1997), 644–646.

[46]  M.K. Ramasubramanian, K. Tiruthani: A capacitive displacement sensing technique for early detection of unbalanced loads in a washing machine, Sensors, 9 (2009) 9559–9571, doi:10.3390/s91209559I.

[47]  P. Bord, Tardy, F. Ménil: Temperature influence on a differential capacitive rain sensor performances, Sens. Actuators B, 125 (2007), 262–267.

[48]  V. Pavlov, H. Ruser, H.-R. Tränkler, J. Böttcher: Multi-electrode capacitive person detector using wavelet denoising; IEEE Conference on Multisensor Fusion and Integration (MFI'06), Heidelberg, Germany, 3–6 September 2006.

[49]  B. George, H. Zangl, T. Bretterklieber, G. Brasseur: Seat occupancy detection based on capacitive sensing, IEEE Trans. Instrum. Meas., 58(5) (2009), 1487A–1494A.

[50]  A. Satz, D. Hammerschmidt, D. Tumpold: Capacitive passenger detection utilizing dielectric dispersion in human tissues, Sens. Actuators A, 152 (2009), 1–4.

[51]  R.N. Aguilar Cardenas, H.M.M. Kerkvliet, G.C.M. Meijer: Design and empirical investigation of capacitive human detectors with opened electrodes, Meas. Sci. Technol., 21 (2010) doi: 10.1088/0957-0233/21/1/015802, 015802, 8 pp.

[52]  H. Fuchs, G. Zangl, E.M. Brasseur: Petriu: flow-velocity measurement for bulk granular solids in pneumatic conveyor pipes using random-data correlator architecture, IEEE Trans. Instrum. Meas., 55(4) (2006), 1228–1234.

[53]  S. Zurek, T. Meydan: A novel capacitive flux density sensor, Sens. Actuators A, 129 (2006), 121–125.

# Literature for Further Reading

[1] L.K. Baxter, R.J. Herrick: Capacitive sensors: design and applications; IEEE Press Series on Electronics Technology, 1996; ISBN 0780311302.

[2] G.A. Bertone, Z.H. Meiksin, N.L Carroll: Investigation of a capacitance-based displacement transducer, IEEE Trans. Instrum. Meas., IM-39(2) (1990), 424–427.

[3] A. Fertner, A. Sjölund: Analysis of the performance of the capacitive displacement transducer, IEEE Trans. Instrum. Meas., IM-38(4) (1989), 870–875.

[4] G.T. Jankauskas, J.R. LaCourse, D.E. Limbert: Optimization and analysis of a capacitive contactless angular transducer, IEEE Trans. Instrum. Meas., IM-41(2) (1992), 311–315.

# 6 Inductive and Magnetic Sensors

Inductive sensors employ variables and parameters like magnetic induction $B$, magnetic flux $\Phi$, self-inductance $L$, mutual inductance $M$ or magnetic resistance $R_\mathrm{m}$. By a particular construction of the device, these quantities are made dependent on an applied displacement or force. First we review various magnetic quantities and their relations. Next the operation and the specifications of the major types of magnetic and inductive sensors are reviewed. Special attention is given to transformer-type sensors. The chapter concludes with a section on applications.

## 6.1 Magnetic and Electromagnetic Quantities

### 6.1.1 Magnetic Field Strength, Magnetic Induction and Flux

The magnetic field strength $H$, generated by a flow of charged particles, is defined according to

$$I = \oint_C H \cdot \mathrm{d}l \tag{6.1}$$

where $I$ is the current passing through a closed contour $C$ (Figure 6.1A). The quantities $H$ (A/m) and $\mathrm{d}l$ (m) are vectors. For each configuration of conductors carrying an electric current the field strength in any point of the surrounding space can be calculated by solving the integral equation (6.1). A current $I$ through a long, straight wire produces a magnetic field with strength $H = I/2\pi r$ at a distance $r$ from the wire. So the field strength is inversely proportional to the distance from the wire. The field lines form concentric circles around the wire so the vector direction is tangent to these circles (Figure 6.1B).

Only for structures with a strong symmetry, simple analytical solutions can be obtained. Magnetic fields of practical devices and shapes are studied using FEM (finite element method). The space is subdivided in small (triangular) areas and for each area the equations are solved numerically. FEM programmes calculate the field strength and direction over the entire region of interest; the results are visualized in colour or grey-tone pictures or with 'field lines'.

As an illustration Figure 6.2 shows the magnetic field of a permanent magnet and of a wire loop with a DC current. The FEM programme also shows the numerical values of the field quantity in each point of the space enclosed by a specified boundary.

Sensors for Mechatronics. DOI: 10.1016/B978-0-12-391497-2.00006-6
© 2012 Elsevier Inc. All rights reserved.

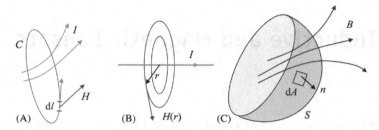

**Figure 6.1** (A) Magnetic field generated by current $I$, (B) calculation of field strength due to a straight wire carrying a current $I$ and (C) calculation of magnetic flux.

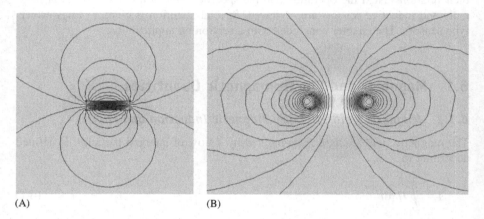

(A)                                                    (B)

**Figure 6.2** Examples of magnetic fields obtained by FEM analysis: (A) a permanent magnet and (B) a single turn with DC current 1 A.

Evidently a stronger field strength can be obtained by increasing the current. However a more efficient method is to make multiple turns of the wire. Each turn carries the current $I$, thus contributing to the field strength. For example the field inside a coil with $n$ turns is proportional to the product of $n$ and $I$. The product $n \cdot I$ (expressed as ampere-turns) is a measure for the strength of such a magnetic source.

Other quantities that describe magnetic and induction phenomena are the magnetic induction $B$ (unit Tesla, T, kg/As$^2$) and the magnetic flux $\Phi$ (unit Weber, Wb = kg m$^2$/As$^2$ = Tm$^2$). By definition the flux is:

$$\Phi = \iint_S B \cdot dA \tag{6.2}$$

In words the flux is the inner vector product of the magnetic induction vector and a surface patch $dA$, integrated over the total area $S$ for which the flux is calculated (Figure 6.1C). In particular for a homogeneous field that makes an angle $\alpha$ with the normal on a flat surface $A$, the flux is:

$$\Phi = B \cdot A \cdot \cos \alpha \tag{6.3}$$

When the surface is in parallel to the field ($\alpha = \pi/2$), there is no flux through that surface. The flux is maximal through a surface perpendicular to the field ($\alpha = 0$).

The magnetic field strength of (permanent) magnets is expressed in terms of magnetic induction $B$ (so in Tesla) rather than of $H$. For instance the Earth's magnetic field strength is about 60 $\mu$T, and the strengths of permanent magnets range from 0.01 to 1 T.

Free charges moving in a magnetic field experience a Lorentz force, driving them into a direction according to the well-known right-hand rule:

$$F_l = q(v \times B) \tag{6.4}$$

This happens also with the free electrons in a conductor that moves in a magnetic field. The movement results in a potential difference, the induction voltage, across the conductor and satisfies the equation

$$V_{ind} = -\frac{d\Phi}{dt} \tag{6.5}$$

which is the induction law of Faraday. In an open wire loop, moving in a magnetic field, the induction voltage appears between both ends of the wire; the current through the wire is zero. In a closed loop, the induced voltage causes a current equal to $V_{ind}/R$, where $R$ is the resistance of the wire loop. From Eq. (6.5) it follows that the induced voltage differs from zero only when the flux changes with time. At constant flux, the induced voltage is zero. The definition of flux in Eq. (6.5) links the units volt (V) and Weber (Wb).

### 6.1.2 Permeability

The magnetic quantities $H$ and $B$ are related by the equation

$$B = \mu H = \mu_0 \mu_r H \tag{6.6}$$

The quantity $\mu$ is the (magnetic) permeability; $\mu_0$, the permeability of free space, equals $4\pi \cdot 10^{-7}$ Vs/Am by definition. The relative permeability $\mu_r$ is a material property (compare $\varepsilon_r$ for a dielectric material). For vacuum $\mu_r = 1$, for gases and many nonferrous materials it is very close to 1. The permeability of ferromagnetic materials is much higher, but strongly non-linear; at higher values of $H$ the material shows saturation and hysteresis. Table 6.1 shows some values of the static permeability for several materials, as well as values for the saturation induction.

**Table 6.1** Permeability of Various Construction Materials [1,2]

| Material | $\mu_r$ (max) | $B_{sat}$ (T) |
|---|---|---|
| Vacuum | 1 | |
| Pure iron | 5,000 | 2.2 |
| Transformer steel | 15,000 | 2.0 |
| Mumetal ($Fe_{17}Ni_{56}Cu_5Cr_2$) | 100,000 | 0.9 |
| Supermalloy ($Fe_{16}Ni_{79}Mo_5$) | 1,000000 | 0.8 |

Mumetal is often used for shielding system parts that are sensitive to magnetic fields. The non-linearity of $\mu_r$ is employed in, for instance, fluxgate sensors (Section 6.2.3).

### 6.1.3 Eddy Currents

Any conductor in a non-stationary induction field experiences induction voltages. This holds not only for wires (where it is used for the generation of electric currents) but also for bulk material (as the iron cores of transformers and electric machines). The induced currents through such material follow more or less circular paths; therefore, they are called eddy currents. They produce unwanted heat, so normally they are minimized by, for instance, increasing the resistance of the bulk material. In constructions with a lot of iron (e.g. transformers and electric machines), this is accomplished by laminating the material: instead of massive material the construction is built up of a pile of thin iron plates (lamellae) packed firmly together. Eddy current can only flow in the plane of the plates, but cannot cross the boundary between two adjacent plates. A useful application of eddy currents for sensors is described in Section 6.3.3, the eddy current proximity sensor.

### 6.1.4 Magnetic Resistance (Reluctance) and Self-Inductance

The analogy between the description of magnetic circuits and electrical circuits is demonstrated by the equations in Table 6.2.

Equation (6.7) links the intrinsic and extrinsic field variables in the electrical and magnetic domain, respectively (Chapter 2). The electrical conductivity $\sigma = 1/\rho$ opposes the magnetic permeability $\mu$. With Eq. (6.8) the field quantities $E$ and $H$ are converted to the circuit quantities $V$ and $I$. Equation (6.9) defines density properties, and Eq. (6.10) defines electrical and magnetic resistances, respectively. The latter is also called *reluctance*. Finally, Eq. (6.11) expresses the electric and magnetic resistances in terms of material properties and shape parameters: $l$ is the length of a device with constant cross section and $A$ its cross-section area.

In an electric circuit consisting of a series of elements, the current through each of these elements is the same. Analogously the flux through a series of magnetic

**Table 6.2** Comparison Between the Electrical and the Magnetic Domain

| Electrical Domain | Magnetic Domain | |
|---|---|---|
| $E = \dfrac{1}{\sigma} \cdot J$ | $H = \dfrac{1}{\mu} \cdot B$ | (6.7) |
| $V = \displaystyle\int E \cdot \mathrm{d}l$ | $n \cdot I = \displaystyle\int H \cdot \mathrm{d}l$ | (6.8) |
| $I = \displaystyle\iint J \cdot \mathrm{d}A$ | $\Phi = \displaystyle\iint B \cdot \mathrm{d}A$ | (6.9) |
| $V = R_e \cdot I$ | $n \cdot I = R_m \cdot \Phi$ | (6.10) |
| $R_e = \dfrac{1}{\sigma} \cdot \dfrac{l}{A}$ | $R_m = \dfrac{1}{\mu} \cdot \dfrac{l}{A}$ | (6.11) |

elements is the same. So the resistances (reluctances) of these elements can simply be summed to find the total reluctance of the series circuit.

The *self-inductance* of a magnetic circuit with coupled flux is found as follows. The induced voltage equals $V_{\mathrm{ind}} = n \cdot (\mathrm{d}\Phi/\mathrm{d}t)$ (when there are $n$ turns). Substitution of $\Phi$ using Eq. (6.9) yields $V_{\mathrm{ind}} = (n^2/R_m) \cdot (\mathrm{d}I/\mathrm{d}t)$ and since $V = L(\mathrm{d}I/\mathrm{d}t)$ the self-inductance is:

$$L = \frac{n^2}{R_m} = n^2 \cdot \frac{\mu A}{l} \tag{6.12}$$

So the coefficient of self-inductance (unit Henry, H = Wb/A) is proportional to the square of the number of turns and inversely proportional to the reluctance. Several sensors, based on a change in self-inductance and reluctance, will be further discussed in this chapter.

## 6.1.5 Magnetostriction

All ferromagnetic materials exhibit the magnetostrictive effect. Basically it is the change in outer dimensions of the material when subjected to an external magnetic field. In the absence of an external field the magnetic domains (elementary magnetic dipoles) are randomly oriented. When a magnetic field is applied, these domains tend to line up with the field, up to the point of saturation. The effect is not strong: materials with a large magnetostriction (for instance Terfenol-D) show a sensitivity of about 5 μstrain per kA/m, with a maximum strain between 1200 and 1600 μstrain at saturation [3].

The inverse magnetostrictive effect is called the Villari effect: a change in magnetization when the material is stressed. This effect is used in magnetostrictive force sensors, discussed in Section 6.3.6.


## 6.2  Magnetic Field Sensors

This section presents various sensors for the measurement of magnetic field strength or magnetic induction. In most mechatronic applications the magnetic field is not the primary measurement: combined with a magnetic source (e.g. permanent magnet and coil) they are used to measure displacement quantities and (with an elastic element) force quantities. Such sensors are discussed in Section 6.3. The sensors described in this section are: coil, Hall sensors, fluxgate sensors and magnetostrictive sensors. One of the most sensitive magnetic field sensors is the Superconducting QUantum Interference Device (SQUID). This sensor operates at cryogenic temperature (liquid helium or liquid nitrogen) and is used mainly in medical applications and for material research. They are rarely used in mechatronics.

Most research on magnetic field sensors is focussed on Hall sensors and fluxgate sensors, in particular to reduce dimensions and fabrication costs, by applying MEMS technology and integration with interface electronics (see further Section 6.4). Sometimes innovative concepts are introduced, but the application to mechatronic systems requires further development. An example of such a new principle is given in Ref. [4]. The sensor proposed herein consists of two thin flexible plates (cantilevers); when magnetized by an external field, the repulsive force causes a displacement of one of the plates relative to the other, similar to the classic gold-leaf electrometer. In force equilibrium the displacement is a measure for the external magnetic field.

Combining Eqs (6.3) and (6.6) gives for the flux through a magnetic circuit with $n$ windings and area $A$

$$\Phi(t) = n \cdot A(t) \cdot \mu(t) \cdot H(t) \tag{6.13}$$

where all parameters may vary with time due to a time-varying quantity. A magnetic or inductive sensor can be based on a change in each of these parameters, resulting in an induction voltage equal to

$$V(t) = \frac{d\Phi(t)}{dt} = n \cdot \left\{ \mu H \frac{\partial A(t)}{\partial t} + AH \frac{\partial \mu(t)}{\partial t} + \mu A \frac{\partial H(t)}{\partial t} \right\} \tag{6.14}$$

The remainder of this section deals with three types of magnetic field sensors: the coil, Hall sensors, fluxgate sensors and magnetostrictive sensors. Their principle of operation and performance are essentially based on this equation.

### 6.2.1 Coil

The most straightforward method for the transduction from magnetic field to an electric voltage is a coil: Eq. (6.5) relates the induced voltage in a coil to the magnetic flux. At first sight, only AC fields can be measured in this way since the induced voltage is proportional to the rate of change in flux. Static fields can nevertheless be measured, just by rotating the coil. Let the surface area of the coil be $A$ and the frequency of rotation $\omega$, then for a homogeneous induction field $B$, the induced voltage equals:

$$V(t) = -B \cdot A \cdot \frac{d \sin \alpha(t)}{dt} = -B \cdot A \cdot \omega \cdot \cos \omega t \tag{6.15}$$

With a rotating coil very small induction fields can be measured. Disadvantages of the method are movable parts, the need for brushes to make electrical connection to the rotating coil and for an actuator to procure rotation.

### 6.2.2 Hall Plate Sensors

The Hall plate is based on the magnetoresistive effect. In 1856 W. Thomson (Lord Kelvin) discovered that a magnetic field influences the resistivity of a current-conducting wire (see also Section 4.5 on magnetoresistive sensors). Later this effect was named the Gauss effect. Only after the discovery of the Hall effect in 1879, by the American physicist E.F. Hall, could the Gauss effect be explained. Both the Gauss and the Hall effects are remarkably stronger in semiconductors, so they became important for measurement science only after the development of semiconductor technology.

The Hall effect is caused by the Lorentz forces on moving charge carriers in a solid conductor or semiconductor, when placed in a magnetic field (Figure 6.3). The force $F_1$ on a particle with charge $q$ and velocity $v$ equals:

$$F_1 = q(v \times B) \tag{6.16}$$

The direction of this force is perpendicular to both $B$ and $v$ (right-hand rule). As a result the flow of charges in the material is deflected and an electric field $E$ is built up, perpendicular to both $I$ and $B$. The charge carriers experience an electric

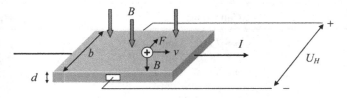

**Figure 6.3** Principle of the Hall sensor.

force $F_e = qE$ that, in the steady state, counterbalances the Lorentz force: $F_e = F_l$. Hence:

$$E = v \times B \tag{6.17}$$

Assuming all charge carriers have the same velocity, the current density $J$ equals $n \cdot q \cdot v$ with $n$ the particle density. When $B$ is homogeneous and perpendicular to $v$ (as in Figure 6.3), the electric field equals simply $E = JB/nq$. Finally with $I = b \cdot d \cdot J$ and $V = E \cdot b$, the voltage across the Hall sensor becomes:

$$V = \frac{1}{nq} \cdot \frac{IB}{d} = R_H \cdot \frac{IB}{d} \tag{6.18}$$

The factor $1/nq$ is called the Hall coefficient, symbolized by $R_H$. In p-type semiconductors holes are majority carriers, so $q$ is positive. Obviously the Hall voltage is inversely proportional to the thickness $d$. A Hall sensor, therefore, has often the shape of a plate (as in Figure 6.3), explaining the name 'Hall plate' for this type of sensor. The Hall voltage has a polarity as indicated in Figure 6.3. For n-type semiconductors, the charge carriers (electrons) are negative thus the polarity is just the inverse. The Hall coefficient is large in semiconductors, because the charge density is much lower compared to that in conductors.

The assumption of equal velocity for the charge carriers is only an approximation: the velocity distribution within the material in not completely uniform. This results in a deviation of the Hall coefficient from that given in Eq. (6.18). Practical values range from 0.8 to 1.2 times the theoretical values. Table 6.3 shows typical values of the Hall coefficient for various semiconductor compounds as used in Hall plates.

Hall plates can also be implemented in silicon. The thin 'plate' consists of the epitaxial layer of a silicon chip (see also Chapter 4). Currents that are generated in this layer (by on-chip current sources) deflect in a magnetic field, resulting in a voltage difference between two lateral points on the chip. This voltage difference is measured by some integrated electronic interface. The Hall coefficient can be tuned to a desired value by the doping concentration (see some examples in Table 6.3). Here too the measured value of $R_H$ deviates strongly from the value according to

**Table 6.3** Hall Coefficient and Resistivity of Various Semiconductors Used for Hall Plates and Integrated Silicon Hall Sensors [5,6]

| Material | Conditions | $R_H$ (m³/C) |
|----------|------------|--------------|
| GaAs | $\rho = 2 \cdot 10^{-3}$ $\Omega$m; $T = 25°C$ | $-1.7 \cdot 10^{-3}$ |
| InAs | $\rho = 1 \cdot 10^{-3}$ $\Omega$m; $T = 25°C$ | $-3.7 \cdot 10^{-3}$ |
| InSb | $\rho = 5 \cdot 10^{-5}$ $\Omega$m; $T = 25°C$ | $-3.8 \cdot 10^{-4}$ |
| p-Si | $n = 0.3 \cdot 10^{10}$ cm$^{-3}$ | 50 |
| n-Si | $n = 1.5 \cdot 10^{10}$ cm$^{-3}$ | $-150$ |

Eq. (6.18) because of the non-homogeneous velocity distribution of the charge carriers. Silicon technology offers the possibility of 2D and even 3D directivity. Such silicon Hall devices can be realized in bipolar, JFET as well in CMOS processes [6−9]. Obviously these technologies allow integration of all necessary circuitry, resulting in a very compact device.

When applying Hall plates for the measurement of magnetic fields, some shortcomings should be considered. First the construction is never fully symmetric: the voltage contacts are not precisely positioned perpendicular to the main axis of the plate. This results in an *offset voltage*: the Hall voltage is not zero at zero field. This error is called the resistive zero error and is proportional to the current through the plate. It can be compensated for by using a compensation voltage derived from the current. Uncompensated InSb and InGaAs Hall sensors have offsets in the order of 0.1 and 1 mT, respectively. Silicon Hall devices suffer from an even larger offset, arising from layout tolerances and material inhomogeneity. Various methods have been proposed to eliminate this offset (see Section 3.3), of which the spinning method is rather popular [10]. Typical offset values range from 10 to 50 mT for standard silicon Hall-effect devices down to 10 μT for offset-compensated designs.

In an AC magnetic field, the output voltage of a Hall plate is not zero, even at zero current. The changing flux induces a voltage in the loop consisting of the connection wires and the voltage contacts. This is called the inductive zero error. Such errors are small in integrated Hall-effect sensors because of the small dimensions of such devices.

As shown in Section 3.3, *modulation* (and in particular chopping) is an effective way to circumvent offset. Chopping has been demonstrated to be feasible for magnetic fields as well [11]. The chopping is realized by surrounding the Hall sensor by a torroidal coil with a ferromagnetic core of high permeability, as a magnetic shield. When the coil is not activated, the core acts as a magnetic shield, and the external magnetic field cannot reach the sensor. When the coil is activated such that the core is saturated, the permeability is low, and hence the shielding effect is also low; the sensor receives almost the total external field to be measured.

Another way to improve the performance of a magnetic field measurement (and in particular for Hall sensors) is the application of so-called *flux concentrators*. The sensor is provided with a piece of material with high permeability; the magnetic field to be measured is locally concentrated in this material and the Hall sensor is positioned in that region of concentrated field lines. The ferromagnetic material acts as a field amplifier, so very weak fields can be measured in this way. Depending on the material and the shape of the concentrators, a gain of more than 6000 can be achieved [12]. Figure 6.4 illustrates the effect of flux concentration. The piece of iron located at the right side of the permanent magnet constricts the field lines of the otherwise symmetric field pattern, due to its lower reluctance compared to air (as on the left side). The method is also applicable in CMOS-type Hall sensors, by deposition of thin layers of high permeability material nearby and even on top of the chip [13]. Condition for an accurate measurement is that the field gain has a known value.

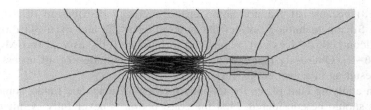

**Figure 6.4** Effect of flux concentrator; the rectangular piece of iron at the right of the permanent magnet deforms the original field pattern.

Traditional Hall sensors are sensitive in one direction only, perpendicular to the plate or chip. A triaxial sensor can simply be created by mounting three of these sensors orthogonally on a common base. This solution is chosen in Ref. [14] for the realization of a magnetodosimeter for monitoring exposure to strong magnetic fields. To achieve a low offset, spinning current Hall devices (manufactured in CMOS technology) are used.

Hall sensors are still a subject of research, aiming at higher sensitivity, lower offset and reduced manufacturing costs. Special technological knacks make it possible to create 2D and 3D Hall sensors in silicon (see for instance Refs [15,16]). The Hall effect in standard silicon Hall sensors is located in a thin plate-like layer close to the surface of the chip. To obtain sensitivity in the third direction, perpendicular to the plate, sensors are constructed with a vertical sensitivity [17].

Not only crystalline silicon but also other materials are investigated for their applicability to Hall sensors. Polycrystalline Hall sensors have been reported with a sensitivity of 19 mV/T [18], or a plasma as sensitive material, resulting in a 10 mV/Gauss sensitivity [19]. Hall sensors with a self-calibration functionality have been reported as well. The sensor in Ref. [20] has a built-in reference (a coil) to perform self-calibration; all circuit parts are implemented in CMOS technology.

### 6.2.3  Fluxgate Sensors

Like the Hall plate, the fluxgate sensor measures magnetic field strength and is therefore a suitable device in mechatronic systems, for instance as part of a displacement sensor (combined with a permanent magnet or an activated coil). Basically the fluxgate sensor (or saturable-core magnetometer) consists of a core from soft magnetic material and two coils: an excitation coil and a sense coil (Figure 6.5C).

The excitation coil supplies an AC current that periodically brings the core into saturation. Hence the permeability of the core material changes with twice the excitation frequency, between values corresponding to the unsaturated and the (positive and negative) saturated states. An external magnetic field $H$ produces an additional induction field $B$ in the core of the sensor, according to Eq. (6.6). Since the permeability varies periodically, so does this added field: it is modulated by the varying

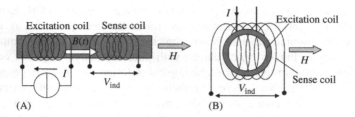

**Figure 6.5** Structure of a fluxgate sensor: (A) basic structure and (B) ring shaped.

permeability. In the sense coil a voltage is induced according to Eq. (6.5). Assuming homogeneous and uniform fields, $\Phi = n \cdot \mu \cdot H \cdot A$, with $A$ the cross section of the core and $n$ the number of turns. The voltage of the sense coil due to the external field only becomes:

$$-V_{ind} = \frac{d\Phi}{dt} = \frac{d\{nA\mu_0\mu_r(t)H\}}{dt} = nA\mu_0 H \frac{d\mu_r(t)}{dt} \tag{6.19}$$

where $H$ is assumed to be a static external field. Hence the output voltage of the sensor is periodic, with an amplitude proportional to the magnetic field strength to be measured. Since the relation between $B$ and $H$ is strongly non-linear, the shape of the output voltage is a distorted sine wave, almost pulse like. Frequency selective detection of the output (synchronous demodulation on the second harmonic of the excitation frequency, see Chapter 3) makes the measurement highly insensitive to interference. A more detailed description of the operating principle is given in (for instance) [21].

Since the actuating current produces an AC induction field $B(t)$ and hence an output voltage in the sense coil, the output is not zero at zero external field. To counterbalance this voltage, a second excitating coil could be added. A more elegant way is to employ a ring-shaped core (Figure 6.5B). The excitation coil is wound toroidally around the ring, whereas the sense coil fully encloses the ring, with its radial axis in the plane of the ring. At $H = 0$, the induced voltage is zero because of ring symmetry: the sensor is in balance. A field along the axis of the sense coil disturbs this symmetry, resulting in an output voltage proportional to the field strength. The sensitivity strongly depends on the core material, the coil configurations and the interfacing (e.g. excitation and sensing).

A further improvement of the sensor performance is obtained by the feedback principle (see Chapter 3). The output of the phase-sensitive detector is amplified and activates a feedback coil around the sense coil. When the loop gain is sufficiently large the resulting induced field is zero, whereas the amplified sense voltage serves as the sensor output (analogous to the feedback accelerometer in Figure 3.9).

Coils are relatively bulky components, and are not easy to miniaturize. Many attempts have been made to decrease the size of fluxgate sensors without deteriorating its performance. A first step is to create coils on a PCB. Examples of flat core designs for fluxgate sensors are found in Refs [22−25]. A next step towards further miniaturization is to implement the coils in IC-compatible or micromachining technologies, combined with microstructured magnetic core material. Various configurations and technological solutions have been proposed and tested to further reduce size, increase sensitivity, improve directivity, extending the frequency range and create multi-axis sensors, see for instance Refs [26−32]. Interfacing of fluxgate sensors (e.g. the actuation signal and the demodulation method) is an important issue to achieve maximum results [33,34].

Table 6.4 lists some maximum specifications of commercial fluxgate sensors. Table 6.5 compares major properties of various magnetic field sensors discussed so far.

## 6.3 Magnetic and Induction Based Displacement and Force Sensors

In this section we discuss various types of sensors based on induction or a change in magnetic field strength. Most of these sensors are constructed to determine linear or angular position, or their time derivatives (e.g. velocity and acceleration).

**Table 6.4** Specifications of Fluxgate Sensors

| Parameter | Value (typ.) | Unit |
|---|---|---|
| Range | $10^{-4}$ | T |
| Resolution | $10^{-10}$ | T |
| Sensitivity | 10 | $\mu$V/nT |
| Noise | 10 | pT/$\sqrt{\text{Hz}}$ |
| Excitation frequency | 0.4−100 | kHz |
| Bandwidth | 1 | kHz |
| Offset drift | 1 | pT/K |
| Maximum temperature | 70 | °C |

**Table 6.5** Comparison of Various Magnetic Field Sensors

| Property | Fluxgate | AMR | Hall (Si) | Unit |
|---|---|---|---|---|
| Range | $10^{-4}$ | $10^{-3}-10^{-2}$ | 100 | T |
| Sensitivity | $4 \cdot 10^4$ | 5 | 0.1 | V/T |
| t.c. sensitivity | 30 | 600 | | ppm/K |
| Linearity | $10^{-5}$ | $10^{-2}$ | | − |
| Offset | 0.001−0.01 | 10−100 | 1−100 | $\mu$T |
| t.c. offset | 0.03−0.2 | 10−100 | | nT/K |
| Bandwidth | $10-10^4$ | $10^6-5 \times 10^6$ | | Hz |

### 6.3.1 Magnetic Proximity Switches

A simple, versatile magnetic proximity sensor is the reed switch: a magnetically controlled switch consisting of two magnetizable tongues or reeds in an hermetically sealed housing filled with an inert gas (Figure 6.6A).

When exposed to a magnetic field, the reeds become magnetized and attract each other, making electrically contact. When the field disappears, the reeds open by their own elasticity. A proximity sensor is built by combining the reed switch with a movable permanent magnet or coil (Figure 6.6B and C).

Reed switches are available in normally-on and normally-off types, and with built-in switching coils. When the switch is closed, its resistance (the on-resistance) should be as low as possible. To that end, the reeds are covered with a thin layer of ruthenium or an alloy of gold and ruthenium: materials with a very low contact resistance. The on-resistance is in the order of 0.1 $\Omega$. Conversely the isolation resistance or off-resistance should be as high as possible; practical values are in the order of $10^{12}$ $\Omega$.

A reed switch is a mechanical device and, therefore, the switching speed is low compared to electronic switches (typically 0.2 ms turn-on time, 0.02 ms turn-off time). Further, the switch is susceptible to wear (life times in the order of $10^7$ switches). Another disadvantage of the reed switch is the *bouncing* of the reeds during switching, caused by the mass-spring behaviour of the reeds. Finally the switch shows hysteresis: the turn-on value exceeds the turn-off value. Some additional characteristics are given in Table 6.6.

Reed switches are applied in numerous commercial systems: from cars (e.g. monitoring broken lights and level indicators), up to electronic organs (playing contacts), in telecommunication devices and in test and measurement equipment.

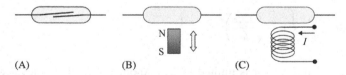

(A)  (B)  (C)

**Figure 6.6** Reed switch (A) structure, activated by (B) a permanent magnet (proximity sensor) and (C) a coil (switch).

**Table 6.6** Characteristics of Reed Switches with Built-In Switching Coil (Small to Large Types)

| Parameter | Typical Values | Unit |
| --- | --- | --- |
| Turn-on value | 8−16 to 50−75 | ampere-turns |
| Turn-off value | 4−14 to 17−30 | ampere-turns |
| Maximum switching power | 10−25 | W |
| Maximum switching current | 750 | mA |
| Maximum switching voltage | 100−250 | V |

### 6.3.2 Inductive Proximity and Displacement Sensors

The general principle of a displacement sensor based on variable self-inductance is depicted in Figure 6.7A. The self-inductance of the configuration is approximately equal to $L = n^2/R_m$, according to Eq. (6.12). A displacement $\Delta x$ of the movable part causes the reluctance $R_m$ to change because the air gap width changes by an amount of $2\Delta x$. With Eq. (6.12), and noting that the iron and the air gaps are in series and hence their reluctances add, the self-inductance modifies according to

$$L(x) = \frac{\mu_0 A n^2}{l_r/\mu_r + 2(x_0 + \Delta x)} = \frac{\mu_0 A n^2}{l_0 + 2\Delta x} = \frac{L_0}{1 + 2\Delta x/l_0} \tag{6.20}$$

where $\mu_r$ is the permeability of the iron core, $l_r$ the path length of the flux through the iron part, $x_0$ the initial air gap, $L_0$ the self-inductance at the reference position $x_0$ and $l_0 = l_r/\mu_r + 2x_0$ the total effective flux path in the initial position. Then at a displacement $\Delta x$ the self-inductance changes by

$$\Delta L = -L_0 \frac{2\Delta x/l_0}{1 + 2\Delta x/l_0} = -L_0 \frac{2\Delta x}{l_0}\left\{1 - \frac{2\Delta x}{l_0} + \frac{1}{2}\left(\frac{2\Delta x}{l_0}\right)^2 - \cdots\right\} \tag{6.21}$$

This result expresses a strongly non-linear sensitivity. A differential configuration gives some improvement (Figure 6.7B). The sensitivity is doubled, whereas the non-linearity is reduced to a second-order effect, as is shown in Eq. (6.22):

$$\Delta L = L_2 - L_1 = \frac{L_0}{1 - 2\Delta x/l_0} - \frac{L_0}{1 + 2\Delta x/l_0} = L_0 \frac{4\Delta x/l_0}{1 - (2\Delta x/l_0)^2}$$

$$= L_0 \frac{4\Delta x}{l_0}\left\{1 + \left(\frac{2\Delta x}{l_0}\right)^2 - \cdots\right\} \tag{6.22}$$

The range of this sensor is limited to $2x_0$. Another variable inductance configuration with a much wider range is shown in Figure 6.8A: a coil with movable core. Since the self-induction varies non-linearly with displacement too, the differential configuration in Figure 6.8B is better because quadratic terms cancel, as was demonstrated in the previous equations.

A common disadvantage of all self-inductance types is the coil itself. The device is not an ideal self-inductance: the wire resistance and the capacitance between the

**Figure 6.7** Displacement sensors based on variable self-inductance: (A) single and (B) differential.

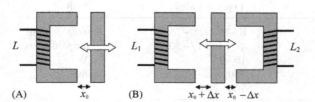

$L$     $L_1$     $L_2$

(A)    $x_0$     (B)     $x_0 + \Delta x$   $x_0 - \Delta x$

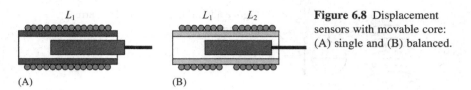

$L_1$          $L_1$    $L_2$

(A)          (B)

**Figure 6.8** Displacement sensors with movable core: (A) single and (B) balanced.

turns cannot be neglected. Further the sensors suffer from iron loss (by eddy currents) and hysteresis. Due to these factors, the sensor impedance deviates remarkably from a pure self-inductance. In Section 6.3.4 a type of inductive displacement sensor will be introduced in which most of these drawbacks do not apply.

### 6.3.3  Eddy Current Displacement Sensors

Eddy currents originate from induction: free-charge carriers (electrons in a metal) experience Lorentz forces in a varying magnetic field, and cause currents to flow in that material (Section 6.1.3). In an eddy current sensor this effect is used to measure displacement.

The self-inductance of an isolated coil (not surrounded by any material) can be approximated by Eq. (6.12). When a conductive or ferromagnetic object approaches the coil, its self-inductance will change. We distinguish two possibilities: a conductive object and a ferromagnetic object.

A conductive object located in the magnetic field of a coil activated with an AC current experiences an induction voltage according to Eq. (6.5), and since the material is conductive, currents will flow through the object (Figure 6.9). They are called *eddy currents* or Foucault currents, because they follow circular paths within the object. These currents produce a magnetic field that counteracts the original field from the coil (Lenz's law). Hence the flux produced by the coil will reduce and, since $L = \Phi/I$, the self-inductance decreases due to the presence of the conductive object. When the object comes closer to the coil (where the magnetic field is stronger), the eddy currents become larger, and so the self-inductance further decreases. This effect is stronger in materials with a low resistivity, because of the higher eddy currents. In conclusion the sensor impedance *decreases* upon the approach of a *conducting, non-ferromagnetic* material.

The situation is different when the object is ferromagnetic. The reluctance of a ferromagnetic material is $R_m = l/\mu_0\mu_r A$, where $l$ is the length of the (closed) flux line in the magnetic structure (composed of the sensor, the target and the surrounding air) and $A$ the cross section. Further the self-inductance equals $L = n^2/R_m$. Hence when a ferromagnetic material approaches a coil, the impedance of the sensor coil *increases*; the effect is stronger for materials with a high permeability $\mu_r$. So the sensitivity of an eddy current sensor depends strongly on the material the movable object is made of. Figure 6.10 shows typical transfer characteristics (impedance change versus target distance) for various materials.

The sensor impedance is a rather complicated function of the distance: the frequency, the nature of the material and the orientation of the object have substantial

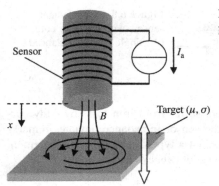

**Figure 6.9** Principle of an eddy current proximity sensor.

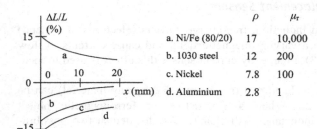

| | $\rho$ | $\mu_r$ |
|---|---|---|
| a. Ni/Fe (80/20) | 17 | 10,000 |
| b. 1030 steel | 12 | 200 |
| c. Nickel | 7.8 | 100 |
| d. Aluminium | 2.8 | 1 |

**Figure 6.10** Relative sensitivity $\Delta L/L$ (%) of an eddy current proximity sensor, for various materials (after Kaman Measuring Systems).

influence on the sensor sensitivity. Other metal or ferromagnetic objects in the neighbourhood of the sensor may reduce the measurement accuracy.

Figure 6.11 shows an electric model of the sensor. The network elements in this model have the following meaning:

- $R_1$: copper losses of the coil (increases with frequency due to the skin effect)
- $R_2$: resistance of the object (depends on the conductivity and the frequency, the latter due to the skin effect and the permeability)
- $L_1$: coil's self-inductance
- $L_2$: measure for the counter-field produced by the eddy currents in the object (depends strongly on the magnetic properties of the material)
- $M$: measure for the coupling between sensor and object (hence dependent on the distance); another measure for the coupling is the coupling factor, defined by $k^2 = M^2/L_1L_2$.

In Ref. [35] more detailed analysis of an eddy current displacement sensor is given. In this approach the target is divided in concentric circles carrying the eddy currents. In Ref. [36] the current density and contours of eddy currents are studied with the Maxwell equations as a starting point.

The measurement frequency of an eddy current sensor is in the order of 1 MHz. Some types contain a second or compensation coil, arranged in a half-bridge or a transformer configuration. More specifications are listed in Table 6.8 at the end of this section. Non-ferrous and isolating objects normally give no response; however, by applying a thin conducting layer, for instance an aluminium foil, they can be

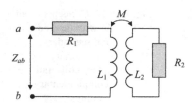

**Figure 6.11** Simplified model of an eddy current sensor (left part) with target object (right part).

detected as well, since eddy currents flow just near the surface of the conductor (*skin effect*).

Two basic eddy current sensor types are distinguished: simple coil (as in Figure 6.9) and differential coil. The latter consists of two adjacent or concentric coils, one of which is activated, the other of which acts as a sensor. The coupling between the two coils is determined not only by the sensor structure but also by eddy currents induced in the object near the coils.

Wire-wound coils are bulky. Special designs have been developed to reduce the dimensions. A first step to a smaller and cost-effective coil is the development of flat coils on a PCB (cf. fluxgate sensors). The coils are either single sided, having spiral conductors just on one side of the PCB [37−39], double sided (half of each turn is deposited on the top side, the other half on the back side, and connected by conducting material in small holes through the PCB [40] or multilayered [41]. A further step is the application of micromachining technology, for instance the LIGA process [42], and the integration of the coils with the read-out electronics [43,44].

### 6.3.4  Variable Differential Transformers

A coil with a movable core as displacement sensor (Figure 6.8) shows strong non-linearity even when configured in balance. A much better linearity is obtained with a transformer with a movable core, the Linear Variable Differential Transformer (LVDT). Figure 6.12A shows an LVDT with one primary and one secondary coil. The coils have tapered windings in opposite directions. Without a ferromagnetic core, the coils are only weekly coupled: a voltage on the primary coil gives only a small output voltage at the secondary coil.

However when a ferromagnetic core (with high permeability) is inserted in the hollow cylinder, the coupling between the coils is much stronger at the position of the core. The transformer ratio varies from high to low when the core moves from left to right, according to the varying winding ratio at the core position. So the output voltage varies with core displacement. Although irregularities in the windings are smoothed out by the core, the linearity is not very high.

A better construction is the one shown in Figure 6.12B. This device consists of one primary coil and two secondary coils, positioned symmetrically with respect to the primary, and wound in an opposing direction. When the core is in its centre position, the voltages over the secondary coils are equal but with opposite polarity (due to the winding directions). The total output voltage of the two coils *in series* is zero. When the core shifts from its centre position, the output voltages are not

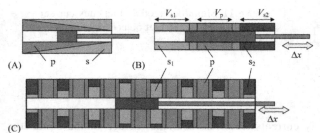

**Figure 6.12** Concepts of an LVDT (A) with tapered primary and secondary, (B) with one primary and two secondary coils and (C) with complementary tapered windings.

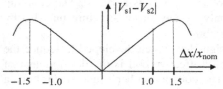

**Figure 6.13** Typical transfer characteristic of an LVDT.

equal anymore: the one becomes higher while the other lower. The difference is a measure for the distance from the centre position (Figure 6.13).

Note that the output amplitude is equal for positive and negative displacements. Information about the direction of displacement is given by the phase of the output signal: the phase with respect to the primary voltage is either 0 or $\pi$, depending on the direction of the displacement.

The stroke of the construction in Figure 6.12B is limited to about 1/3 of the total length. A shorter core would result in a much larger range but at the expense of linearity. The solution is obtained by distribution of the primary and secondary windings over the sensor body as depicted in Figure 6.12C. The distribution is such (tapered) that the output amplitude varies linearly with core displacement, over a range about 80% of the length [45]. End-of-stroke non-linearity is further compensated for by additional turns at the extremities.

The LVDT is operated with an AC voltage; its amplitude and frequency can be chosen by the user within a wide range (typically 1–10 V, 50 Hz to 10 kHz). Its sensitivity is expressed in mV (output) per volt (input) per m (displacement), so in $mV\ V^{-1}\ m^{-1}$. Users who wish to avoid the handling of AC signals (i.e. amplitude and phase measurement) can choose the (more expensive) DC–DC LVDT, with built-in oscillator and phase-sensitive detector (see Chapter 3 for details on modulation and synchronous detection): both input (supply voltage) and output (displacement signal) are DC voltages.

A stainless steel housing guarantees proper shielding against mechanical and chemical influences as well as against electrical and magnetic interference. Special types withstand high pressures (up to 200 bar) or are radiation resistant. To further improve the performance, special constructions and compensation techniques have been developed. A self-compensating arrangement with two sets of secondary coils is presented in Ref. [46]. The two sets are exposed to the same environmental influences and excitation voltages, but they generate two outputs: a sum and a difference voltage. The ratio of the two is independent on common interferences (more or less similar to the post-processing method for a PSD in Chapter 7). A 2D version of the LVDT is presented in Ref. [47], with a range of 30 cm in two directions. A special design for very low temperatures (70 K) is given in Ref. [48].

To illustrate the quality of standard LVDT sensors, Table 6.7 presents selected specifications of a typical commercial type LVDT. The overview Table 6.8 at the end of this section lists some general characteristics of the LVDT.

The rotational version of the LVDT is the RVDT. The principle of operation is the same as for the LVDT. Due to the particular configuration of the coils, the range is restricted to less than 180° rotation.

### 6.3.5 Resolvers and Synchros

Inductive sensors for the full range of $2\pi$ are the *resolver* and the *synchro*. These types are based on a variable transformer as well, with fixed and rotating coils. The coupling between primary and secondary coils depends on the angle between the coils. The resolver (Figure 6.14A) essentially consists of two fixed coils (stator) making angles of $\pi/2$ and one rotating coil, the rotor.

The voltages across the primary coils have a phase difference $\pi/2$ rad: $V_1 = V \cos \omega t$ and $V_2 = V \sin \omega t$, respectively. Both stator coils induce voltages in the rotor coil; the two induced voltages add up in the rotor. Assuming equal amplitudes of the stator voltages, the rotor voltage is:

$$V_3 = a \cdot V \cdot \cos \omega t \cos \alpha + a \cdot V \cdot \sin \omega t \sin \alpha = a \cdot V \cdot \cos(\omega t + \alpha) \qquad (6.23)$$

**Table 6.7** Product Information of an LVDT (Novotechnik, Germany)

| | |
|---|---|
| *Electrical and mechanical data* | |
| Mechanical range | 12 mm |
| Required measuring force | 4 N |
| Weight | 90 g |
| Electrical range | 10 mm |
| Absolute non-linearity | $\pm 0.2\%$ |
| Operating voltage | 18−30 V (DC) |
| Signal out | 4−20 mA |
| Current consumption | 50 mA |
| Load impedance | 0−500 ohm |
| Temperature coefficient | <80 ppm/°C |
| | |
| *Environmental data* | |
| Temperature range | −25 to +70 °C |
| Frequency of operation | maximum 10 Hz at 10 mm measuring stroke |
| Shock | 50 g |
| Mechanical life | $10^8$ operations |

Hence the phase relative to $V_1$ is equal to the geometric angle of the rotor (Figure 6.14B). The coefficient $a$ in Eq. (6.23) accounts for the transfer between stator and rotor.

Instead of supplying the input voltage to the stator coils (so-called stator supply mode), the input voltage can be supplied to the rotor coil as well (rotor supply mode): in this case the stator coils provide two output signals from which the mechanical angle can be derived. Multi-pole resolvers have two stator coils and a $2p$ wound rotor. The output phase at the stator coils varies over a full period when the rotor rotates over an angle of $2\pi/p$, so the resolution is increased by a factor $p$.

Configurations with three stator coils are called synchros. The three coils make geometric angles of $120°$ and are supplied with three sinusoidal voltages which also have phase differences of $120°$. The principle is the same as for the resolver.

Resolvers and synchros have a measurement range of $2\pi$ rad. Measurement frequencies are around 10 kHz. The performance is mainly limited by the read-out electronics, in particular the phase measurement circuitry. The accuracy can be better than 0.001 rad (3 min of arc or minarc), whereas a resolution down to 40 s of arc can be achieved ($1° = 60$ min of arc = 3600 s of arc). An analysis of the various kinds of errors can be found in Ref. [49], together with a compensation method to reduce errors. Other attempts to reduce errors are found in, for instance, [50], using a look-up table (so calibration is required) and [51], in which special electronics perform an accurate conversion from resolver output to a DC signal with an error as low as $0.01°$.

A variation on the resolver is the *inductosyn*®. We discuss only the linear inductosyn® here: the rotational version operates in the same way. The linear inductosyn® (Figure 6.15) consists of a fixed part, the ruler or scale, and a movable part, the slider.

Both the ruler and the slider have meander-shaped conducting strips on an isolating carrier (like a printed circuit board) and isolated from each other by a thin insulating layer. The ruler acts as one elongated coil; its length determines the measurement range. The slider consists of two short coils with the same periodicity as that of the ruler. However, the geometric distance between the two parts of the slider is $(n + ¼)$

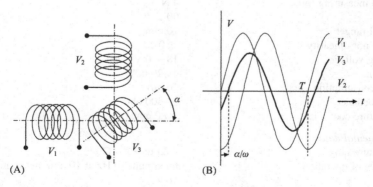

**Figure 6.14** Resolver: (A) operating principle and (B) input and output voltages.

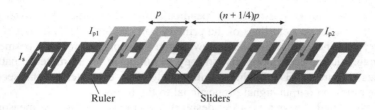

**Figure 6.15** Layout of an inductosyn®.

of a geometric period $p$. The inductosyn® is activated in a similar way as the resolver: the sliders are supplied with AC currents $I_{p1}$ and $I_{p2}$ with $\pi/2$ phase difference. The ruler receives the sum of two induced voltages, according to Eq. (6.23). Again the amplitude of the output signal is almost independent of the slider position; the phase is a measure for the displacement. Unlike the resolver, the inductosyn® has no iron, so the operating frequency can be higher: up to about 150 kHz.

To measure absolute displacement over the total extent of the ruler, the system must keep track of the number of electrical periods that have passed. In this way an almost infinite range can be achieved. The inductosyn® is very suitable for accurate positioning in, for instance, a lathe.

Table 6.8 compares major specifications of the various inductive sensors discussed in this chapter.

### 6.3.6  Magnetostrictive or Elastomagnetic Sensors

A magnetostrictive (or magnetoelastic or elastomagnetic) sensor is based on the change of the magnetic flux in a magnetic material due to deformation. Therefore the effect is mainly used for force sensing. Basically two different magnetoelastic sensor systems can be recognized: uni- and multidirectional. Sensors of the first type consist of a coil with (cylindrical) core from a suitable magnetostrictive material (Figure 6.16A). Upon compression of the core the permeability changes, and so do the reluctance and the self-inductance, according to the empirical and linearized expression:

$$\frac{\Delta L}{L} = -\frac{\Delta R_m}{R_m} = \frac{\Delta \mu_r}{\mu_r} = k \cdot T \tag{6.24}$$

where $k$ is the sensitivity (m²/N) and $T$ is the mechanical stress in the material. The sensitivity is small: about $2.10^{-9}$ m²/N. Furthermore, the output signal depends on the temperature and on the frequency and amplitude of the supply voltage. The overall accuracy is several percent.

Multidirectional devices use the effect that magnetic flux lines change direction when a force is applied. Such sensors can be used for the measurement of force or torque and using special designs for both force and torque simultaneously [52]. Figure 6.16B shows a magnetoelastic force sensor configured as a transformer. The core consists of a pile of ferromagnetic lamellae. Both primary and secondary coils have only a few turns, positioned perpendicularly to each other. When no load is applied, the magnetic field lines of the primary coil (Figure 6.16C) do not cross the secondary coil, hence the transfer is low. Upon an applied force, the permeability of the magnetic material looses its isotropy, resulting in an asymmetric pattern of field lines (Figure 6.16D). In this situation field lines can intersect the secondary coil, generating an output signal proportional to the force.

In the so-called *Torductor*® [53], designed for contact-free torque measurement of a (rotating) axis, this axis itself acts as the magnetostrictive medium. It makes up the magnetic joint between the primary and the secondary coils of a transformer. The primary coil introduces a magnetic field in the axis. The transformer is configured in such a way that at zero torque the transfer is zero too. An applied torque results in torsional stress (cf. Figure 4.14) causing a change in direction of the field lines, according to the Villari effect. This in turn results in an induced voltage in the secondary coil of the transformer, similar to the torque sensor in Figure 6.16B. By a set of magnetoelastic transducers with proper configuring the magnetic cores and the primary and secondary windings, a combined force and torque sensor can be realized, with reasonably small crosstalk between the two measurands [52].

An alternative for the magnetostrictive metal is stress-sensitive amorphous ribbon [54]. When taped on the curved surface of an axis the ribbon is stressed, becomes anisotropic and exhibits magnetoelastic properties that can be used to measure torque, similar to strain gauges.

Magnetostriction can also be used for linear displacement sensing, through use of a special design. One such design consists of a rod made up of a magnetostrictive material (the fixed part, as the body of a linear potentiometer). A movable coil surrounding the rod is connected to the part whose position or displacement has to be measured (cf. the wiper of the potentiometer). The coil is activated by a short pulse, which generates an ultrasonic wave in the rod. The wave travels from the coil towards each of the ends of the rod, where ultrasonic receivers measure the travel times. Clearly the position of the coil follows directly from the travelled path of the two waves. Magnetostrictive-related hysteresis limits the accuracy, but this can be reduced (up to a factor of 10) by a proper biasing of the magnetic material [55].

This section concludes with a linear induction-type position sensor of which the concept is shown in Figure 6.17. The structure is similar to the magnetostrictive sensor from the previous paragraph. It consists of an elongated soft magnetic core

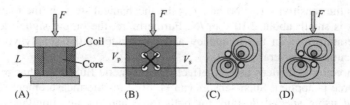

**Figure 6.16** Magnetostrictive force sensors: (A) unidirectional sensor, (B) transformer configuration, (C) field lines at zero force and (D) field under stress [53].

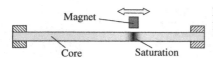

**Figure 6.17** Linear displacement sensor based on local saturation.

**Table 6.8** Overview of Characteristics of Various Inductive Sensors

| Type | Range[a] | Sensitivity; Resolution | $T_{max}$ (°C) |
|------|-------|------------------------|----------------|
| Differential coil | 0.1−200 mm | 10−100 mV/(mm V) | |
| Eddy current | 0.1−60 mm | 0.1−5 V/mm | 300 |
| LVDT | ±1 mm to ±50 cm | 1−500 mV/(mm V) | 500 |
| RVDT | ±40 to 360 | 1−10 mV/(°V) | 500 |
| Resolver | $2\pi$ | $2\pi$/rev | |
| Inductosyn®, linear | Several m | 0.1 μm | |
| Inductosyn®, rotational | $2\pi$ | 0.05 arcsec | |
| Coded magnetic strip | 10−50 cm | − | 60 |
| Steel band | 3 m to ∞ | 50 μm | 50 |
| Magnetostrictive | 10−$10^3$ N m | $10^{-9}$ m²/N | 120 |
| Proximity | 1−100 mm | 1% FS reproducibility | 70 |

[a]Range in FS, so from 0 up to the indicated value, for various types.

and a magnet (or activated coil); the latter moves along the core similar to the wiper of a potentiometer.

At the position of the magnet, the core is locally saturated, so the permeability of the core at that position has a much lower value compared to the remainder of it. The total reluctance of the structure depends now on the position where the core is saturated: the magnet acts as a magnetic shortcut of the core. The position-dependent reluctance can be measured in various ways. In Ref. [56] this is accomplished by an LVDT-like configuration with primary and secondary coils, and in Ref. [57] by a self-inductance measurement in differential configuration. In both these cases the prototypes have a measurement stroke of about 20 cm.

# 6.4 Applications

Inductive and magnetic sensors cover a wide application area. In mechatronics they can be used to measure many quantities: linear and angular displacement and their time derivatives, force, torque and material properties. Eddy current sensing is also used to detect defects in metal structures or for quality monitoring in production.

It is important to note the difference between the various sensor groups. Figure 6.18 presents an overview of the various sensing principles using magnetic or inductive sensing:

- *Magnetic type*: the *output* of the magnetic sensor varies with distance between sensor and magnet.

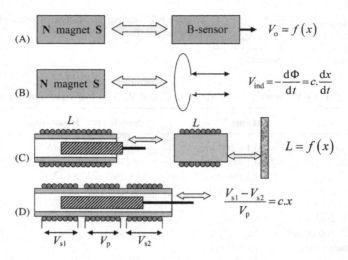

**Figure 6.18** Various sensor principles: (A) magnetic, (B) inductive, (C) inductance and (D) transformer type.

- *Induction type*: the moving magnet induces a *voltage* in a coil; the output is proportional to the (relative) velocity.
- *Inductance type*: the *self-inductance* of the coil(s) varies with the position of the moving core.
- *Transformer type*: the *transformation ratio* varies with the position of the moving core.

### 6.4.1 *Interfacing Inductive Sensors*

According to the type of sensor, an interface circuit has to be designed for the conversion of its output parameter into a voltage or current. Modern integrated magnetic field sensors have a built-in interface: the output is a voltage or sometimes a current or a frequency proportional to the applied magnetic field. The output of an induction-type sensor is a voltage, which only needs to be amplified if the voltage level is too low. It should be pointed out that though the overall sensitivity is increased by the gain of the amplifier, the S/N ratio is decreased, or at best remains the same when the amplifier noise is small compared to the noise from the sensor.

The output quantity of an inductance sensor is a change in self-inductance. The conversion of this quantity into an electrical signal can be accomplished in similar ways as for capacitive sensors (Figure 5.11). However while capacitors behave as an almost ideal capacitance, this is not the case for coils: the wire resistance and the capacitances between the turns are usually not negligible, making the impedance of a coil differ from just a pure inductance.

Most integrated interface circuits are based on a relaxation oscillator because of circuit simplicity. The relaxation oscillator generates a pulse or square-shaped periodical signal with a frequency determined by the value of the self-inductance of the sensor. Since the method is essentially analogue, a high resolution can be achieved, down to 50 nm [58].

A completely different approach is given in Ref. [59], where the frequency-dependent transfer characteristic of a low-pass filter (including the sensor impedance) is used to convert a change in self-inductance into a change in the output amplitude of the filter.

### 6.4.2  Contact-Free Sensing Using Magnetic and Inductive Sensors

The magnetic displacement sensors in this section are based on a position-dependent magnetic field strength. The strength of a magnetic field, originating from a permanent magnet or an activated coil, depends on the distance from the magnetic source [compare Eq. (6.1)]. Generally the strength decreases with increasing distance from the source, in a way that depends on the pattern of magnetic field lines (as represented in Figure 6.2).

Magnetic field strength can be measured by any magnetic sensor, for instance a Hall plate, a fluxgate sensor or a detection coil. The latter responds to fluctuating fields only, so in this case the magnetic field must be generated by an AC-activated coil.

Displacement measurements using a magnet−sensor pair have a limited range, a strong non-linearity (depending on the relative movement, Figure 6.19), and are sensitive to the orientation of the source relative to the sensor. Ferromagnetic construction parts in the neighbourhood of the sensing system may disturb the field pattern. The major application, therefore, is as *proximity sensor* or as *switch*.

An important parameter is the switching distance, which depends on the mechanical layout of the sensor but also on the dimensions and material of the moving part. In commercial systems the switching point can be adjusted over a limited range by varying the sensitivity of the electronic interface (see Table 6.8).

Long range displacement measurements can be realized using permanently magnetized strips. Figure 6.20 shows some possible configurations. One option is a strip that is magnetized with alternating north and south poles (Figure 6.20A).

A magnetic field sensor moves along the strip, like the slider in a potentiometer. Starting from a reference position, an electronic counter keeps track of the number of transitions from negative to positive sensor output, similar to the optical incremental encoder that will be discussed in Chapter 7. With two sensors spaced $(n + \frac{1}{4})$ of a magnetic period (which can be as small as 200 μm), the direction of movement can be determined as well. The reference marks are not susceptible to

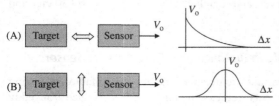

**Figure 6.19** Two general configurations for magnetic displacement sensors; the block arrows show relative movement between magnetic source and sensor.

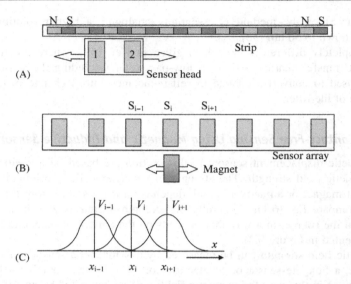

**Figure 6.20** Long range displacement sensing with magnetic sensors: (A) array of magnets and single sensor, (B) array of sensors and single magnet and (C) sensitivity characteristics.

magnetic fields. Advantages of this type of sensor are its robustness and corrosion resistance (strips of stainless steel). The measurement range is unlimited. It is possible to attach the strip to curved surfaces as well. Attainable resolution is 0.5 μm (over a limited range); maximum velocity of the sensor head is about 20 m/s. Opposed to this multi-magnet single-sensor system stands the multi-sensor single-magnet system, based on the configuration of Figure 6.19B. When the sensors are positioned in such a way that the sensitivity characteristics overlap, unique position measurement over a wide range is achieved (Figure 6.20C). Interpolation between two adjacent sensor positions provides accurate distance data. Absolute linear encoders using a special code and read-out procedure are available, providing a measurement accuracy of ± 5 μm and a resolution down to 1 nm, over a range of over 25 m.

The self-inductance of a coil not only depends on the lift-off (distance to a conducting target) but also on the conductivity of the target (see Figure 6.10). This property is exploited in Ref. [60] to measure liquid level and liquid conductivity at the same time, using two coaxially coils located above the liquid. A model-based approach is followed to infer both quantities from the frequency-dependent imaginary part of the impedance. Level and conductivity ranges are 0−30 cm and 4−10 S/m, respectively.

### 6.4.3 Applications of Variable Reluctance and Eddy Current Sensors

Eddy current sensors and other variable reluctance sensors are suitable for testing the surface quality of metal objects (cracks increase the resistance for eddy currents), object classification based on material properties (the sensor impedance depends on the resistivity, Figure 6.10), contact-free distance measurement and event detection (e.g. holes, edges and profiles, in metal objects).

## Angular Displacement

An inductive *tachometer* consists of a toothed wheel fixed on the rotating shaft and an inductive measuring head positioned close to the circumference of the wheel. Each time a tooth passes the sensor head, the interface electronics gets a count pulse; the pulse rate is related to the rotational speed of the shaft. Inductive tachometers are very robust and often applied in hazardous environments. A miniaturized version of this tachometer is described in Ref. [44]. The (flat) coil measures $1 \times 1$ mm, and has 10 turns only. A CMOS interface circuit converts a change in inductance (due to a passing tooth of a gear wheel) to a frequency change from about 9 to 11.6 MHz. This sensor also allows analogue distance measurements with 50 nm resolution over a 0.5 mm range [43].

Although the working range of eddy current sensors is limited, the measurement range can be enlarged by mechanical reduction of the movement to be measured. As an example, Figure 6.21 shows a configuration with an eccentric and two (balanced) eddy current sensors to measure *rotation*. In the neutral position (left) distances $x_1$ and $x_2$ are equal; when the bar rotates clockwise $x_1$ decreases whereas $x_2$ increases. Both the amount and the direction of the rotation are measured using very simple and robust sensors.

In the eddy current angle sensor described in Ref. [61], the large range (30 degrees) is obtained by a special mechanical construction: the rotating part has a semi-cylindrical shape, turning in a coil that makes an angle with the axis.

## Linear Displacement and Velocity

When a conductor moves in a magnetic field, eddy currents are induced, affecting the self-inductance of the coil that produces the magnetic field. This is the basic principle for a contact-free *velocity* or *acceleration* sensor. A detailed analysis of this concept is given in Ref. [62].

An alternative approach to measure velocity using eddy current sensors is described in Ref. [63], applied to rail vehicles. This method is based on correlation (the same as described in Chapter 7 on optical sensors and Chapter 9 on acoustic

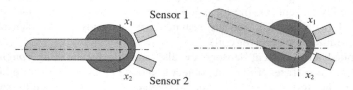

**Figure 6.21** Rotation measurement using eddy current sensors; in the left position sensors 1 and 2 have the same output.

sensors). In this case two eddy current sensors positioned about 10 cm above the rail track and with a fixed distance $d$ apart pick up the irregularities of the rail (the rail clamps are actually detected). The cross-correlation function of the two signals shows a peak for a delay time equal to $d/v$. In this case the correlation function is periodic but a special correlator is able to find the correct maximum value. In practice a rail vehicle also makes common (vertical and sideways) movements. A differential configuration of eddy current sensor is insensitive to such movements.

The presence of capacitances and the skin effect make the impedance of an eddy current sensor frequency dependent. So the transfer characteristic contains not only information about the distance between the sensor and the target (the 'lift-off'), but also on the properties of the target: thickness and material. In general the sensor can be modelled as a second-order system, of which the resonance frequency and the quality factor (or damping) are determined by the material properties, the thickness and the lift-off. This makes eddy current sensing suitable for *thickness* measurements (for instance of a non-conductive coating on a metal background or on conductive sheets [64]), for measuring conductivity and permeability [65] and for material identification [66], to mention just a few examples.

*Pressure* is usually measured by an intermediate elastic element: its deformation is proportional to the applied force or pressure. This deformation can be measured by an eddy current or variable reluctance sensor. In Ref. [67] this principle is demonstrated, where the elastic element is a flat membrane. The deflection is sensed by a flat coil (made on a PCB) and by force feedback using a second coil the deflection is compensated, according to the general feedback principle explained in Chapter 3.

## Non-Destructive Testing and Material Characterization

Eddy current sensors are widely used for *non-destructive testing* (NTD) of all kind of products and materials. In this application the object is scanned by moving one or more eddy current sensors over the area of interest. Surface irregularities are detected as an unwanted variation in distance to the object. Material defects (e.g. cracks and corrosion) are detected based on a change in local conductivity or permeability. The most simple sensor is the flat coil, which is adequate for, for instance, profile measurements [68]. With an array of sensors the scanning can be limited to just one direction, a useful solution when the resolution requirements are not very high. For the detection of more complicated structures like pipes, various other possible configurations have been proposed [41,69]. A particular example is the use of eddy current sensing for metal tag recognition [70]. The tags consist of a specific flat metal pattern, and are scanned with an eddy current sensor (in this case a transmitter and receiver coil).

Flat coiled eddy current sensors are also suitable for imaging *roughness* and defects in metal surfaces. Either the object is moved along an array of eddy current sensors or the test surface is scanned mechanically by a single sensor (cf. Section 1.1.4). The technique provides images with a spatial resolution down to 1 mm (see for instance Ref. [41]).

Another method to detect defects in conductive structures is magnetic flux leakage (MFL). The principle of this technique is shown in Figure 6.22. The object under test is magnetized by an external magnet. In the ideal case the magnetic field is mainly located inside the object because of its high permeability compared to the surrounding air. In case of a surface defect (e.g. hole and crack) the permeability at that location is low, and this causes the magnetic flux to bulge out of the material. This 'leakage' of the magnetic field can be measured by a magnetic sensor located just above the object's surface. Note that the defects create field lines perpendicular to the surface, so sensing in this direction is recommended. With sensitive magnetic sensors it is possible to detect defects in the order of micrometers, and even defects at the bottom side of (relatively thin, several mm) plates or pipe walls.

The MFL technique is still widely studied, aiming at better resolution and a more accurate localization and characterization of defects. A promising technique is the pulsed method that employs the dynamic behaviour of the pulse response for the interpretation of the defects. For instance in Ref. [71] it is shown that surface and sub-surface defects give distinctive responses, and from the measured time-to-peak the depth of the defects can be reconstructed.

A completely different application of a magnetic field sensor is found in Ref. [72]. Goal is to create an image of steel objects embedded in concrete structures. An excitation coil supplied by a DC current generates a static magnetic field over the area of interest. The pattern of the magnetic field reflects the shape of an embedded steel object. An image is obtained by a 2D scanning sequence over the test area by a sensitive magnetic sensor. An optimized layout of the sensor setup and advanced signal processing are required to obtain an image of reasonable quality, but the method enables the visualization of objects at a depth up to10 cm.

### 6.4.4  Applications of Other Inductive Sensors

*Velocity* and *acceleration* can be derived from position by taking the first and second derivative of the displacement signal. Real accelerometers make use of a seismic mass: when accelerated, the inertial force causes a deformation of an elastic element connected to the mass. The *accelerometer* described in Ref. [73] is based on an LVDT with a magnetic fluid as movable core. The position of the

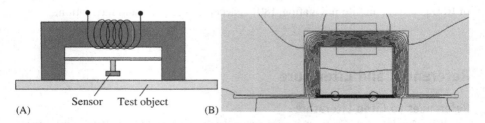

(A)  Sensor   Test object   (B)

**Figure 6.22** Principle of magnetic flux leakage for NDT: (A) measurement setup and (B) simulation example with two defects in the plate under test: left at the top side, right at the bottom side.

ferrofluidic sphere is measured in the same way as is the core in a normal LVDT. In this sensor the inertial force on the fluid is in equilibrium with the magnetic force.

A magnetic fluid is also used in the *tilt sensor* presented in Ref. [74]. Here the magnetic fluid acts as the support of a permanent magnet moving between two magnets located at both sides of the sensor housing. When tilted the magnetic core is displaced over a distance determined by the tilt angle and the magnetic forces. The position is measured by a differential pair of coils around the construction.

*Pressure* and *force* are likewise determined by an elastic intermediate. The deformation or displacement of that element can be measured by an induction method and is accomplished in various ways. For instance the bending of a conductive membrane affects the reluctance of a magnetic circuit composed of this membrane and one or more coils. If the membrane is not conductive or ferromagnetic, small ferrite objects can be attached to both sides of the membrane [75], together with two coils forming a differential variable reluctance circuit. A strain sensor in which the coil itself acts as a deforming element is presented in Ref. [76]. The coil is embedded in a flexible epoxy, and the change in coil shape causes its self-inductance to change.

The combination of a directional sensitive magnetic field sensor and a permanent magnet that can rotate with respect to the sensor makes the basic structure of a magnetic *angle sensor*. The magnetic sensor can be of any type. An angular position sensor over the full 360° based on a 3D integrated Hall sensor is presented in Ref. [77]. Although a 2D sensor would suffice in this application (giving the sine and cosine of the angle), the third sensitivity axis (measuring tilt angle) is used to compensate for the misalignment of the sensor relative to the magnet.

The magnetic angular sensor presented in Ref. [78] is based on a different approach. The (integrated) device comprises four Hall sensors, a flux concentrator and an analogue signal processing circuit. By a special configuration of the Hall sensors and a ratiometric principle an almost linear output (1%) is realized over an angular range of 120°.

Fluxgate sensors have a high sensitivity, and are therefore suitable for measuring the Earth's magnetic field (e.g. as part of a compass). They are widely used for navigation purposes. Research is aiming at the increase of overall performance, a lower price and reduced dimensions. Examples of such research work are presented in Ref. [79] for satellite navigation, [80] spacecraft and [81] land navigation.

# References and Literature

## References to Cited Literature

[1]  P.T. Moseley, A.J. Crocker: Sensor materials; Institute of Physics Publishing, Bristol and Philadelphia. 1996; ISBN 0-7503-0015-9.

[2]  P. Ripka (ed.): Magnetic sensors and magnetometers; Artech House Publishers, Boston, London. 2000; ISBN 1-58053-057-5. p. 30.

[3]  G.P. McKnight, G.P. Carman: Large magnetostriction in terfenol-D particulate compo-
     sites with preferred [112] orientation, smart structures and materials 2001: active mate-
     rials: behavior and mechanics; in: Christopher S. Lynch (ed.), Proceedings of SPIE,
     USA. Vol. 4333, 2001, pp. 178–183.
[4]  D. Ciudad, C. Aroca, M.C. Sánchez, E. Lopez, P. Sánchez: Modeling and fabrication
     of a MEMS magnetostatic magnetic sensor, Sens. Actuators A, 115 (2004), 408–416.
[5]  G. Asch: Les capteurs en instrumentation industrielle; Dunod, Paris. 1983; ISBN 2-04-
     015635-6.
[6]  S. Middelhoek, S.A. Audet: Silicon sensors; Academic Press, London, San Diego,
     New York, Berkeley, Boston, Sydney, Tokyo, Toronto. 1989; ISBN 0-12-495051-5.
[7]  S. Kordic: Integrated 3-D magnetic sensor based on an n-p-n transistor, IEEE Electron
     Device Lett., EDL7 (1986), 196–198.
[8]  D. Misra, M. Zhang, Z. Cheng: A novel 3-D magnetic field sensor in standard CMOS
     technology, Sens. Actuators A, 84 (1992), 67–75.
[9]  F. Burger, P.-A. Besse, R.S. Popovic: New fully integrated 3-D silicon Hall sensor for
     precise angular-position measurement, Sens. Actuators A, 67 (1998), 72–76.
[10] A.A. Bellekom: Origin of offset in conventional and spinning-current Hall plates, PhD
     thesis, Delft University of Technology, Delft, the Netherlands, 1998, ISBN 90-407-1722-2.
[11] L. Chiesi, K. Haroud, J.A. Flanagan, R.S. Popovic: Chopping of a weak magnetic field
     by a saturable magnetic shield; Eurosensors X, Leuven, Belgium, 8–11 September
     1996, pp. 123–126.
[12] P. Leroy, C. Coillot, A. Roux, G. Chanteur: High magnetic field amplification for
     improving the sensitivity of Hall sensors; Institut Pierre-Simon Laplace, CETP Velizy,
     France, 2004; Note no. 49.
[13] P.M. Drljaca, V. Schlageter, F. Vincent, R.S. Popovic: High sensitivity Hall magnetic
     sensors using planar micro and macro flux concentrators; Eurosensors XV, Munich,
     Germany, 10–14 June 2001.
[14] P. Malcovati, F. Maloberti: An integrated microsystem for 3D magnetic field measure-
     ments, IEEE Trans. Instrum. Meas., 49(2) (2000), 341–345.
[15] R.S. Popovic: Not-plate-like Hall magnetic sensors and their applications, Sens.
     Actuators A, 85 (2000), 9–17.
[16] E. Schurig, M. Demierre, C. Schott, R.S. Popovic: A vertical Hall device in CMOS
     high-voltage technology; Eurosensors XV, Munich, Germany, 10–14 June 2001.
[17] Z. Randjelovic, Y. Haddab, A. Pauchard, R.S. Popovic: A novel non-plate like Hall
     sensor; in: N.M. White (ed.), Eurosensors XII, 13–16 September 1998, Southampton,
     UK; IOP Publishing, Bristol; ISBN 0 7503 0536 3, pp. 971–974.
[18] F. Le Bihan, E. Carvou, B. Fortin, R. Rogel, A.C. Salaün, O. Bonnaud:
     Realization of polycrystalline silicon magnetic sensors, Sens. Actuators A, 88 (2001),
     133–138.
[19] Y.H. Seo, K.H. Han, Y.H. Cho: A new magnetic sensor based on plasma Hall effect;
     Eurosensors XIV, Copenhagen, Denmark, 27–30 August 2000, pp. 695–698.
[20] M. Demierre, S. Pesenti, J. Frounchi, P.A. Besse, R.S. Popovic: Reference magnetic
     actuator for self-calibration of a very small Hall sensor array; Eurosensors XV,
     Munich, Germany, 10–14 June 2001.
[21] P. Ripka: Review of fluxgate sensors, Sens. Actuators A, 33 (1992), 129–141.
[22] O. Dezuari, E. Belloy, S.E. Gilbert, M.A.M. Gijs: Printed circuit board integrated flux-
     gate sensor, Sens. Actuators A, 81 (2000), 200–203.
[23] T. O'Donnell, A. Tipek, A. Connell, P. McCloskey, S.C. O'Mathuna: Planar fluxgate
     current sensor integrated in printed circuit board, Sens. Actuators A, 129 (2006), 20–24.

[24]  A. Baschirotto, E. Dallago, P. Malcovati, M. Marchesi, G. Venchi: A fluxgate mag-
      netic sensor: from PCB to micro-integrated technology, IEEE Trans. Instrum. Meas.,
      56(1) (2007), 25−31.
[25]  B. Andò, S. Baglio, V. Sacco, A.R. Bulsara, V. In: PCB fluxgate magnetometers with
      a residence times difference readout strategy: the effects of noise, IEEE Trans.
      Instrum. Meas., 57(1) (2008), 19−24.
[26]  H.S. Park, J.S. Hwang, W.Y. Choi, D.S. Shim, K.W. Na, S.O. Choi: Development of
      micro-fluxgate sensors with electroplated magnetic cores for electronic compass, Sens.
      Actuators A, 114 (2004), 224−229.
[27]  S. Moskowicz: Fluxgate sensor with a special ring-core; Proceedings XVIII IMEKO
      World Congress, Rio de Janeiro, Brazil, 17−22 September 2006.
[28]  H. Grüger, R. Gottfried-Gottfried: Performance and applications of a two axes fluxgate
      magnetic field sensor fabricated by a CMOS process, Sens. Actuators A, 92 (2001),
      61−64.
[29]  R.A. Rub, S. Gupta, C.H. Ahn: High directional sensitivity of micromachined magnetic
      fluxgate sensors; Proceedings Eurosensors XV, Munich, Germany, 10−14 June 2001.
[30]  P. Ripka, S.O. Choi, A. Tipek, S. Kawahito, M. Ishida: Symmetrical core improves
      micro-fluxgate sensors, Sens. Actuators A, 92 (2001), 30−36.
[31]  P. Ripka, S. Kawahito, S.O. Choi, A. Tipek, M. Ishida: Micro-fluxgate sensor with
      closed core, Sens. Actuators A, 91 (2001), 65−69.
[32]  P.M. Drlača, P. Kejik, F. Vincent, D. Piguet, F. Gueissaz, R.S. Popović: Single core
      fully integrated CMOS micro-fluxgate magnetometer, Sens. Actuators A, 110 (2004),
      236−241.
[33]  S. Kawahito, A. Cerman, K. Aramaki, Y. Tadokoro: A weak magnetic field measure-
      ment system using micro-fluxgate sensors and delta−sigma interface, IEEE Trans.
      Instrum. Meas., 52(1) (2003), 103−110.
[34]  S.C. Tang, M.C. Duffy, P. Ripka, W.G. Hurley: Excitation circuit for fluxgate sensor
      using saturable inductor, Sens. Actuators A, 113 (2004), 156−165.
[35]  D. Vyroubal: Impedance of the eddy-current displacement probe: the transformer
      model, IEEE Trans. Instrum Meas., 53(2) (2004), 384−391.
[36]  M.T. Restivo: A case study of induced eddy currents, Sens. Actuators A, 51 (1996),
      203−210.
[37]  Ph.A. Passeraub, G. Rey-Mermet, P.A. Besse, H. Lorenz, R.S. Popovic: Inductive
      proximity sensor with a flat coil and a new differential relaxation oscillator;
      Eurosensors X, Leuven, Belgium, 8−11 September 1996, pp. 375−337.
[38]  H. Fenniri, A. Moineau, G. Delaunay: Profile imagery using a flat eddy-current prox-
      imity sensor, Sens. Actuators A, 45 (1994), 183−190.
[39]  C. Bartoletti, R. Buonanni, L.G. Fantasia, R. Frulla, W. Gaggioli, G. Sacerdoti: The
      design of a proximity inductive sensor, Meas. Sci. Technol., 9 (1998), 1180−1190.
[40]  O. Dezuari, E. Belloy, S.E. Gilbert, M.A.M. Gijs: Printed circuit board integrated flux-
      gate sensor, Sens. Actuators A, 81 (2000), 200−203.
[41]  Y. Hamasaki, T. Ide: A multilayer eddy current microsensor for non-destructive
      inspection of small diameter pipes; Transducers '95 − Eurosensors IX, Stockholm,
      Sweden, 25−29 June 1995, pp. 136−139.
[42]  P. Kejík, C. Kluser, R. Bischofberger, R.S. Popović: A low-cost inductive proximity
      sensor for industrial applications, Sens. Actuators A, 110 (2004), 93−97.
[43]  Ph.A. Passeraub, P.A. Besse, S. Hediger, Ch. De Raad, R.S. Popovic: High-resolution
      miniaturized inductive proximity sensor: characterization and application for step-
      motor control, Sens. Actuators A, 68 (1998), 257−262.

[44] Ph.A. Passeraub, P.A. Besse, A. Bayadroun, S. Hediger, E. Bernasconi, R.S. Popovic: First integrated inductive proximity sensor with on-chip CMOS readout circuit and electrodeposited 1 mm flat coil, Sens. Actuators A, 76 (1999), 273−278.

[45] E.E. Herceg: Evolution of the linear variable differential transformer; in: Handbook of measurement and control, Chapter 3; Schaevitz Engineering, Pennsauken NJ, USA, 1972.

[46] S.C. Saxena, S.B.L. Seksena: A self-compensated smart LVDT transducer, IEEE Trans. Instrum. Meas., 38(3) (1989), 748−753.

[47] Y. Kano, S. Hasebe, C. Huang, T. Yamada: New type linear variable differential transformer position transducer, IEEE Trans. Instrum. Meas., 38(2) (April 1989), 407−409.

[48] P.C. McDonald, C. Iosifescu: Use of a LVDT displacement transducer in measurements at low temperatures, Meas. Sci. Technol., 9 (1998), 563−569.

[49] L. Mingji, Y. Yu, Z. Jibin, L. Yongping, N. Livingstone, G. Wenxue: Error analysis and compensation of multipole resolvers, Meas. Sci. Technol., 10 (1999), 1292−1295.

[50] S.K. Kaul, R. Koul, C.L. Bhat, I.K. Kaul, A.K. Tickoo: Use of a 'look-up' table improves the accuracy of a low-cost resolver-based absolute shaft encoder, Meas. Sci. Technol., 8 (1997), 329−331.

[51] M. Benammar, L. Ben-Brahim, M.A. Alhamadi, A High: Precision resolver-to-DC converter, IEEE Trans. Instrum. Meas., 54(6) (2005), 2289−2296.

[52] J. Zakrzewski: Combined magnetoelastic transducer for torque and force measurement, IEEE Trans. Instrum. Meas., 46(4) (1997), 807−810.

[53] O. Dahle: The Torductor and the Pressductor; ASEA Res. Lab, Sweden (undated), orig.: O. Dahle: Der Preßduktor, ein Hochleistungs-Druckmeßgerät für die Schwerindustrie, ISA J., 5 (1959), 32−37.

[54] M. Hardcastle, T. Meydan: Magnetic domain behaviour in magnetostrictive torque sensors, Sens. Actuators A, 81 (2000), 121−125.

[55] F. Seco, J.M. Martín, J.L. Pons, A.R. Jéminez: Hysteresis compensation in a magnetostrictive linear position sensor, Sens. Actuators A, 110 (2004), 247−253.

[56] O. Erb, G. Hinz, N. Preusse: PLCD, a novel magnetic displacement transducer, Sens. Actuators A, 25−27 (1991), 277−282.

[57] B. Legrand, Y. Dordet, J.-Y. Voyant, J.-P. Yonnet: Contactless position sensor using magnetic saturation, Sens. Actuators A, 106 (2003), 149−154.

[58] Ph.A. Passeraub, P.A. Besse, S. Hediger, Ch. De Raad, R.S. Popovic: High-resolution miniaturized inductive proximity sensor: characterization and application for stepmotor control, Sens. Actuators A, 68 (1998), 257−262.

[59] D. de Cos, A. García-Arribas, J.M. Barandiarán: Simplified electronic interfaces for sensors based on inductance changes, Sens. Actuators A, 112 (2004), 302−307.

[60] W. Yin, A.J. Peyton, G. Zysko, R. Denno: Simultaneous noncontact measurement of water level and conductivity, Trans. Instrum. Meas., 57(11) (2008), 2665−2669.

[61] J. Cui, F. Ding, Y. Li, Q. Li: A novel eddy current angle sensor for electrohydraulic rotary valves, Meas. Sci. Technol., 19 (2008) doi:10.1088/0957-0233/19/1/015205.

[62] M.T. Restivo, F. Gomes de Almeida: The use of eddy currents on the measurement of relative acceleration, Sens. Actuators A, 113 (2004), 181−188.

[63] T. Engelberg: Design of a correlation system for speed measurement of rail vehicles, Measurement, 29 (2001), 157−161.

[64] W.L. Yin, A.J. Peyton, S.J. Dickinson: Simultaneous measurement of distance and thickness of a thin metal plate with an electromagnetic sensor using a simplified model, IEEE Trans. Instrum. Meas., 53(4) (2004), 1335−1338.

[65]  Y. Nonaka, H. Nakane, T. Maeda, K. Hasuike: Simultaneous measurement of the resis-
      tivity and permeability of a film sample with double coil, IEEE Trans. Instrum. Meas.,
      44(3) (1995), 679–682.
[66]  N.H. Kroupnova, Z. Houkes, P.P.L. Regtien: Application of eddy-current imaging in
      multi-sensor waste separation system; Proceedings of the International Conference &
      Exhibition on Electronic Measurement & Instrumentation, ICEMI'95, Shanghai,
      China, 3 January 1996, pp. 196–199.
[67]  B.M. Dutoit, P.-A. Besse, A.P. Friedrich, R.S. Popovic: Demonstration of a new princi-
      ple for an active electromagnetic pressure sensor, Sens. Actuators A, 81 (2000),
      328–331.
[68]  H. Fenniri, A. Moineau, G. Delaunay: Profile imagery using a flat eddy-current prox-
      imity sensor, Sens. Actuators A, 45 (1994), 183–190.
[69]  P.-Y. Joubert, Y. Le Bihan: Multisensor probe and defect classification in eddy current
      tubing inspection, Sens. Actuators A, 129 (2006), 10–14.
[70]  F. Belloir, R. Huez, A. Billat: A smart flat-coil eddy-current sensor for metal-tag rec-
      ognition, Meas. Sci. Technol., 11 (2000), 367–374.
[71]  John W. Wilson, Gui Yun Tian: Pulsed electromagnetic methods for defect detection
      and characterization, NDT&E Int., 40 (2007), 275–283.
[72]  D.S. Benitez, S. Quek, P. Gaydecki, V. Torres: A preliminary magnetoinductive sensor
      system for real-time imaging of steel reinforcing bars embedded within concrete,
      Trans. Instrum. Meas., 57(11) (2008), 2437–2442.
[73]  S. Baglio, P. Barrera, N. Savalli: Ferrofluidic accelerometers; Eurosensors XIX,
      Barcelona, Spain, 11–14 September 2005.
[74]  R. Olaru, D.D. Dragoi: Inductive tilt sensor with magnets and magnetic fluid, Sens.
      Actuators A, 120 (2005), 424–428.
[75]  H.A. Ashworth, J.R. Milch: Force measurement using inductively coupled sensor, Rev.
      Sci. Instrum., 49(11) (1978), 1600–1601.
[76]  J.C. Butler, A.J. Vigliotti, F.W. Verdi, S.M. Walsh: Wireless, passive, resonant-circuit,
      inductively coupled, inductive strain sensor, Sens. Actuators A, 102 (2002), 61–66.
[77]  Ph.A. Passeraub, P.A. Besse, S. Hediger, Ch. De Raad, R.S. Popovic: High-resolution
      miniaturized inductive proximity sensor: characterization and application for step-
      motor control, Sens. Actuators A, 68 (1998), 257–262.
[78]  C. Schott, R. Racz: Novel analog magnetic angle sensor with linear output, Sens.
      Actuators A, 132 (2006), 165–170.
[79]  J.M.G. Merayoa, P. Brauera, F. Primdahla: Triaxial fluxgate gradiometer of high sta-
      bility and linearity, Sens. Actuators A, 120 (2005), 71–77.
[80]  Å. Forslund, S. Belyayev, N. Ivchenko, G. Olsson, T. Edberg, A. Marusenkov:
      Miniaturized digital fluxgate magnetometer for small spacecraft applications, Meas.
      Sci. Technol., 19 (2008) doi:10.1088/0957-0233/19/1/015202.
[81]  K.M. Lee, Y.H. Kim, J.M. Yun, J.M. Lee: Magnetic-interference-free dual-electric
      compass, Sens. Actuators A, 120 (2005), 441–450.

## Literature for Further Reading

### Books

[1]  P. Ripka (ed.): Magnetic sensors and magnetometers; Artech House Publishers, Boston,
     London. 2000; ISBN 1-58053-057-5.

[2] P. Ripka, A. Tipek (eds.): Chapter 10: magnetic sensors; in: Modern sensors handbook; Wiley-ISTE, 2007; ISBN 978-1-905209-66-8.

[3] R.S. Popovic: Hall effect devices; Institute of Physics Publishing, Bristol, Philadelphia. 1991; ISBN 0.7503.0096.5.

## Review Papers

[4] P. Ripka: Review of fluxgate sensors, Sens. Actuators A, 33 (1992), 129−141.

[5] M. Vopálenský, P. Ripka, A. Platil: Precise magnetic sensors, Sens. Actuators A, 106 (2003), 38−42.

[6] G. Boero, M. Demierre, P.-A. Besse, R.S. Popovic: Micro-Hall devices: performance, technologies and applications, Sens. Actuators A, 106 (2003), 314−320.

[7] D.C. Jiles, C.C.H. Lo: The role of new materials in the development of magnetic sensors and actuators, Sens. Actuators A, 106 (2003), 3−7.

[8] V. Korepanov, R. Berkman, L. Rakhlin, Ye. Klymovych, A. Prystai, A. Marussenkov, M. Afassenko: Advanced field magnetometers comparative study, Measurement, 29 (2001), 137−146.

# 7 Optical Sensors

Most displacement and force sensing systems operating on optical principles consist of three basic parts: a light source, a light sensor and the transmitting medium. Often auxiliaries are required, such as lenses, optical wave guides, mirrors, filters, polarizers, diaphragms and choppers. An optical displacement sensor is constructed in such a manner that a change in distance between two sensor parts or between one sensor part and a moving object results in a change in transmission, reflection, absorption, scattering or diffraction of a beam of light. The majority of optical displacement sensors are based on variable reflection (continuous) or transmission (usually on-off). Other modalities are not often encountered in mechatronics. Optical force sensors are either derived from displacement sensors by adding one or more spring elements, or based on a change in parameters of suitable optical materials (in particular fibres). Since light is a wave, one can also make use of travel time (or ToF), phase shift and interference.

Optical fibres gain increasing interest as sensors, next to their role as data transport medium. The transmission properties of such fibres can be modulated by various physical properties, for instance temperature, humidity, bending, strain and concentration of chemicals put in contact with the fibre. The reader is referred to the literature on this specific subject.

First we discuss in this chapter the major optical variables and parameters. Next we review the properties of some electro-optical and optoelectric components commonly used as measurement tools. The subsequent section deals with methods on how to build up sensing systems for the measurement of position and displacement, using such components. This section also includes a discussion on optical encoders (linear or angular sensors with binary output). Section 7.3 deals with some basic electronic interface circuits. Finally Section 7.4 is devoted to the application of optical sensing systems in mechatronics.

Appendix A.4 lists various optical quantities with their definitions and units, as used in this chapter.

## 7.1 Electro-Optical Components

Most optical sensors applied in mechatronic systems are used for the measurement of displacement or distance between (moving) parts of the construction. Their output is used to control proper operation of the system, for instance to perform a specified action (i.e. movement and rotation) or to maintain a specified condition

**Sensors for Mechatronics. DOI: 10.1016/B978-0-12-391497-2.00007-8**
© 2012 Elsevier Inc. All rights reserved.

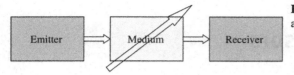

**Figure 7.1** Basic configuration of an optical sensing system.

**Table 7.1** Overview of Optical Emitters and Receivers, Modulation Types and Possible Accessories

| Emitters | Modulation | Receivers | Accessories |
|----------|------------|-----------|-------------|
| Light bulb | Transmission | Photo resistor | Lenses |
| (Gas) Laser | Reflection | Photovoltaic cell | Mirrors |
| Laser diode | Absorption | Photo diode | Reflectors |
| LED | Scattering | Photo transistor | Filters |
| | Refraction | PSD | Fibres |
| | Polarization | Line scan camera | Collimators |
| | Dispersion | Matrix camera | |

(i.e. position and force). A general setup of an optical sensing system is given in Figure 7.1. It consists of a light emitter (light source) and a light receiver (light detector). The light travels through a medium (usually air but may include parts of the construction) from transmitter to receiver. The transmission path is influenced (modulated) by the quantity that has to be measured. Various properties of the light can be affected by the measurand. This can be done in various ways, as shown in Table 7.1. The same table also gives various types of emitters, receivers and possible accessories that may build up the optical sensing system. Clearly the large number of combinations gives the designer much freedom in choosing the best solution for the intended application.

Clearly the application of an optical sensing system requires the incorporation of optical components in the mechatronic construction; they make up an integral part of it. Therefore it is important to include the (optical and any other) sensing system from the very beginning in the design process of the mechatronic system.

Since most optical sensing systems comprise a combination of a light emitter and a light detector, we give a review of some of these electro-optical devices and their properties.

### 7.1.1 Light Emitters

The next list gives commonly used emitters for the application in displacement and force transducers, with a general characterization. For detailed information the reader is referred to the data sheets and catalogues of the manufacturers, most of them freely available on the Internet.

The major types of light emitters are:

*Thermal light bulb*: A very cheap and easy to get optical source, also available in small dimensions (for instance a bicycle taillight). Major characteristics:
- wide optical spectrum (almost white light)
- wide beam angle (almost omnidirectional)
- slow; small bandwidth: due to the thermal operation the intensity cannot change very fast; not suitable in modulated systems.

*LED (light-emitting diode)*: A cheap and widely used light source for numerous applications (e.g. illumination, displays and remote control). Main characteristics:
- materials: GaAs and other 3−5 compounds
- relatively narrow emission spectrum (tens of nm)
- covering all colours in the visible spectrum and IR
- electrically controllable intensity (so suitable in modulated systems)
- moderately focussed beam (with built-in lens).

*Semiconductor laser (laser diode)*: Low power types having appearance and dimensions similar to LEDs, but whose physical operation is different; well known from CD and DVD players, laser pointers and laser printers. Main characteristics:
- materials: GaAs and other 3−5 compounds
- laser stimulation by injection of charged particles (current)
- electrically controllable intensity
- monochromatic (very narrow spectrum, a few nm)
- covering most colours in the visible spectrum, IR and near UV (based on GaN)
- moderately focussed beam (same as LED).

*Solid-state laser*: Much more expensive device, suitable for large optical power. Major characteristics:
- materials: $Nd^{3+}$:$Y_3Al_5O_{12}$ (YAG: yttrium aluminium with neodymium ion impurity); ruby ($Cr^{3+}$:$Al_2O_3$)
- laser stimulation by optical means (light pulses)
- monochromatic (very narrow optical spectrum) in visible spectrum and infrared
- narrow beam
- intensity control only possible with special components.

*Gas laser*: Much more expensive device, suitable for large optical power. Major characteristics:
- materials: He−Ne, Ar gas
- laser stimulation by gas discharge
- monochromatic (very narrow optical spectrum) in visible spectrum and IR
- narrow beam
- intensity control only possible with special components.

Many other types of lasers exist, mainly intended for other applications, such as cutting, drilling and welding. For mechatronic applications the LED and the laser diode are recommendable devices, because of their small size, simple interfacing and easy to modulate. Important parameters to be considered when designing a sensor system are the output intensity or power versus input current (sensitivity), output intensity versus wavelength (spectral sensitivity) and output intensity versus angle of divergence (directivity diagram).

There are a vast number of LEDs and laser diodes on the market. Some manufacturers offer the possibility to search in their database by particular specifications as required by the designer, for instance by power range or wavelength. For each product type, main specifications are tabulated, but not always in a structured, consistent way. Finding exactly what is needed for a particular application is still time consuming.

Graphical data provide better information compared to numerical data. Figure 7.2A−D shows examples of such graphs: the output characteristics for a LED and a laser diode, the optical spectrum and the directivity.

*Intensity versus forward current*: Figure 7.2A shows that the intensity and power of a LED is about proportional to the forward current. The maximal intensity of a LED ranges from a few mW/sr up to several W/sr, depending on the type. The intensity drops with increasing temperature, at a rate of about 1% per Kelvin. A laser diode (Figure 7.2B) starts to laser from a specified forward current, typically between 50 and 100 mA. Here, too, the intensity drops with increasing temperature, mainly because of the decreasing quantum efficiency at higher temperatures.

*Wavelength*: A typical plot of the spectral emission is given in Figure 7.2C. The peak wavelength is determined by the semiconductor material type (Table 7.2). In-between values are also available; these colours are obtained by small additives of other materials. The optical bandwidth is the wavelength interval at 50% relative intensity. LEDs have optical bandwidths in the order of 20−50 nm; a laser diode is much more monochromatic, with optical bandwidths of only 1 nm.

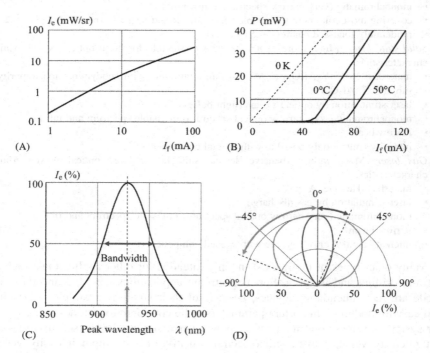

**Figure 7.2** Typical characteristics of light-emitting devices: (A) axial radiant intensity versus forward current (LED), (B) output power versus forward current (laser diode), (C) relative radiant intensity versus wavelength (IR LED) and (D) directivity diagrams (LED).

**Table 7.2** Peak Wavelength of Some LED Types

| Composition | Peak Wavelength (nm) | Colour |
|---|---|---|
| InGaN | 470 | Blue |
| GaP | 555–590 | Green-yellow |
| GaAsP | 630 | Orange |
| GaAlAs | 650 | Red |
| GaAs | 930 | Infrared |
| InGaAsP | 1300 | Infrared |

*Directivity diagram*: Figure 7.2D is an example of the directivity diagram. The half angle (denoting the beam width) is defined as the direction in which the intensity has dropped by 50% relative to the intensity in the direction of the main axis (0°). The diagram combines the directivity of two types: a narrow beam and a wide beam type. Usually a small angle is obtained by a lens at the front. Note that laser diodes have also a relatively wide beam angle (due to the short optical path the light travels in the device).

Some types of laser diodes contain an additional photo diode within the housing, suited to monitor and possibly control the output intensity. Other important parameters to consider when choosing a light source are the temperature sensitivity of the peak wavelength (about 0.1 nm/K for a LED), the rise and fall times (when used as a switching device) and the heat resistance (when used near maximal power).

## 7.1.2 Light Sensors

Commonly used optical sensors and their general characteristics are:

- *Photoresistor*: small; cheap; sensitive to visible and IR light; no directional sensitivity; slow (large response time, see Chapter 4).
- *Photo diode*: small; cheap; sensitive to visible and IR light; moderate directional sensitivity; fast response time; also available as quadrant detector (four diodes in one casing).
- *Photo transistor*: similar to photo diode, somewhat slower.
- *Diode array*: array of photo diodes: from a few (in a single casing) up to over 4000 elements (on a chip); line rate up to 100 kHz; grey-tone and colour.
- *Position sensitive diode (PSD)*: radiometric properties as photo diode; 1D and 2D position sensitive (see Section 7.1.3).
- *CCD camera (area scan)*: resolution up to 2048 × 2048 elements; frame rate up to 1000/s for low resolution types; grey-tone and colour.
- *CMOS camera (area scan)*: higher resolution compared to CCD camera; somewhat worse noise performance.

Line and area scan cameras are excellent devices to perform imaging tasks, and hence play an important role in mechatronics and in particular robotics. The market offers a gradually increasing number of cameras with built-in image pre-processing. Pixel size of both line and scan cameras is in the order of 5 × 5 μm. Current research aims at increasing the resolution (number of pixels) and the speed (measured in either pixel acquisition time or frame rate). Chips with over 60 million pixels are feasible. CMOS cameras have higher resolution because there are no charge buffers on the chip as with CCD. Moreover camera and system clocks are synchronized, eliminating the need for a frame grabber and thus increasing the overall frame rate. Cameras will not be discussed further in this book: the reader is referred to textbooks on this subject and the information provided by the manufacturers.

Figure 7.3A–D shows typical characteristics of photo diodes: voltage–current characteristics, sensitivity and temperature characteristics and the spectral sensitivity. The directional sensitivity is given in a similar way as with emitters; half angles range from 10° to 60°, depending on the presence of a built-in lens.

*Voltage–current characteristic*: Figure 7.3A depicts four curves for different values of the incident light intensity. In the absence of light, the voltage–current relation (uppermost curve) is the same as for a normal pn-diode: an exponentially increasing current with forward voltage and a small leakage current (the dark current) when reversed biased. When light falls on the pn-junction, the leakage current increases because photons generate additional electron–hole pairs which may result in an external current. As can be seen in Figure 7.3A, the short-circuit current (represented by the bullets on the $I$-axis) increases in proportion to the incident light. The open voltage (represented by the circles on the $V$-axis) increases as well with irradiance but in a non-linear way.

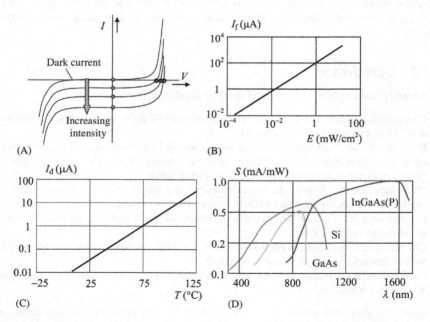

**Figure 7.3** Typical characteristics of photo diodes: (A) current–voltage characteristic, (B) sensitivity characteristic of a silicon photo diode, (C) dark current versus temperature and (D) spectral sensitivity of a Si, a GaAs and an InGaAsP diode.

*Sensitivity characteristic*: The photo current is proportional to the incident light over more than six decades (Figure 7.3B). This is the main reason why the current mode (or conductivity mode) is preferred over the photovoltaic mode of operation. Current readout is easily accomplished using an operational amplifier in current-to-voltage converter arrangement (Section 7.3.1). Usually a reverse voltage is applied to ensure the diode always operates in the flat part of the characteristics in Figure 7.3A.

*Dark current*: Not only light (optical energy) but also thermal energy gives rise to the generation of additional charge carriers around the pn-junction. This is exactly why the reverse current of a pn-diode is not zero. Roughly the thermally induced leakage current doubles each 7°C. At room temperature the leakage current of a silicon diode is very small, some 10 pA, but since a photo diode has an essentially larger active area, its dark current is also much larger, and moreover, it increases exponentially with temperature (Figure 7.3C).

*Spectral sensitivity*: Assuming a quantum efficiency of 100%, each incident photon creates one electron−hole pair, if the photon energy is larger than the bandgap energy $E_g$ (to generate a free electron):

$$E_g = h \cdot v \tag{7.1}$$

where $v$ is the frequency and $h$ Planck's constant. So $E_g$ of the material determines the maximum wavelength. Silicon has a bandgap equal to 1.12 eV, corresponding to a limiting wavelength of about 1100 nm. The lower boundary of the spectrum is set by the absorption of the light before it has reached the pn-junction. In general the absorption coefficient of the materials involved grow with reduced wavelength. The result is a spectral band that strongly depends on the materials. Figure 7.3D gives the spectral sensitivities of three photo diodes: Si, GaAs (both sensitive in the visible part of the spectrum) and a special diode for large wavelengths. Detailed information on device characteristics and performance of other types can be found in data books of the manufacturers.

### 7.1.3  Position Sensitive Diode

A PSD is a light sensitive diode which is not only sensitive to the intensity of the incident light but also to the position where the incoming light beam hits the diode surface. Figure 7.4 shows, schematically, the configuration of a lateral silicon PSD.

The device consists of an elongated pn-photo diode with two connections at its extremities and one common connection at the substrate. An incident light beam can penetrate through the top layer, down to the depleted region around the pn-junction where it generates electron−hole pairs. As in a normal photo diode, the junction is reversed biased. Due to the electric field across the junction, the electrons are driven to the positive side, the holes move to the negative side of the junction. The electrons arriving in the n-region flow through the common contact to ground; the holes in the p-layer split up between the two upper contacts, producing two external photo currents $I_a$ and $I_b$.

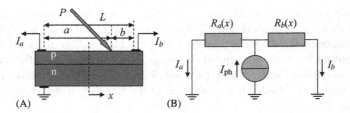

**Figure 7.4** PSD: (A) principle of operation and (B) current division.

Figure 7.4B shows an electric model of the PSD. It is used to explain how the current division depends on the position of the light beam. The photo current (modelled as a current source $I_{ph}$) originates from the point where the beam hits the diode. The division over the two contacts is determined by the resistances from this point on to the two end points ($R_a$ and $R_b$). Applying Kirchhoff's current law yields for the two output currents:

$$I_a = \frac{R_b}{R_a + R_b} I_{ph}$$
$$I_b = \frac{R_a}{R_a + R_b} I_{ph} \qquad (7.2)$$

Assuming a homogeneous resistance over the whole length of the PSD, Eq. (7.2) can be rewritten as:

$$I_a = \frac{b}{a + b} I_{ph}$$
$$I_b = \frac{a}{a + b} I_{ph} \qquad (7.3)$$

Generally the origin of the co-ordinate system of the PSD is chosen in its centre. Thus $x$ is the position of the light beam relative to the centre ($x = 0$). Further when the total length of the PSD is $L$, the distance $a$ in Figure 7.4A equals $L/2 + x$ and $b$ equals $L - a$, so

$$I_a = \left(\frac{1}{2} - \frac{x}{L}\right) I_{ph}$$
$$I_b = \left(\frac{1}{2} + \frac{x}{L}\right) I_{ph} \qquad (7.4)$$

Apparently the difference between the two output currents, $I_a - I_b$ equals $-2xI_{ph}/L$ and is linearly dependent of $x$ and $I_{ph}$. However it is preferred to have an output signal that only depends on the position $x$ and not on the light intensity of the incident beam. In order to obtain such an output, $I_a - I_b$ is divided by $I_a + I_b = I_{ph}$, resulting in

$$\frac{I_a - I_b}{I_a + I_b} = \frac{2x}{L} \tag{7.5}$$

So to perform an intensity-independent position measurement, the PSD should be connected to an electronic interface for three signal operations: addition, subtraction and division (see Section 7.3.2). A PSD is essentially an analogue device: the photo currents vary analogously with the position of the incident light beam. This is one of the reasons why researchers have designed PSDs with on-chip (analogue) signal processing to accomplish intensity-independent output signals according to Eq. (7.5) [1,2].

With the PSD in Figure 7.4A the position of the light beam can be determined only in one direction. For a 2D measurement, a 2D-PSD is required: this type of PSD has a square-shaped sensitive surface and two pairs of current contacts, one pair for each direction. Its operating principle is exactly the same as for the 1D-PSD.

The linearity of a PSD depends largely on the homogeneity of the resistance in the top layer. In the central part (about 70% of the total length or area) the linearity is better than 1%, whereas at the edges the linearity error rises significantly: this should be considered when operating the device in that area.

The range of a linear PSD is almost equal to its length: from several mm up to 10 cm; the width is a few mm. A 2D-PSD measures $1 \times 1$ to $4 \times 4$ cm. The total resistance between two opposing contacts is several k$\Omega$. Devices with larger size are difficult to realize in silicon. Other materials are being investigated to increase the lateral dimensions. For instance PSDs made from hydrogenated amorphous silicon (a-Si:H) have been realized with a range of $10 \times 10$ cm and comparable sensitivity [3]. Research is going on to further improve the detectivity of PSDs and to reduce the effects of background light [4].

Most PSDs are made of silicon, hence their spectral sensitivity is comparable with normal silicon photo diodes: most LEDs and laser diodes match well in this respect with the PSD.

A PSD responds to the *optical centre of intensity* of the incident light. Non-homogeneous background light causes a shift of this point, resulting in a measurement error. Note that homogeneous background light changes the scale factor of the system, as can be seen from Eq. (7.5), where an equal contribution to the currents cancel in the numerator but add in the denominator. A more in-depth analysis of the effect of environmental illumination sources is given in Ref. [5] for a 2D-PSD.

In general there is just one light spot (from a laser) of which the position can be detected. In case of multiple light spots, the beams can be distinguished using modulated light on different carriers [6]. The PSD can be used for contact-free displacement measurements. The way this is accomplished will be discussed in Section 7.2.2 on triangulation.

## 7.2    Optical Displacement Sensors

### 7.2.1    Intensity Measurement

A simple though rather inaccurate method to measure displacement is based on the inverse quadratic law: the light intensity from a point source decreases with the square of the distance (see Appendix A.4.1). We distinguish two possible modes: the direct and the indirect mode. Figures 7.5A and B show the general setup according to the direct mode.

In the direct mode the transmitter (LED or laser diode) is mounted on a fixed part of the construction while the sensor is positioned on the moving part or target (or vice versa). Comparing Figures 7.5A and B makes clear that the light intensity (in $Wm^{-2}$) drops with the square of the distance (see Appendix A.4.3 for a detailed analysis). Assuming a homogeneous light beam over the complete beam angle in the configuration of Figure 7.5A, the sensor output $I_o$ is inversely proportional to the square of the distance, so

$$I_o \approx \frac{1}{x^2} \tag{7.6}$$

Actually the inverse square law is just an approximation, since the emission is not homogeneous over the whole cone, as can be seen in the directivity diagrams in Figure 7.2D.

Evidently the sensor output not only depends on distance but also on the source intensity and the sensitivity of the sensor. These parameters should have sufficient stability to be able to distinguish their changes from displacement-induced intensity changes. The stability requirements can be weakened by applying a second sensor, with fixed position relative to the source, as a reference. The output quantity is the difference or (better but more complicated to implement) the *ratio* of both sensor

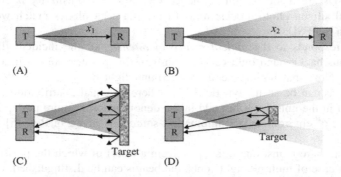

**Figure 7.5** Displacement sensing by variation of intensity (A), (B) direct mode, (C) indirect mode with large target and (D) indirect mode, with small target; T = transmitter (source), R = receiver (sensor).

signals. If both sensors have stable, identical sensitivity, source instability can be completely eliminated. The final accuracy is limited by changes in the difference of sensor parameters.

Figure 7.5C shows the indirect mode, suitable for contact-free distance measurements. Source and sensor are mounted both on a fixed part of the construction. The source casts a light beam on the target which scatters this light in all directions (diffuse reflection; we disregard here specular reflection for simplicity). Most surfaces behave as a Lambertian surface, that is, they scatter the incident light in all directions equally (see Appendix A.4.2). Actually the object acts as a secondary light source, in the same way we see the moon by the sunlight scattered from its surface. The intensity of the detected light depends on the distance from the emitter−receiver system to the reflecting target hence is a measure for the position of the moving part.

The inverse square law (7.6) also applies for this configuration, as long as the target intercepts the full beam. If, on the other hand, the target is smaller (as shown in Figure 7.5D) the reflecting surface area is constant (that of the target). The irradiance (incident power per unit area) of the target decreases with the square of the distance from the source (because the surface area is constant). The illuminated target acts as a secondary light source, the light of which is measured by the sensor. For this part the inverse square law also applies, so in total the sensor output drops with the *fourth* power of the distance between source and target (see Appendix A.4.3 for a more general derivation):

$$I_0 \approx \frac{1}{x^4} \tag{7.7}$$

Note that the output depends also on the angle of the object's surface relative to the optical axis of the source. The fourth-power dependence results in an even smaller measurement range. Clearly the design of a proximity sensor must be such that the object always intersects the whole beam, in other words, a narrow beam is preferred.

Like Eq. (7.6), also Eq. (7.7) is an approximation since it does not account for the directivity of the source and the sensor, or an inhomogeneous light beam. Nevertheless they are useful for a quick check during the design process. Appendix A.4 gives a more general proof of these relationships.

For contact-free distance measurements according to Figures 7.5C and D special devices are available, in which the emitter and receiver are mounted in a single housing. When configured for small distances, this sensor system is referred to as an *optical proximity sensor*. The optical axes make a fixed angle; as can be seen in Figure 7.6A, light can only travel from transmitter to receiver when the reflecting target falls in the area where the directivity curves of the transmitter and receiver overlap. The useful range of the sensing system is somewhat smaller: both the lower and the higher limits of the detection range are set by the configuration geometry and by the directivity characteristics of the devices: $x_a < x < x_b$ (Figure 7.6B).

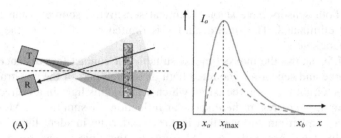

**Figure 7.6** Optical proximity sensing: (A) useful range and (B) transfer characteristics.

All considerations given for this type of linear displacement sensors hold for angular sensors too. Angular sensitivity is obtained using the direction sensitivity of the source or sensor.

The contact-free distance sensors discussed here are simple and cheap. Principally due to the analogue nature, with the reflection method a very high resolution (down to 0.01 μm) can be obtained [7]. However the accuracy is low: since the output is an intensity signal, intensity variations due to other causes will be taken as a displacement. For instance changes in the target reflectivity, the intensity of the emitter and the sensitivity of the receiver all contribute to a change in the output signal. The effect of different reflection coefficients is depicted in Figure 7.6B, where the upper curve refers to a surface with a relatively high reflectivity and the lower (dotted) one to a more absorbing surface.

Compensation, balancing and feedback make the output more tolerant to changes in these parameters (see the general principles in Chapter 3). Instead of a single receiver, a pair or a quad (four diodes arranged in a $2 \times 2$ matrix) is used. The position of the incident light beam is controlled in such a way that all sensors have the same output: this guarantees a fixed position of the beam — exactly between the diode pair or in the centre of the quad. Displacement information is derived from the actuators controlling the beam position. In this way accurate position information is required with rather inaccurate optical sensors. Similar concepts are used in CD-players to keep the reading head exactly on the right track on the disc.

In mechatronics and many other applications it is not always of primary importance to find the exact distance: the detection of an occurrence would be sufficient in many cases (door open or close, presence of a person or object). For this purpose special optical sensors have been constructed with a wide range of specifications. Systems in which the sensor faces the source (direct mode) have the widest range: up to 35 m. Indirect mode systems have a smaller range, because only part of the scattered light will return to the sensor. With special reflectors (retroreflectors) the maximum range is about 10 m; when only the reflection of the object itself is used the distance range is several metres.

Up to here only sensors in reflection mode have been discussed. Measuring (linear and angular) displacement can also be performed by an intensity method in transmission mode. A simple configuration is given in Figure 7.7.

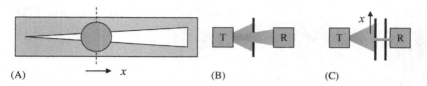

**Figure 7.7** Examples of displacement sensors using intensity variation in transmission mode: (A) long-range front view and (B) side view (C) short-range side view.

Light transmission is modulated by the tapered slot in the moving mask: a very cheap and simple linear displacement sensor suitable for applications with low demand on accuracy. A rotational version can be made in a similar way, with a tapered slot in a circular segment. The dimensions of the slot are chosen such that the whole displacement range is covered from 0% to 100% transmission. The principle is also suitable for measuring very small movements (sub-millimetre). A high sensitivity (but small range) is obtained by a configuration with a narrow slot in the moving mask and a fixed slot in front of the receiver (Figure 7.7C). By shaping the slot in a particular way, an arbitrary transfer function can be obtained, for instance a logarithmic or a parabolic relationship between displacement and transmission. For each application optimal dimensions can easily be found by inspection of the amplitude of the movement to be monitored.

### 7.2.2 Triangulation

A triangle is completely determined by three parameters (sizes or angles), a property that is used to find an unknown distance. Assume $x$ in Figure 7.8 is an unknown distance.

When one side $a$ and the angles $\alpha$ and $\beta$ are known, the other sides $x$ and $y$ follow from

$$x = a\frac{\sin\beta}{\sin(\alpha + \beta)}; \quad y = a\frac{\sin\alpha}{\sin(\alpha + \beta)} \tag{7.8}$$

This principle is used to measure displacement or distances, in a variety of situations. Figure 7.9A shows a configuration using the direct mode.

With one fixed distance $d$, one fixed angle $\alpha$ and one measured distance $y$ (on a PSD or diode array) the unknown distance can be calculated. In this configuration for 1D displacement measurement, the position $x$ of the target has to be measured while moving along the $x$-axis. It carries an omnidirectional light transmitter, viewed by a PSD (or a line camera). The sensor is positioned in the focal point of the lens, with focal distance $f$. This creates a sharp image on the PSD only if the spot is far away (which in most practical situations is indeed the case). In general the optimal position (and orientation) of the PSD should fulfil the Scheimpflug condition, in order to have the spot always in focus on the PSD, irrespective of the distance [8]. From Figure 7.9A it follows that the measurement range (travel from

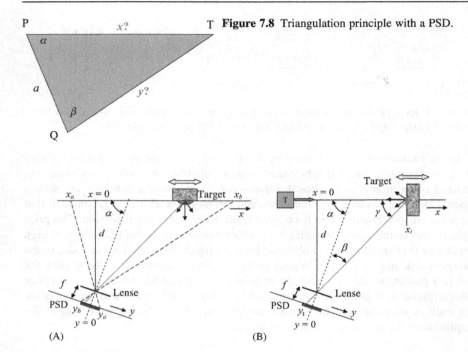

P         $x?$       T **Figure 7.8** Triangulation principle with a PSD.

**Figure 7.9** Triangulation applied to contact-free distance measurement: (A) direct mode and (B) indirect mode.

$x_a$ to $x_b$) depends on the size of the PSD (from $y_a$ to $y_b$), and the focal length $f$ and can further be adjusted with the parameters $d$ and $\alpha$.

Figure 7.9B gives a configuration in indirect mode. A light beam from a laser source casts a spot on a suitable, well-defined point at the movable target. Part of the scattered light is projected onto the PSD. A displacement of the target causes a displacement of the projected light spot on the sensor. The x-position of the laser source is irrelevant here, as long as the spot intensity is large enough to create a bright spot on the PSD. Clearly each target position $x_t$ corresponds to a spot position $y_t$ on the PSD. This relation depends on the geometry and the focal distance of the lens and can be found using the goniometric relation:

$$\tan \alpha = \tan(\beta + \gamma) = \frac{\tan \beta + \tan \gamma}{1 - \tan \beta \tan \gamma} \tag{7.9}$$

With $y/f = \tan \beta$ and $a/x = \tan \gamma$ (see Figure 7.9B) we arrive at

$$y = f \cdot \frac{x \tan \alpha - d}{x + d \tan \alpha} \quad \text{or} \quad x = d \cdot \frac{y \tan \alpha + f}{f \tan \alpha - y} \tag{7.10}$$

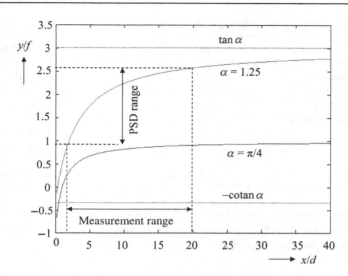

**Figure 7.10** Relation between distance *x* to be measured and position *y* of the light spot on the PSD, normalized to *x/d* and *y/f*; measurement range and PSD range are given for $\alpha = 1.25$.

Figure 7.10 shows the effect of the parameter $\alpha$ on the range and on the transfer characteristic. The horizontal lines − cotan$\alpha$ and tan$\alpha$ represent the values for $x = 0$ and $x \to \infty$, respectively, in this case for $\alpha = 1.25$ rad. At a fixed length of the PSD, the measurement range increases for increasing $d$ and decreasing $f$. In general the transfer is non-linear; only for $\alpha = \pi/2$ (the optical axes of the source and the sensor are perpendicular) the transfer is linear: $y/f = x/d$; the range is rather small. For $\alpha = 0$ (parallel optical axes) the transfer is simply the inverse: $y/f = -d/x$. In practice a compromise must be made with respect to range and inaccuracy by non-linearity. The range can be enlarged using rotational mirrors, a technique that will be discussed in Section 7.4 on applications.

The size of the light spot image on the sensor depends on the beam diameter and divergence, and may be substantially larger than the required resolution of the sensor. For a PSD this is not very relevant because it responds to the optical centre of intensity, which lies on the main axis of the light beam. Obviously the spot size should not exceed the *width* of the PSD. A more serious problem is the low S/N ratio due to poorly reflecting surface materials, a slanting angle of incidence or simply a too large distance. Using multi-sensor techniques and signal averaging the accuracy can be improved over a factor of 10 [9].

When using a diode array instead of a PSD, the light spot may simultaneously activate several adjacent elements of the array. Therefore the element in the centre of all activated elements (or the median) should be selected as being the best position of the spot. While designing a displacement sensing system often a trade-off has to be made between a (1D or 2D) PSD and a (line or matrix) camera. Table 7.3

**Table 7.3** Comparison Between PSD and Camera for Position Measurements

| Item | PSD | Camera (b/w) |
|---|---|---|
| Position information | Analogue | Quantized |
| Number of positions | 1 | #pixels |
| Position resolution | 1 ppm | 4096 pixels |
| Dynamic position range | $1:10^6$ | $1:10^3$ |
| Accuracy | High | Calibration required |
| Speed | 30 kHz per point | 80 MHz pixel rate (20 kHz line rate) |
| Light source | LED, laser | Environmental |

presents some basic differences between these two devices which could be helpful in making a proper choice. Note that numerical data are typical; higher values for resolution and speed are available.

### 7.2.3 Optical Encoders

The optical sensors described in this section are digital in nature. They convert, through an optical intermediate, the measurement quantity into a binary signal, representing a binary coded measurement value. Sensor types that belong to this category are optical encoders, designed for measuring linear and angular displacement, optical tachometers (measuring angular speed or the number of revolutions per unit of time) and optical bar code systems for identification purposes.

Optical encoders are composed of a light source, a light sensor and a coding device (much the same as the general setup in Figure 7.1). The coding device consists of a flat strip for linear displacement or disc for angular displacement, containing a pattern of alternating opaque and transparent segments (the transmission mode) or alternating reflective and absorbing segments (the reflection mode). Both cases are illustrated in Figure 7.11.

The coding device can move relatively to the assembly of transmitter and receiver causing the radiant transfer between them switch between a high and a low value. In the transmission mode the encoder consists either of a translucent material (e.g. glass, plastic and mylar), covered with a pattern of an opaque material (for instance a metallization), or just the reverse, for instance a metal plate with slots or holes.

Two basic encoder types are distinguished: absolute and incremental encoders. An *absolute encoder* gives instantaneous information about the absolute displacement or the angular position. Figure 7.12 gives examples of absolute encoder devices in transmission mode.

Each (discrete) position corresponds to a unique code, which is obtained by an optical readout system that is basically a multiple version of Figure 7.11A. The acquisition of absolute position with a resolution of $n$ bits requires at least $n$ optical tracks on the encoder and $n$ separate sensors.

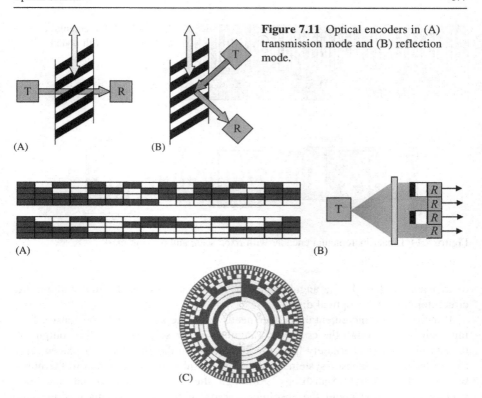

**Figure 7.11** Optical encoders in (A) transmission mode and (B) reflection mode.

**Figure 7.12** Absolute encoders: (A) linear encoders with 4 bit dual code and with 4 bit Grey code, (B) readout system with collimator and (C) angular encoder disc.

Obviously the maximum resolution of an absolute encoder is set by that of the least significant bit (LSB) track; evidently it is the outer track of an angular encoder. Practical limitation of the resolution is limited by the dimensions of the detectors. Special integrated circuits including photo diodes and built-in compensation have been developed to maintain the performance also with high resolution encoder discs [10].

The natural binary code is polystrophic which means that two or more code bits may change for a 1-LSB displacement (see the linear encoder of Figure 7.12A, upper track). Mechanical tolerances will introduce error codes between two adjacent binary words. For this reason most encoders use the Grey code, which requires also the minimum number of tracks at a given resolution but is monostrophic: at each LSB displacement only one bit will change at a time, as can be seen from the linear encoder in the lower track of Figure 7.12A.

Serious disadvantages of an absolute encoder are its limited resolution, the large number of output sensors and the consequently high price. A much better resolution is achieved by the *incremental encoder*, of which Figure 7.13 shows a typical example for linear displacement. It has only one track consisting of a large number

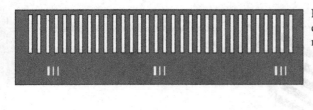

**Figure 7.13** Incremental linear encoder scale with coded markers.

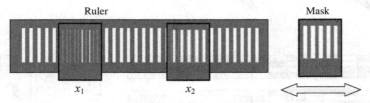

**Figure 7.14** Linear incremental encoder with fixed scale and moving mask.

of transparent slots. In the angular type these slots are arranged radially along the circumference of the optical disc.

Principally the incremental encoder needs only one source and one sensor, facing each other through the code disc (angular) or the scale (linear). The output of the sensor changes alternately from low to high when the encoder is displaced relative to the source—sensor system, generating a number of pulses equal to the number of light-dark-light transitions. Although the actual output signal does not contain information about the absolute position of the encoder, this position can nevertheless be found by counting the number of output transitions or pulses, starting from a certain reference position. The reference position is marked by an extra slot in the disc or scale. Once the position information is lost (for instance due to a power interruption), the system should go to the nearest reference marker. At very large scales, this could take too much time; therefore, long scales have multiple markers, which are coded to make them distinguishable from each other. In Figure 7.13 the markers are identical but positioned on different distances. The number of lines between two adjacent markers is a unique code for the marker position.

If the incremental encoder is used in the same way as the absolute encoder (one sensor for each line) its resolution would not be much better: the size of the sensor should not exceed the width of one slot. However, the combination with a fixed mask positioned against the moving disc or scale gives a much higher resolution, as is illustrated in Figure 7.14. The mask has an extension of several line widths and the same pitch as that of the scale.

The light transfer changes periodically from a minimum to a maximum value and back for each displacement over the pitch of the optical pattern. If the scale and the mask are perfectly matched and aligned, the maximum light transfer is 50% (position $x_2$ in Figure 7.14), whereas the minimum light transfer is zero over the whole area of the mask (position $x_1$). So the sensor size is not limitative

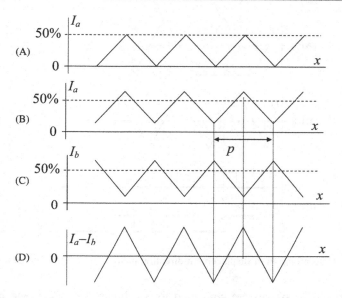

**Figure 7.15** Output currents versus displacement (A) without offset, (B) and (C) transfer of each mask with offset and (D) difference between the two signals.

anymore to the resolution. However due to mechanical tolerances and not perfectly collimated light the minimum transfer is not completely zero. These effects become more significant at decreasing line width.

In the ideal case the sensor output changes linearly with the displacement or rotation over half a grating period. At continuous displacement, the output changes periodically as indicated with the triangular characteristic in Figure 7.15. One period of this triangle corresponds to a displacement over one grating period $p$.

Unfortunately real encoders show some deviations from the ideal behaviour. For instance the base line of the triangle $I_a$ in Figure 7.15A can be shifted upwards or the amplitude decreased, due to misalignment of the scale and the masks, resulting in a position as in Figure 7.15B. Adding a second moving grating and sensor offers some improvement. When the gratings are positioned $(n + \frac{1}{2})p$ ($n$ real) apart, the outputs $I_a$ and $I_b$ are in antiphase (Figure 7.15B and C). The difference between both outputs is a triangular signal that has a more stable relation with displacement (Figure 7.15D): the (common) offset is eliminated and zero crossings define dark–light transitions in a more reproducible manner.

The incremental encoders discussed so far do not provide information about the *direction* of displacement or rotation. They are useful for unidirectional displacements and as (absolute) velocity sensors, but fail when position information is required after displacement in an arbitrary direction. Figure 7.16 clarifies how to solve this problem. Two pairs of masks are now positioned at a distance roughly $(n + \frac{1}{4})$ of the pitch. A displacement results in two triangular signals according to Figure 7.15C, $\pi/2$ rad out of phase (Figure 7.16A and B).

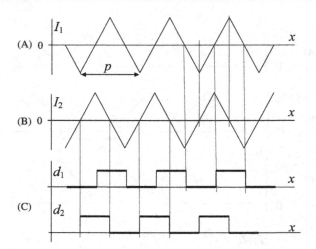

**Figure 7.16** Determination of displacement direction using two optical encoders.

These triangular signals are converted to two binary signals $d_1$ and $d_2$, having also a phase difference $\pi/2$ (Figure 7.16C). Such a conversion can easily be implemented using comparators that switch their output just on the zero crossings of the triangular signal. From these two signals $d_1$ and $d_2$ the direction of the movement can simply be deduced. A movement in the positive $x$-direction results in a binary word sequence $d_1d_2$ of $00-01-11-10-00$, etc. So code 11 is followed by 10. When the direction of the movement is reversed, code 11 is followed by 01. By driving a pair of flipflops, information about the direction is easily achieved.

The resolution of an incremental encoder can further be increased as follows. When the relation between displacement and output is known, this information can be used to find intermediate positions between the maximum and the minimum values of the triangle. Hence the incremental encoder offers the possibility to interpolate between successive slots, increasing substantially the resolution of the sensor. In practice the output versus displacement is not a nice triangle as given in Figure 7.15: the tops may be rounded or the shape is more trapezium like. Better interpolation accuracy is obtained when the output is first converted into a sinusoidal shape. This results in one or two pairs of signals in quadrature ($\pi/2$ phase differences), from which the 'electrical' angle can be reconstructed (compare the resolver signal processing in Chapter 6). Another way is the arctan method: the quotient of two quadrature signals is a tangential function, so by taking the arctan of this function the displacement is obtained. Obviously errors in the amplitude, offset, and phase of these signals introduce position errors. A relation between signal errors and resulting position errors is given in Ref. [11]. A method to further reduce the interpolation error is the generation of higher harmonics as 'in-between' quadrature signals [12].

Note that the measurement head of commercial incremental encoders may contain four masks to obtain direction sensitivity (according to Figure 7.16),

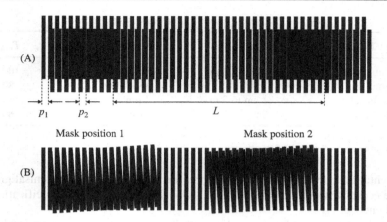

**Figure 7.17.** Incremental encoder employing (A) the Vernier effect and (B) the Moiré effect for increased resolution.

zero-crossing accuracy (Figure 7.15) and interpolation all in one. A fifth mask serves for detection of the reference markers.

The intrinsic resolution of an incremental encoder is set by the number of slots per unit distance or rotation. Apart from interpolation the resolution can further be increased by employing the *Vernier effect* or the *Moiré effect*. Two identical scales with patterns having a slightly different pitch cause, when superimposed, alternately light and dark zones along the encoder scale, with a synthetic period $L$ (Figure 7.17A).

Similarly two scales with the same pitch but making a small angle cause also white and dark zones (Figure 7.17B). When the scales shift a distance of one pitch relative to each other, the dark zone shifts over a complete synthetic period. The Vernier period equals $L = |p_1 p_2/(p_1 - p_2)|$. The configuration behaves as an optical displacement amplifier. Its amplification factor is inversely proportional to the pitch difference and can be very large for pitches having slightly different values. The zones have an extension that allows the use of normally sized sensors.

Direction sensitivity is obtained in the same way as is illustrated in Figure 7.16, using two sets of source—sensor pairs. Here, too, the system allows interpolation between two adjacent positions with maximum and minimum output, similar to the normal encoder. A special design based on this concept is described in Ref. [13], where both shifted and rotated masks are studied.

The Vernier concept can also be applied in absolute encoders, to enhance their intrinsically lower resolution. Details are presented in Ref. [14], dealing with an absolute angular encoder used to control a servo motor. With two discs comprising only three binary tracks each, a resolution of 1.6° can be achieved, instead of 45° with a conventional three-bit code disc.

The main errors of optical encoders are the quantization error $\frac{1}{2}n$ (where $n$ is the number of binary tracks in an absolute encoder or the number of lines in the

**Table 7.4** Maximum Specifications of Optical Encoders

| Encoder Type | Range (FS) | Resolution | $T_{max}$ (°C) |
|---|---|---|---|
| Incremental linear[a] | 1 cm to 3 m[b] | 80 lines/mm | 100 |
| Incremental angular[a] | $2\pi$ | 18,000 lines/$2\pi$ | 80 |
| Absolute linear | 1 cm to 3 m | 1 μm | |
| Absolute angular | $2\pi$ | 14 bit | 80 |

[a]Without interpolation.
[b]Larger range encoders are constructed upon request.

incremental encoder without interpolation), the mask tolerances, the misalignment and the eccentricity of the code patterns. Table 7.4 shows ultimate specifications of several encoder types.

So far we have considered encoders with linear, regular code patterns and static light beams. Many other code patterns have been proposed to reduce instrument complexity, increase accuracy or enhance applicability (see further section 7.4 on applications). Instead of a collimated light beam (to fully cover the coded area), a scanning light beam (as used for reading bar codes) is a useful alternative [15]. This approach allows the use of just a single detector and makes additional detectors for determining the direction of movement unnecessary. It is also easier to read codes with a non-regular format, for instance pseudo-random codes. Such codes may yield higher accuracy because they are less noise sensitive. The scanning may slow down the system, but for many applications, such as navigation, this is not a serious limitation.

## 7.2.4  Interferometry

The concept of interferometry is based on the phenomenon of (optical) interference, occurring when two waves with equal frequency coincide. The resulting amplitude (or intensity) varies with the phase difference between the two waves. In an interferometric sensor such a phase difference occurs due to a path length difference between the two light beams, one of which usually follows a reference path, the other a path of which the length is modulated by a movable part of the construction.

At equal amplitudes of the individual waves, the total intensity doubles when the waves are in phase (*constructive interference*), and drops to zero when in antiphase (*destructive interference*). If the wave amplitudes are not the same, the minimum intensity is not zero and the interference effect becomes less pronounced.

The wave form of monochrome light (i.e. light with just one wavelength) is described by:

$$A(x, t) = A_o \cos(\omega t - kx) \tag{7.11}$$

where $A_o$ is the wave amplitude (for both the electric and the magnetic field components), $\omega = 2\pi f$ is the angular frequency of the wave, $k = 2\pi n/\lambda$ is the wave number and $x$ the co-ordinate in the direction of propagation. Remember that the refractive index $n$ of a medium is defined as the ratio of the propagation speed of light in a vacuum $c$ and in the medium $v$, so $n = c/v = \lambda_0/\lambda$. In the remainder of the text we assume $n = 1$, because most applications are in air having a refraction index very close to 1.

When two waves with equal frequency (wavelength) travel distances $x_1$ and $x_2$, respectively, their wave forms are described as $A_1 \cos(\omega t - kx_1)$ and $A_2 \cos (\omega t - kx_2)$. Both waves fall simultaneously on the same sensor, so the wave functions are added:

$$A_{\text{tot}} = A_1 \cos(\omega t - kx_1) + A_2 \cos(\omega t - kx_2) = A(x) \cos\{\omega t - \phi(x)\} \tag{7.12}$$

where

$$A(x) = \sqrt{A_1^2 + A_2^2 + 2A_1 A_2 \cos k(x_1 - x_2)} \tag{7.13}$$

and

$$\tan \phi(x) = \frac{A_2 \sin k}{A_1 + A_2 \cos k(x_1 - x_2)} \tag{7.14}$$

Equation (7.12) describes a periodic signal with radial frequency $\omega$ and in which amplitude and phase both vary with $k(x_1 - x_2)$; hence with the difference between the travelled distances.

A flat wave transfers optical (electromagnetic) energy into the direction of propagation $x$ that amounts to $S_x = E_y \cdot H_z$, a quantity that is proportional to the square of the wave function. Since a photo-sensor responds to the irradiance — that is the average wave energy per unit area — its output is also proportional to the square of the wave function amplitude $A(x)$ in Eq. (7.13). So the sensor output satisfies:

$$I_d = |A(x)|^2 = A_1^2 + A_2^2 + 2A_1 A_2 \cos k(x_1 - x_2) = I_o(1 + m \cos k\Delta x) \tag{7.15}$$

with

$$m = \frac{2A_1 A_2}{A_1^2 + A_2^2} \tag{7.16}$$

Apparently the sensor output changes sinusoidally with the path difference $\Delta x$. The differential sensitivity is at maximum for odd multiples of $\pi/2$ (the so-called *quadrature point*), and minimal at multiples of $\pi$ (Figure 7.18). Only when both light wave amplitudes are equal, $m = 1$, resulting in an optimal interference effect. This is why the factor $m$ is called the interference visibility.

In an interferometer the two waves travel along separate paths, one with a fixed length (the reference) and the other having a length that is set by a movable object

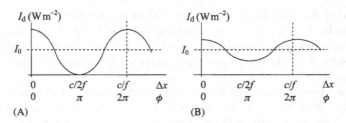

**Figure 7.18** Sensor output as a function of the relative displacement: (A) $m = 1$, (B) $m = 0.5$.

whose displacement has to be measured. The two waves fall on the same detector where they are effectively added up. According to Eq. (7.15) this results in an output signal proportional to the intensity, which varies with the optical path difference hence with the displacement of the object.

Figure 7.18 shows that the output varies periodically with displacement; the unambiguous measurement range is just half a wavelength (for red light about 300 nm). To cover a wider range, there are two possibilities. The first is to count the number of dark−light transitions during the movement starting from some reference position, as with the incremental encoder. Another method makes use of an artificially enlarged wavelength, accomplished by mixing two waves of slightly different frequencies $\omega_1$ and $\omega_2$ (or wavelengths $\lambda_1$ and $\lambda_2$):

$$A_{\text{mix}} = A\{\cos(\omega_1 t - k_1 x) + \cos(\omega_2 t - k_2 x)\}$$

$$= 2A \cos\frac{1}{2}\{(\omega_1 - \omega_2)t - (k_1 - k_2)x\} \cdot \cos\frac{1}{2}\{(\omega_1 + \omega_2)t - (k_1 + k_2)x\}$$

$$= 2A \cos\{\omega_{\text{mix}}t + \phi_{\text{d}}(x)\} \cdot \cos\{\omega_{\text{s}}t + \phi_{\text{s}}(x)\} \tag{7.17}$$

This is an amplitude modulated wave, of which the envelope (see Section 3.2.4) has a frequency $\omega_{\text{mix}}$ equal to half the difference of the two frequencies $\omega_1$ and $\omega_2$. This corresponds to a wavelength $\lambda_{\text{mix}}$ equal to

$$\lambda_{\text{mix}} = \frac{2\lambda_1\lambda_2}{\lambda_1 - \lambda_2} \tag{7.18}$$

also called the *synthetic wavelength*. For instance when the two wavelengths are 650 and 652 nm, the synthetic wavelength amounts 0.424 mm, resulting in an unambiguous displacement range of 0.2 mm (where it was originally 325 nm).

Interference is only visible with two monochromatic, coherent light beams. The creation of such a pair of coherent beams starts with a single, monochrome light source (for instance a laser or laser diode). The emitted light beam is split up into two separate beams, for instance using a semi-transparent mirror. Figure 7.19 shows such a system, known as the classical interferometer or Michelson interferometer.

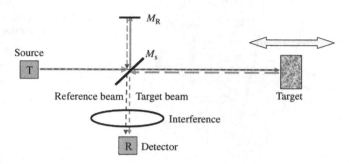

**Figure 7.19** Classical interferometer configuration.

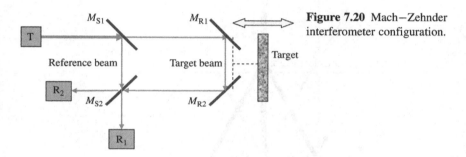

**Figure 7.20** Mach–Zehnder interferometer configuration.

The beam splitter $M_S$ is a semi-transparent mirror: its transmission and reflection coefficients are (about) 50%. The two light beams arrive at the sensor via different paths: one runs from the source, via $M_S$ and $M_R$ and back through $M_S$ to the sensor; it has a fixed length. A second beam travels via the (movable) target and ends at the sensor as well: its total path length varies with the position of the target. Evidently all optical components should be accurately aligned to achieve a proper operation.

A major disadvantage of the interferometer configuration of Figure 7.19 is the optical feedback towards the source, which may introduce instability of the output frequency. In the configuration of Figure 7.20 such feedback is eliminated by using an extra beam splitter. This type of interferometer is referred to as the Mach–Zehnder interferometer. Originally designed for the investigation of the properties of transparent samples, it can also be used for displacement measurements. Again two coherent light waves are involved: the reference wave travels over a fixed distance via semi-transparent mirrors $M_{S1}$ and $M_{S2}$, the other reflects at a set of mirrors $M_{R1}$ and $M_{R2}$ fixed on a movable target.

There are two receivers, which can act in balance mode: when one has maximum output, the other has minimum output. The disadvantage of this interferometer is the necessity to connect mirrors to the moving target.

The simplest interferometer is the Fabry–Perot configuration (Figure 7.21). It is essentially an optical cavity made up from two semi-transparent mirrors in parallel.

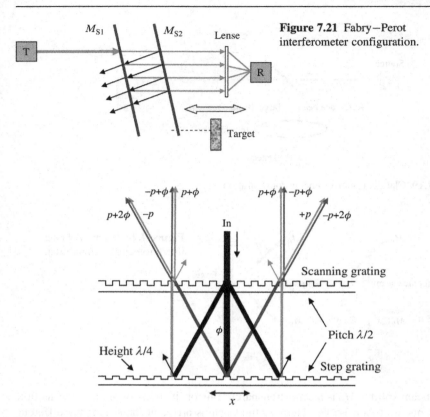

**Figure 7.21** Fabry–Perot interferometer configuration.

**Figure 7.22** Interferometric scale (after Heidenhain).

The moving object is fixed to one of these mirrors. The sensor receives light that has travelled a number of path lengths in the cavity.

Suppose the path length between the mirrors is $\frac{1}{2}\lambda$, then the waves leaving the moving mirror have phase differences of $\lambda$ and are added up (constructive interference); when the path length is just $\frac{1}{4}\lambda$, the waves have phase differences of $\frac{1}{2}\lambda$ resulting in destructive interference.

When the pitch of an incremental encoder is made smaller, in the order of the wavelength, the light is refracted at the grating. This effect is used to further increase the resolution of a linear encoder, by combining it with interference. Figure 7.22 shows the basic principle of this *interferometric encoder*. The instrument comprises two scales moving relatively to each other, at about 1 mm apart. One is transparent, the other not. The non-transparent scale has a pattern of small grooves, with pitch $p$; this is the step grating. The transparent one is the scanning grating, with the same pitch. Light that falls on such a grating will be partly diffracted, resulting in diffracted beams of order 0, ±1, ±2 …. In this case we only consider diffraction of the orders −1, 0 and +1; the order 0 corresponds to the non-diffracted beam.

The 0th-order wave front passing the transparent grating hits the step grating, where it is diffracted again in three directions 0, +1 and −1. When the gratings are in line, the wave fronts have all the same phase. However when the step grating is shifted over a distance $x$, the diffracted wave front shows a phase shift equal to $\phi = 2\pi np/\lambda$ ($n$ is the order of diffraction). The same holds for the two diffracted beams from the scanning scale: when they hit the step scale, three diffracted beams result, with the same phase shift. The design is made in such a way that the diffracted beams interfere when entering back to the scanning scale: for each of the three directions the phase difference of the interfering waves is $2p$. Each time the wave front passes the scanning grating, it experiences an additional constant phase shift; by design this phase shift is $\phi$ for the 0th-order diffracted waves and 0 for the other waves.

Interferometry is mainly applied for linear displacement measurements. However the interferometric sensing principle can also be used to create angular displacement and velocity sensors. Rotation measurements by interferometry require additional optics, for instance a prism assembly [16,17]. Since the resolution is high, interferometry is also applied in sensor systems where the measurand introduce only small displacements of a sensitive part, for instance pressure, vibration and inertial acceleration sensors. In these cases the small displacements of a membrane [18], a cantilever [19] and a seismic mass [20], respectively, are measured by an interferometric principle. In all these examples light is guided through fibres to and from the sensor element, allowing the construction of very small sensing devices.

As an interferometer has a very high sensitivity to displacement parameters, it requires a rigid setup and careful adjustment of the optical components. Furthermore a frequency shift $\Delta f$ induces an output change equivalent to a phase shift of $4\pi\Delta f/c$, where $c$ is the speed of light. An interferometer, therefore, needs a very stable source, to minimize phase noise in the output signal. Altogether an interferometer is quite an expensive instrument.

The range of an interferometer runs from several cm up to several tens of metres and depends on the maximal number of subsequent fringes (light−dark transitions) that can be counted. The resolution is set by that of the interference pattern measurement (either in time or in the spatial domain). The accuracy depends largely on the accuracy and stability of the wavelength (the displacement between two consecutive counts). The wavelength in air varies (non-linearly) with the index of refraction, and hence with temperature, pressure and air humidity. For small variations of these parameters, an approximation for the wavelength change is [21]:

$$\frac{\Delta\lambda}{\lambda} = 0.9\cdot 10^{-6}\cdot\Delta t - 0.3\cdot 10^{-6}\Delta p + 10^{-8}\Delta H \qquad (7.19)$$

denoting the relative change in wavelength around atmospheric conditions. In this expression $\Delta t$, $\Delta p$ and $\Delta H$ are the changes in temperature (°C), pressure (hPa) and relative humidity (%), respectively. An example: the wavelength of a helium−neon laser in vacuum is $\lambda_0 = 0.6329914$ μm. In air with temperature $t = 20°C$, pressure $p = 1013$ hPa and 50% humidity $\lambda = 0.632820$ μm.

The relative accuracy of a non-compensated interferometer in a conditioned room is not much better than 10 μm per metre (hence $10^{-5}$), mainly due to pressure dependency. When compensated for environmental parameters (which is never perfect), an accuracy of 3 μm/m can be obtained, within a temperature range from 0°C to 40°C. Some types of errors can be reduced using two waves with different wavelengths and assuming the errors are common to both waves [22].

### 7.2.5  Time-of-Flight

Light travels through a medium with a fixed speed $v$ determined by its optical properties, so the elapsed time $t$ for covering a certain distance $x$ is proportional to that distance: $t = x \cdot v$. The ToF method is based on measuring the time interval between the emission of a light pulse and the arrival time of the reflected pulse. The speed of light is high: about 300.000 km/s in air. So the ToF of a light pulse is very short: roughly 3 ns for each metre, or 6 ns for a distance of 1 m to a target and back.

The measurement of short distances calls for fast electronics to obtain a reasonable resolution. For the long range resolution is not a major problem; the boundary is mainly set by the decreasing intensity of the reflected light at increasing distance, hence the reduced S/N ratio. When the incident laser light makes an angle with respect to the target's normal, the light is reflected away from the receiver. When such a situation is expected to occur, the target can be provided with a retroreflecting foil: the reflected beam makes the same angle as the incident beam and most light will be captured by the detector. The same can be done for an absorbing target surface.

## 7.3  Interfacing

### 7.3.1  LEDs and Photo Diodes

LEDs and photo diodes are the most frequently used electro-optical components in mechatronic applications. Therefore we give some basic electronic interface circuits for these devices. The simplest circuits are with discrete resistors and transistors, but they have rather poor stability. Figure 7.23A presents an interface circuit for controlling a LED.

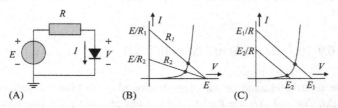

**Figure 7.23** Interface circuit for a LED, with resistor and voltage source: (A) circuit, (B) adjustment by resistor, (C) adjustment by a voltage; dots represent the bias points.

With either $E$ or $R$ the current through the LED, and hence the output power, can be adjusted to a proper value. The exact value follows from the voltage-to-current characteristic of the device and the load line:

$$I = (E - V)/R \qquad\qquad (7.20)$$

Figure 7.23B shows two load lines corresponding to two different values for $R$; in Figure 7.23C the resistance $R$ is fixed, and the current is adjusted by the voltage $E$. In both cases the bias point is set by the intersection of the characteristic and the load line.

In Figure 7.3A two possible readout approaches for a photo diode are indicated: current and voltage readout. In the first case the circuit is loaded by a system with zero input resistance (in Figure 7.24A modelled by a resistance $R = 0$); in the latter case the measuring system has an infinite input resistance, $R = \infty$. For a finite value of the resistance, the output voltage is given by the intersection of the diode characteristic and the line $V = I_p \cdot R$, as shown in Figure 7.24B.

Clearly only current readout provides an output voltage that varies linearly with the photo current. Similarly only current control of a LED provides a well-defined optical output. These conditions cannot be met easily just by a resistor. Using operational amplifiers (Appendix C), interface circuits for both source and receiver can be designed with much better performance. Figure 7.25 shows the basic configurations for each of these components.

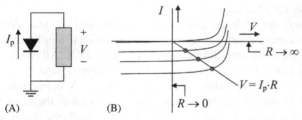

(A)          (B)

**Figure 7.24** Interface circuit for a photo diode, with a resistor: (A) circuit, (B) current, voltage and mixed readout; dots represent output voltage at different input power.

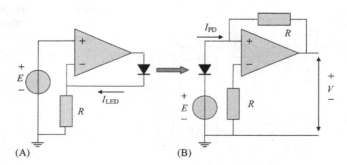

(A)                                  (B)

**Figure 7.25** Interface circuits with operational amplifiers for (A) a LED and (B) photo diode.

In both circuits the operational amplifiers are configured in feedback mode; this makes the voltage difference between the two input terminals zero. Hence the voltage at the positive input terminal in Figure 7.25A is equal to $E$, forcing a current $E/R$ flowing through resistor $R$. Since no current flows in the input of the (ideal) operational amplifier, this current must also flow through the LED, generating an associated radiant flux. By varying $E$ the radiant flux varies accordingly.

In Figure 7.25B the operational amplifier is configured as a current-to-voltage converter. The photo current generated by the photo diode is forced to flow through feedback resistor $R$, producing an output voltage $V = -R \cdot I_{PD}$. The resistor $R$ in series with the positive input terminal serves for compensation of the bias current of the operational amplifier, which flows also through the feedback resistor (see Appendix C).

An effective way to eliminate the influence of environmental light is intensity modulation. This can easily be achieved using the circuit of Figure 7.25A: the optical output power is proportional to the input voltage $E$, so when $E$ varies, the intensity is modulated by this voltage. The modulation frequency (the carrier) should be chosen such that interference signals can easily be separated from the signal by filtering (Appendix C).

## 7.3.2 Interfacing PSDs

A proper signal processing makes the output of a PSD independent of the intensity of the light beam, as shown in Section 7.1.3. Figure 7.26 shows a combined analogue/digital processing circuit for PSD signals.

The currents $I_a$ and $I_b$ from the PSD are converted into voltages $V_a$ and $V_b$, respectively, using simple I−V converters configured with operational amplifiers. Next the output voltages are converted into digital signals by two ADCs. All mathematical operations according to Eq. (7.5) are performed by a microprocessor.

The best performance is obtained when the output range of the I−V converters matches with the input range of the ADCs. In that case the system takes the most advantage from the full resolution of the ADCs. However the irradiance varies

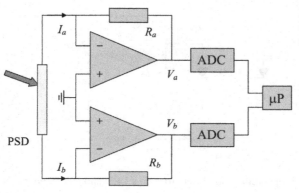

**Figure 7.26** Signal processing circuit for a PSD triangulation system.

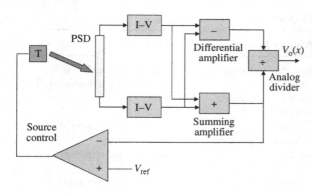

**Figure 7.27** PSD interface (analogue) with stabilized input radiant power.

strongly with distance, as shown in Section 7.2.1. At large distances the output signal may drop below the acceptable noise level. On the other hand at close distances where the irradiance is high, the interface electronics may run into overload.

To extend the measurement range, the source intensity could be controlled in such a way that the PSD receives the highest possible radiant power, irrespective of the distance. The sum of both output currents of a PSD is proportional to the irradiance; hence, this sum is a suitable quantity for controlling the emitted radiant power of the source. Figure 7.27 gives an example of such a control circuit; here, the signal processing is performed by analogue circuits.

The output of the summing amplifier is compared with a reference voltage $V_{ref}$. The amplified difference controls the current through the light source in such a way that the sum voltage remains $V_{ref}$. Within the operating range of the control circuit, the summing voltage equals the reference voltage. The measurement range can be set by adjusting the reference voltage $V_{ref}$. Since the intensity of the incident beam is fixed by this feedback, the output signal is that of the differential amplifier, and the (analogue) division can be omitted, which is an important advantage of this approach.

# 7.4  Applications

This section presents various applications of optical sensors in mechatronics. Line and area *cameras* are often encountered in robotics for the purpose of object recognition, environmental exploring, navigation and (robot) calibration. However we will not discuss these applications further here; the reader is referred to the extensive literature on computer vision. Colour information can be very useful in a number of applications, such as for painting robots, object sorting (for recycling purposes) and so on. A colour camera is a suitable transducer, but when no spatial information but just colour is needed, a simple set of three diodes with colour filters will do as well.

The applications discussed here are categorized in Table 7.5.

**Table 7.5** Applications of Optical Sensing

| Application | Primary Quantity | Methods |
|---|---|---|
| Linear distance (1D) | Medium- and long-range distance | Intensity; triangulation; encoder; interferometry; ToF |
| Proximity (1D) | Short-range distance | Fibre optic |
| Roughness, flatness | Distance | 2D scanning |
| Linear velocity | Velocity | Doppler; correlation |
| Linear acceleration | Velocity; displacement | Differentiation |
| Rotation; angle; angular rate | Angle | Encoder; interferometry |
| Object shape (2D, 3D) | 3D co-ordinates | Probing; triangulation |
| Ranging (2.5D) | Distance | Passive |
| Navigation (2D) | Distance | Active/passive |
| Object tracking | Distance | 3 modi; active |
| Strain, force, torque | Deformation | Elastic element; interferometry |

## 7.4.1   Linear Displacement Sensing

Major measurable attributes of linear displacement are (relative) position, velocity and acceleration. Optical sensing allows the measurement of these quantities without making contact with the moving object.

### Linear Distance

Optical sensing systems are widely used for contact-free *distance measurement* in mechatronics. The major principles are, in the order of increased accuracy, reflectance, triangulation and interferometry. The reflectance method according to Section 7.2.1 is very useful for low-accuracy applications and binary sensors. The small-sized electro-optical components can easily be mounted on or integrated into the mechatronic system. If there is not enough space to mount the sensors at the measurement location, the light can be transported to and from the proper positions using optical fibres.

Since *reflectance* depends on the reflectivity of the object, operation is restricted to environments with invariant objects. In applications with a variety of possible objects, accurate distance data can only be achieved after a calibration with those objects. This problem was recognized in the early days of robotics when designing a controlled gripping systems (see for instance Ref. [23]). To become independent of the reflectivity and the orientation of the objects to be handled, special algorithms and sensor configurations have been developed [24]. These problems still exist, for instance for medical robots and for robots for (unmanned) inspection and exploration of unknown areas. If objects are completely unknown, as is the case of navigation in an unknown environment, sufficient tolerances should be allowed in the obtained distance data using reflection methods.

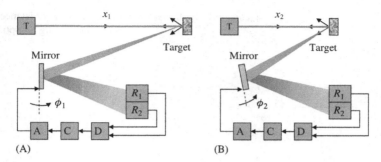

**Figure 7.28** Optical position measurement in feedback configuration; (A) far and (B) close by.

To eliminate the influence of changes in source intensity and other unwanted interference in an optical displacement system based on intensity modulation, multiple sensing and feedback are effective solutions. Adding one or more receivers to compensate for the effect of unwanted changes makes reflection-based systems less sensitive to the surface properties and may even allow uncalibrated operation [25,26]. Another solution is presented in Ref. [27], where the relative position of transmitter and receiver is adjusted such that the maximum intensity is received. The distance is found irrespective of the surface characteristics. Feedback is an even more powerful solution but has higher system complexity. Figure 7.28 shows an example of such a system.

The light reflected by the (movable) target is projected onto two photo-sensors via a rotating mirror. The position of this mirror is controlled such that the two sensors receive equal radiation (position $x_1$ in Figure 7.28A). The corresponding angle of the mirror is $\phi_1$. Upon displacement of the object to position $x_2$ the light beam shifts from the centre of the dual sensor, resulting in different photo currents. This difference is used to drive the mirror to rotate to an angle $\phi_2$ where the photo currents become equal again. The control signal or the angle is a measure for the displacement. By feedback the requirements to the source and the sensors are reduced significantly, as explained in Chapter 2. For 2D displacements the moving mirror has 2 d.o.f. and the detector comprises four receivers arranged in a square (a quad).

The principle is applied in for instance *automatic focussing* systems in certain camera types. Another application is found in contact-free measurement of *profiles*. Commercial instruments for that purpose are available with a resolution down to 50 nm, and a range of about 1 m.

All aforementioned examples use the change in irradiance with distance between source and sensor (usually a $1/x^2$ relation). In a light absorbing (or attenuating) medium the intensity drops exponentially, according to Lambert−Beer's law:

$$I(x) = I(0)e^{-\alpha x} \tag{7.21}$$

where $I(0)$ is the intensity at a point $x = 0$, $I(x)$ the intensity at the position $x$ and $\alpha$ is the absorption coefficient of the medium. The absorption coefficient of air is rather low; some materials show a much larger absorption and can be used to create

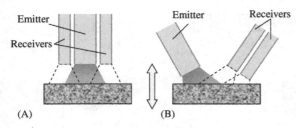

**Figure 7.29** Short distance measurement using an optic fibre.

a distance sensor based on light attenuation. A design with a fluorescence optical fibre is described in Ref. [28]. A light source injects light into the fibre at a particular position along its length. The intensity of the light that arrives at both ends of the fibre is measured by two photo diodes. According to Eq. (7.21) the intensity at each fibre end depends on the distance between the position of the light source and the photo diodes. Similar to the PSD the difference of the photo currents is a measure for the position of the light source. The distance range equals about the length of the fibre (about 88 cm in the experimental setup).

Distance measurements over a very short range can be performed using optic fibres. Two different arrangements are depicted in Figure 7.29.

Light from the emitting fibre reflects at the target and is intercepted by the receiving fibre. The light diverges within a short distance from the exit end, so the irradiance at the entrance face of the receiving fibre depends on the distance $x$, in a similar way as in the arrangements of Figures 7.5 and 7.6. With two receiving fibres in differential mode, most common mode interference (for instance the intensity of the emitter) is suppressed. The sensitivity characteristic depends on the angle of divergence, the stand-off distance and the angles between the fibre ends and the flat target. For a more detailed analysis see for instance Ref. [7]. Since the output is an analogue signal, the resolution of the distance measurement is mainly noise limited. Using modulation techniques and proper filtering the equivalent noise can be as low as $1 \text{ nm}/\sqrt{\text{Hz}}$ [29]. A method to significantly extend the measurement range is based on two wavelengths, as described in Ref. [30], for example. At the expense of higher system complexity the range can be extended up to 40 mm.

Distance measurements based on reflectance have a limited range, due to the inverse power laws given by Eqs (7.6) and (7.7). *Triangulation* systems are less sensitive to the intensity of the reflected beam, if properly interfaced. In the triangulation system shown in Figure 7.8 the PSD may be replaced by a diode array, which simplifies the interfacing, but the resolution is restricted to that of the array.

Commercial triangulation systems are available in a compact housing that includes laser, one or two PSDs, optical components and interface electronics to generate an output independent of the intensity, according to Eqs (7.5) and (7.10), (Figure 7.30). The instrument either contains a display or provides an electrical output. Measurement ranges run from a few mm up to several m; reproducibility is in the range $\pm 1$ to $\pm 10 \ \mu\text{m}$.

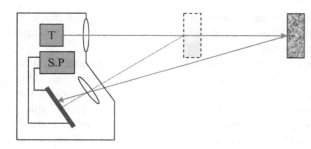

**Figure 7.30** General layout of a commercial triangulation distance sensor.

Size reduction of the optical system is another issue to increase the versatility of distance measurement systems based on triangulation. An example of a technological solution to miniaturization is given in Ref. [31], where the LIGA process (German: Lithographie, Galvanoformung und Abformung) is applied to realize the required optical structure. The total device measures $7 \times 7 \times 3$ mm; obviously the measurement range is small too — about 1 mm. However the repeatability error is less than 3 μm.

An example of a displacement sensor using interferometry in combination with the feedback principle of Chapter 2 can be found in Ref. [32]. The system uses the speckle effect of a laser beam falling on a non-specular surface. Speckles are randomly distributed light and dark areas on the surface; this speckle pattern changes with the distance from the laser to the surface. The feedback system described in Ref. [32] uses this dependency. With a set of piezoelectric actuators the light spot on the target is controlled in such a way that it follows a bright speckle when the target is moving. In this way the movement of the target is transformed to that of the actuator. An essential element of the system is a self-mixing interferometer (this will be explained in the section on velocimeters). The system enables displacement measurements over a range of 500 mm with sub-micron resolution.

Commercial sensing systems based on the *ToF* principle cover ranges from about 0.5 m up to 10 km, with an inaccuracy ranging from a few mm for short-range types to a few cm for the long-range types. The resolution is typically one order in magnitude better. Besides simple distance measurements, the principle is also applicable for range finding and imaging; such systems comprise a rotating scanning mirror for 1D or 2D scanning of the reflective object.

## Surface Properties: Roughness, Flatness, Thickness

Triangulation is applied in a multitude of applications where accurate contact-free distance measurement is required. Very high resolution is needed in applications like roughness measurements and profile measurements. In such applications 1D laser triangulation is combined with a (2D) scanning mechanism.

In Ref. [33] a triangulation system is described that performs the simultaneous measurement of surface *roughness* and surface height (displacement in the normal direction). The roughness measurement is based on the scattering of a light beam incident normal to the surface; the height measurement follows the standard triangulation method. A linear photo diode array is used to determine the position of the reflected light. With this system roughness can be measured over a range from 5 to 100 nm; the height range amounts to $\pm 300$ μm.

Since the triangulation principle allows contact-free distance measurements, it is pre-eminently suitable for measuring of objects having a high temperature. An example of this application is found in Ref. [34] where a system is described for measuring the flatness of hot steel plates just after being pressed by a pair of rolls. In the application a light strip is projected under an angle of 30° onto the sheet (moving at a maximum speed of 15 m/s). A line scan camera with 2048 pixels views the strip in perpendicular direction, so it receives a light spot of about 10 pixels in width. Using a special algorithm the resolution is increased to an equivalent of 0.1 mm in height of the steel plate.

Position or displacement sensors which are intended for a very particular application may be redesigned to serve other applications in mechatronics and quality inspection. Examples are the optical mouse and the DVD pick-up system. In Ref. [35] an optical mouse, consisting of a LED and a CMOS camera, is used for measuring the viscoelastic properties of polyethylene. Experiments showed the feasibility (and limitations) of this application. A commercial DVD pick-up head appears also suitable as a basis for an optical accelerometer [36] and a touch trigger probe [37].

A particular application area is the exploration of an (unknown) environment or object reconstruction for recognition purposes. Triangulation methods used for this specific application field are further discussed in Section 7.4.3.

## Linear Velocity and Acceleration

Velocity can be derived from distance information by taking the time derivative of the displacement signal. Likewise acceleration is obtained by differentiation of the velocity signal. So, in principle, any displacement sensor can be taken as the starting point of a velocity or acceleration measurement. Since any measurement signal contains noise, differentiation will increase the noise level of the velocity signal obtained in this way. A better option is to select a measurement principle that produces an output signal that is directly related to the speed or acceleration of the object under test, without the need of a differentiating operation.

The velocity of a vehicle (e.g. car, train and mobile robot) is often determined from the rotational speed of the wheels (obtained by a tachometer, for instance):

$$v_{lin} = 2\pi R \cdot n / 60 \tag{7.22}$$

where $n$ is the number of revolutions per minute (rpm) and $R$ the circumference of the wheel. The method is rather inaccurate, because slip and wear of the wheels introduce measurement errors. Alternative approaches are Doppler measurement and correlation, both to be implemented by optical sensing. The basic operation of the correlation method using optical sensors is presented in Figure 7.31.

Two sets of a transmitter−receiver pair in reflection mode are mounted at the moving vehicle, a distance $d$ apart in the direction of the movement. When the

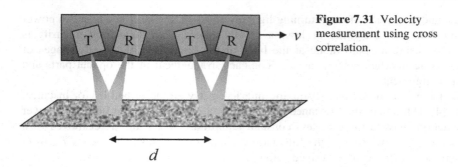

**Figure 7.31** Velocity measurement using cross correlation.

surface is rough, the reflected signals depend somewhat on the particular structure at the point of impact. While moving the output signals will show a random component according to the varying scattering of the surface. Both output signals are (about) the same but displaced over a time interval $d/v$. The time delay between these signals can easily be derived by cross correlation: the cross-correlation function shows a maximum for a time difference equal to $d/v$, from which the velocity is calculated.

Although the implementation seems straightforward, a proper result is not evident. The S/N ratio is adversely affected by vibrations or other movements (i.e. roll and pitch) of the vehicle, misalignment of the optics and surface characteristics. Moreover a high speed requires a high sampling rate to obtain a reliable (digital) correlation function, whereas at low speed the delay time is long, which necessitates longer processing time to avoid large measurement errors. Various solutions have been proposed to overcome such difficulties. In Ref. [38] a system is presented that makes use of optical fibres transmitting the light from two laser diodes towards the moving surface. The laser heads are 30 cm apart, and the distance to the surface is about 1 m. Using a special correlation technique, the speed of the moving surface was measured up to 30 km/h, with an error of 9%.

Many alternative methods for contact-free velocimetry using optical means have been proposed. We mention just some of them here. Apart from time filtering of the reflected signals, spatial filtering is another useful approach. The moving object is illuminated by environmental lighting (so not necessarily coherent light). The illuminated area is viewed by a photo detector through a grating with small periodicity $p$. When the object moves with velocity $v$ in the plane of the grating, the output of the detector contains a frequency component equal to $kv/p$, with $k$ the optical magnification. In Ref. [39] a velocimeter is described that uses this principle; the grating has been replaced by a CCD camera, acting simultaneously as detector and spatial filter device. Information about the direction of the movement is obtained by creating a quadrature signal, which is simply accomplished by a modified CCD readout sequence. Experiments revealed velocity errors of less than 0.1% up to a velocity of 20 m/s.

Various optical velocimeters are based on *self-mixing interferometry*. Self-mixing is obtained by optical feedback: part of the laser light is fed back via an external reflecting surface into the laser cavity. When particular conditions with respect to phase and amplitude of the returning light are fulfilled, the laser beam shows power fluctuations that are more or less saw-tooth shaped. The fluctuating intensity is observed using a photo diode at the back of the laser. One of the advantages of self-mixing interferometry is a more compact arrangement of the optical parts and simple alignment.

Velocimeters based on self-mixing interferometry are described in, for instance, Refs [40,41]. At constant distance between the target and the laser head, the power fluctuations have a more or less constant amplitude. When the target moves with velocity $v$, the spectrum of the fluctuations contain the Doppler frequency $f_D$, from which the velocity can be derived using:

$$v = \frac{\lambda}{2} \cdot f_D \cdot \frac{1}{\sin \phi} \qquad (7.23)$$

with $\lambda$ the wavelength of the emitted laser light and $\phi$ the angle of incidence. At moderate velocities the Doppler frequency is rather high; for instance, using a laser with wavelength 680 nm and normal incidence ($\phi = \pi/2$), the Doppler frequency at a velocity of 1 m/s equals 2.9 MHz. Accurate determination of this frequency is hampered by the presence of the speckle effect, which introduces multiplicative noise to the signal [40]. Moreover when used in automotive applications, for example, other movements introduce measurement errors. In particular fluctuations in the angle $\phi$ and variations in the height above the ground may give large errors, as can be seen in Eq. (7.23). These errors can (largely) be eliminated using a differential configuration, as described in Ref. [41]: two lasers are pointing to the ground, with incident angles of $\pm 18°$ with respect to the ground normal. The Doppler frequency that can be reconstructed from the two output signals is less sensitive to car movements other than in the driving direction. Experiments with the two-laser configuration showed a mean standard deviation less than 0.04%, at velocities up to 100 km/h.

A well-known application of the optical *ToF* principle (Section 7.2.6) is found in laser guns that measure car speed. Usually these systems emit a series of short pulses; from the ToF of the reflected pulses both distance and velocity of the vehicle can be derived. Typical range and accuracy are $5-500 \pm 2$ km/h.

Inertial-based sensors respond directly to *acceleration*, avoiding the noise-enhancing differentiation of position or velocity signals. Such an accelerometer comprises a mass and a spring element (see also Chapter 8 on piezoelectric sensors). When the mass is accelerated the resulting inertial force causes the elastic element to deform in proportion to the applied acceleration. This deformation can be measured in various ways, for instance by optical means. However the displacement is generally very small, and proper measurement requires a sensor with high resolution in terms of displacement. Configurations with optical fibres in reflection mode (Figure 7.29) are suitable candidates for this task.

A particular solution is proposed in Ref. [42]. Here the (small) displacement is measured by an incremental encoder using the Moiree effect (according to Figure 7.17B). One grating is fixed to the housing; the other grating, making a small angle with respect to the first, is connected to the seismic mass. Light is conducted to and from the gratings through a pair of optical fibres (transmitter and receiver), providing the measurement of the transmission, which is modulated by the movement of the mass. Due to the small angle even small displacements of the mass result in a measurable displacement of the Moiree pattern. Actually there must be two pairs of transmitter–receiver fibres to enable measurement of the direction of displacement according to the principle given in Figure 7.16. The system is used for measuring vibrations of large civil structures and has a useful bandwidth of 5 Hz. The resonance frequency of the mass-spring combination (here 20 Hz) limits the bandwidth, as is further explained in Chapter 8 on piezoelectric sensors.

### 7.4.2  Angular Displacement Sensing

Angle measurements can be categorized in angular displacement ($\phi$), angular speed or rate ($\dot{\phi}$) and angular acceleration ($\ddot{\phi}$). Accurate measurement of *angular displacement* is mostly performed by encoders, either absolute or incremental (Section 7.2.4). Angular movement over a very small range can also be detected using the configuration of Figure 7.29A: the transmission from transmitting to receiving fibre is modulated not only by a vertical displacement but also by rotation of the target as well.

A particular application is the *inclinometer* or *tilt sensor*: it measures the angle with respect to the earth normal and is a useful device for walking robots to control movement and equilibrium. The market offers various ready-to-use inclinometers, covering ranges from ±1° to ±90°, with excellent linearity. The mechanical types are based on a free-moving pendulum that always points downwards, also when the housing makes an angle $\phi$ with the earth normal. At the pivot of the pendulum some means of displacement sensing is connected, which gives an output signal that is related to $\phi$. This can be a simple optical displacement sensor in reflection or transmission mode, by an encoder or a vane connected to the other end of the pendulum. Application of a feedback loop (Chapter 3) reduces errors due to, for instance, a strong non-linear relationship between angle and measured displacement.

The proximity sensor in Figure 7.29 is also sensitive to angular displacement of the reflecting surface. This concept is further detailed in Ref. [43]. The fibre-optic angular sensor is composed of a rotatable mirror, one central input fibre and four output fibres. The configuration combines a high angular sensitivity and a negligible sensitivity to normal displacements.

Fluidic inclinometers use a liquid with air bubble to measure tilt [44]. A small hemispherical compartment with the liquid is mounted on top of a silicon wafer in which a quad diode structure is deposited. The bubble always resides at the upper position of the compartment. Light from an LED, mounted on top of the structure, travels through the liquid with the bubble and arrives at the diode structure, at a position that depends on the place of the bubble and hence on the tilt angle. With four photo diodes the inclination angle in two directions is obtained.

*Angular rate* information could be obtained by counting the number of zero transitions in the output signal of an incremental encoder (Figure 7.15) over a specified time interval. However incremental encoders have a limited slew rate (maximum speed), above which the transitions cannot be detected correctly. Besides this the high resolution of most incremental encoders is not needed to achieve rate information with reasonable accuracy: the number of revolutions per second (rps) or rpm will do in most applications.

Low-cost angular rate sensors or *tachometers* are realized simply by mounting one or more segments of a reflective material, such as aluminium foil, along the circumference of the rotating shaft or on the head of the shaft (Figure 7.32). In case of a reflective shaft one should take an absorbing material, like black paint or black paper. The receiver output signal is a pulse-shaped signal, with pulse frequency $f = np/2$ Hz, $n$ being the number of rps and $p$ the number of transitions between reflective and non-reflective material along the circumference. The measured angular speed equals $2\pi n = 4\pi f/p$ rad/s and can be determined by digital processing.

At high speed only one pair of black-and-white areas suffices ($p = 2$). At lower speeds a larger number of intensity changes per revolution is required, to minimize the quantization error, being $1/p$ s$^{-1}$. The performance is further limited by possible environmental light: although this effect can be eliminated by thresholding the detector output signal, illumination other than by the transmitter should be minimized, for instance by optical shielding of the whole system from the environment, whenever possible. Commercial optical tachometers are available in various sizes and ranges, up to 12,000 rpm.

For mechatronics and robotics applications, various mechanical specifications should be considered as well (Table 7.6), for instance the starting torque, the

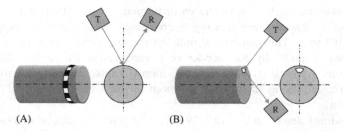

(A)                                          (B)

**Figure 7.32** Simple optical tachometers; reflection from (A) the side and (B) the head of the shaft.

**Table 7.6** Some Mechanical Specifications of Encoders

| Property | Maximum Value |
| --- | --- |
| Starting moment | $10^{-3}$ N m |
| Inertial moment | 10 g cm$^2$ |
| Shaft load − radial | 10−150 N |
| Shaft load − axial | 6−250 N |
| Slewing speed | 5000 rpm |
| Life time | $4.10^8$ rev. |

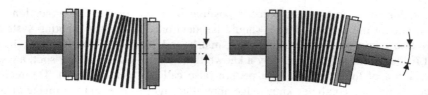

**Figure 7.33** Bellows couplers to obviate parallel and angular misalignment.

inertial moment and the shaft load. As tachometer the maximal rotation (or slewing) speed should be kept in mind. Some manufacturers give an indication of the expected lifetime.

When proper alignment of encoder and rotating shaft is difficult to realize, special accessories can be applied, for instance bellows couplings, available in numerous implementations (angular, axial, parallel; Figure 7.33).

A special *interferometric* arrangement allows the measurement of angular speed, utilizing the *Sagnac effect*. Light is forced to travel a circular path, by means of an optic fibre loop. Actually there are two light waves, one travelling clockwise through the loop, the other counter-clockwise. Both waves arrive on the same detector, where they interfere. The intensity is maximal when the travel times are equal, which is the case in an inertial system in rest. When the loop turns around its axis, the travel times of the two opposite travelling waves differ somewhat, resulting in a phase shift at the detector and hence a change in the interference pattern. The range of such a system is about ±60°, whereas a thermal drift of 10° per hour can be expected.

*Angular acceleration* is obtained by differentiating angular speed. The rotational acceleration sensor described in Ref. [45] uses special encoder discs. One disc is elastically coupled to the other, to transform angular acceleration into angular displacement (according to the law of inertia). The slots of the two discs make small angles, resulting in a radial shift of the overlapping parts and hence a shift of the light spot that passes the two discs. This shift is measured by a PSD.

### 7.4.3 Object Tracking

Measuring the 3D position of a particular (moving) point is a rather common task in mechatronics (for instance the position of the stage in a co-ordinate measurement machine) and robotics (the position of the end-effector or an object in the robot's environment). In many applications a further task is to track the said object or a specific point on that object when it is moving around. In general two different approaches can be considered: imaging (with cameras) and active lighting (such as with triangulation). The latter approach is characterized by the presence of a light source with known position (in the system's co-ordinate frame), and that this position is used to reconstruct the position of the object. Vision systems having environmental lighting do not have that information and require some kind of calibration to reconstruct the object's position from one or two 2D projections of the scene on the camera chip. It should be mentioned that active lighting systems may also comprise a camera (e.g. line, matrix and 2D-PSD). However the object that is looked for is illuminated by a known light source such as a laser spot, a laser line, a pair of lines or a 2D light pattern (also called structured light). The reconstruction process combines knowledge about that light source and the image of the reflected light.

Tracking by vision is a topic on its own and will not be discussed further. The reader is referred to the extensive literature. Table 7.7 compares both concepts with respect to some important design considerations.

Tracking systems using a specified light source can be configured in various ways, depending on such system specifications as power consumption, available space and environmental conditions. Since the transmitter usually takes the most of the available power, the choice between putting the transmitter or the receiver on the (moving) object is an important issue. Further there is the choice between direct mode and indirect or reflection mode. The latter has the advantage that the object to be tracked is completely free from wiring and power sources. Only a distinctive indicator should be present, detectable by the optical system, for instance a light spot, a light pattern, a black/white marker, a (retro)reflector or something similar.

When the design is based on the *direct mode*, again various options are open: transmitter on the moving object and receiver somewhere in the environment or vice versa. Further since the object has to be tracked, either the receiver or the transmitter should be able to explore the environment by some kind of scanning. Of all possible combinations practical applications can be found.

At start-up of the tracking process, the object position is usually not known. Therefore first a search is performed, in which the environment is scanned by either the transmitter or the receiver, until the receiver gets a proper signal back. Once the object has been found, a tracking algorithm controls the transmission path to follow the movement of the object. Further details of the system design strongly depend on the specific application. As an illustration a few examples will be described.

A first application is the 3D tracking of the *end-effector of a robot*. Usually the position and orientation of the tool-centrepoint (TCP) is derived from the position of the joints. However errors in the mechanical construction and the

**Table 7.7** Comparing Vision and Ranging for Object Tracking

| Vision | Active Lighting |
| --- | --- |
| Environmental illumination | Controlled illumination (laser) |
| 2D information (1 camera) | Depth information (by triangulation) |
| Intensity and colour information | No such information |
| No mechanical scanning | Mechanical scanning (mirrors) |
| Feature extraction (image processing) | Predefined point |
| Multiple objects | Single object |

compliancy of the composing parts cause errors in the position of the TCP that may obstruct accurate positioning of the end-effector. Measuring the TCP with respect to the robot co-ordinate frame may provide data for compensation for such errors. A well-known approach uses a light source mounted on a suitable place near the TCP, emitting light in all directions. The light is perceived by a set of cameras (2D-PSD, matrix CCD), positioned at fixed, known locations, together covering the full work space of the robot. By co-ordinate transformation or triangulation the position of the TCP can be reconstructed from the spot positions in the image planes. The working range is set by the viewing field of the cameras.

This basic idea can be implemented and extended in various ways. The principle of feedback may be applied to enhance accuracy and increase the working range. This is achieved for instance by mounting the camera on a rotatable platform; the orientation of the camera is controlled in such a way that the spot is always in the centre of the image. In this application a 2D-PSD camera is a good choice because only a single light spot has to be detected. Since the spot is always in the centre, errors due to non-linearity of the PSD, occurring mainly at the edges of the device, are eliminated. Encoders on the axes of the platform provide the information about the light source position. To find the orientation of a particular part of the robot, two or more light sources may be placed on the construction, viewed by one or more cameras in the world co-ordinate system. In all cases the system, once set up, must be calibrated to obtain the highest accuracy.

In an alternative approach to find the end-effector position the light source and detector from the previous example are interchanged: the sensor (camera) is mounted on the end-effector and a source transmits a light pattern into the direction of the sensor. The pattern can be a light spot, a line (when emitting a light plane), a cross hair or any known pattern. A 2D pattern in combination with a matrix sensor allows the measurement of the position and orientation. An algorithm to obtain the required information is described in Ref. [46], dealing with the measurement of the end-effector position.

Tracking systems are also used in *rehabilitation*, to study the movement of people. Here the reflection mode is common practice to minimize movement restrictions for the test person. A set of (retro)reflecting pads or spheres is positioned on specific locations on the body (e.g. arms, legs, head, shoulders and feet). These markers are illuminated by a specific light source (for instance IR), while the reflected light is viewed by a set of cameras. To increase the S/N ratio, the light is first filtered, so the camera actually sees mainly the reflection from the markers. Since not all markers can be seen at the same time due to occlusion by the test person, and since all markers are identical, special tracking algorithms have been developed to identify the tracked body parts. Accuracy requirements are not severe; processing speed is a more important issue since fast movements must be tracked and analysed.

In high-accuracy applications, for instance *co-ordinate measurement machines*, interferometers can be deployed to accurately measure the position of a machine part. Takatsuji et al. [47] describes a tracking system for the co-ordinates of the stage. It comprises a retroreflector mounted on the stage and four lasers that track the position of the retroreflector. A special algorithm has been developed to reconstruct all three co-ordinates; the fourth, redundant laser is used for self-calibration.

### 7.4.4 Object Shape

The shape of an object is characterized by the co-ordinates of its outer surface. Measuring the shape is useful in lots of applications, for instance localization and orientation of an object in a specified co-ordinate system, identification of objects, quality inspection and parameter extraction. The accuracy requirements are set by the application. The highest accuracy is needed in product inspection, where submicron resolution is often required. Identification and (global) characterization are less demanding applications.

The transmission mode is the most simple and effective way to control production processes and to monitor the dimensions of products. The applicability of the method is almost unlimited. As an example of a highly demanding 1D measurement we refer to Ref. [48]. The aim is the on-line determination of the rotor palette diameter (up to several m) with an error less than 10 μm. Light source and detector (a quad photo diode) are positioned just at the outer edge of the blade. A small change in diameter modulates the transmission, yielding information about the error.

An illustration of object characterization using distance sensing with fibre optics is the (on-line) measurement of *tool wear* [49]. In this intensity-based method, light to and from the tool (here a milling device) is transmitted via a bundle of fibres. A group of fibres in the bundle transmits the light towards the object; another group transmits the reflected light to the sensor unit. Basically the distance to the tool is measured according to the principle shown in Figure 7.29. While rotating an on-line plot of the distance is made; this plot provides information about the wear state of the tool. In this application the intensity method is satisfactory, because the setup is fixed, the distances are short and the environmental conditions are more or less constant.

For more accurate, numerical shape data, and for applications in a less structured environment, the intensity method is usually not adequate. We discuss two other methods: probing and triangulation.

Many 3D co-ordinate measuring instruments are based on a touch probe with which the contours of an object are followed to obtain 3D shape information with high accuracy (also referred to as *dimensional metrology*). The touch probe consists of a ball-shaped touching point with well-known dimensions and which is connected to the end of a stylus. The probe is moved by a set of actuators until it touches the object. Next the movement is controlled in such a way that the point follows the surface of the object, thereby just keeping contact between probe and surface. The measurement quantity can be the force on the ball, the bending of the stylus or the displacement of the ball relative to the base of the system. Shape information is reconstructed from the momentary positions of the stages and possibly corrected for the bending of the stylus.

Various experimental setups of the latter approach using optical sensing are described in Refs [50,51], where the changing position of the ball at the end of the stylus is measured by a couple of optical fibres. The ball is elastically attached to the end of the stylus, so when the probe touches the object, the ball is displaced somewhat in the direction of the reaction force. Further mirrors are connected at the top of the ball; they follow the rotation of the ball when touching the object. A set or a bundle of fibres is fixed to the stylus, with their ends just above these mirrors. According to the principle of Figure 7.29A, the ball movement is measured in two directions normal to the stylus. Ball deflections down to 1 μm (corresponding to a sideways force less than 1 mN) can readily be measured.

In less demanding applications, or when touching is not possible or allowed, triangulation is a suitable method to obtain shape information. For shape information the triangulation configuration from Figure 7.9 should be extended by a scanning mechanism to cover the whole area of the object to be investigated. The method is based on measuring the distance from a specified point to the surface of the object, in a straight line. The method is also referred to as *ranging*. Evidently due to self-occlusion, only that part of the surface that is in direct sight of both the transmitter and the receiver can be investigated.

Optical range finding (some call it 2.5D imaging) is also frequently applied in robotics, for instance to obtain information about the robot's environment, for navigation purposes (automatically guided vehicles) and many other tasks. Navigation by ranging is discussed in the next section.

For shape measurements, the most straightforward configuration is a 2D point scanning (see also Figure 1.3A). The range distance is determined for each individual sample point, by triangulation. Needless to say this is a rather slow acquisition process. A faster solution is line scanning, according to Figure 1.3B. Instead of a point a line is projected onto the object. The line is created either by a special optical lens or by a fast-rotating mirror and is viewed by a matrix camera. The line is (mechanically) scanned in the perpendicular direction to cover the whole surface of the object. If the object itself is moving, as on a conveyer belt, such scanning mechanism may be unnecessary. Figure 7.34 shows the basic setup of a line scan range finder.

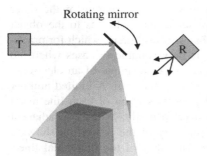

**Figure 7.34** Optical range finder for object recognition using line scanning.

In this simple configuration the line is created by a rotating mirror. The projected line follows the shape of the object; this pattern is viewed by a camera. From the known positions of the mirror, the light plane and the camera, the shape of the intersection between the object and the light plane can be reconstructed. In general not all points of the object are illuminated, and not all illuminated points may be viewed by the camera. Those shadows prohibit capturing the whole cross section. When this is not acceptable for the application, a second triangulation system can be used operating under a different angle. Another option is a set of mirrors through which the scanning plane transmitted by a single source is projected from two sides onto the object; the same mirrors can be used to redirect the light from two sides to the common receiver. This solution is proposed in, for instance, Ref. [52], in which the scanning is performed by acousto-optic deflectors. Evidently the two scans should be complementary to build up a complete image of the object. A similar solution with two mirrors is described in Ref. [53]; in this example, elliptical mirrors are used, which allows dynamical reconfiguration of the geometry. This results in a better depth resolution and reducing occlusion effects. Finally we mention a laser range finder that uses a multifaceted pyramidal mirror. This mirror acts as a double scanner, one for the transmitting beam and one for the reflected beam. The system, of which design details can be found in Ref. [54], features high speed, high resolution and reduction of the shadow effect.

Much research is being done to monitor the condition of buried pipes. In Chapter 6 magnetic methods have been discussed, but optical systems are being developed as well. Pipe robots carrying various sensors travel through the pipes under test, collecting important data about the pipe's condition (i.e. obstacles, corrosion, leaks and mechanical defects). The smaller the diameter, the more severe the requirements with respect to size, power efficiency (when battery powered) and costs. As an example of an optical test system for small pipe diameters we refer to Ref. [55]. This work concerns an optical system that measures the inner surface of a 10 mm pipe. The sensing system consists of a laser, a scanning mirror, a motor and a PSD, axially assembled to fit inside the pipe. The system enables the detection of surface defects as small as 0.1 mm, by means of triangulation.

Mechanical scanning is relatively slow and forms the major limitation of the acquisition time, in particular for a 2D scanning mechanism. Line projection using cylindrical optics considerably reduces the acquisition time and is therefore widespread in range finding systems. One solution to speed up scanning of the other dimension is an array of transmitters sending a set of parallel lines to the object under test. The resolution is limited by the number of transmitters, which for practical reasons cannot be very high. This solution is only suitable in cases where no high resolution in the scanning direction is required. An example of an electronic scanning system for laser range finding is presented in Ref. [56]. A limited number of parallel lines is projected onto the object. The cross section of each line has a bell-shaped intensity distribution and the lines overlap partly. When two adjacent lines are transmitted simultaneously, the total intensity of the projection shows a peak at a position that depends on the relative intensities of the individual lines.

By controlling this ratio, the peak value shifts across the object under test, over a distance about equal to the line separation. The range can be extended simply by adding more lines. Based on this principle an acquisition time of 6 μs was obtained, with a spatial resolution of 1 μm over a scanning area of 22 × 24 mm.

Measuring the shape of very large objects can also be performed by optical ranging. Since the distances involved in such applications are much larger than in the previous examples, triangulation is not a suitable technique. Ranging using optical ToF, as discussed in Section 7.2.6, is a better option. Moreover to keep the sensor dimensions small, a feedback approach (with tracking) is recommended. An example of this combination, ToF and tracking, is extensively described in Ref. [57]. The system has a tracking accuracy better than 1 mm and a measurement resolution in the same order, over a working distance of 12 m and angle range ±45°.

Another application area concerning acquisition of shape information is the determination of a *height profile* (or *depth profile*), for instance in seam tracking for automatic welding. A specific example of a height profile measurement is given in Ref. [58], where methods for recognition of electronic components on a PCB are studied. One of these methods is based on triangulation to obtain 3D shape information about classes of components (e.g. ICs, capacitors, batteries and so on). Also here the object (PCB with components) is illuminated from two sides to eliminate shadow effects. Instead of one line a sequence of line patterns is projected on the PCB, which is viewed by a single camera. From the obtained range data, the shape of the relevant parts in the image is reconstructed and parameterized [59] to match them with the corresponding parameters of standard electronic components.

One of the limitations of the resolution in a structured light system (with lines or other patterns) is caused by the line width of the projected light. Several researchers have proposed methods to reduce the resolution by using special image-processing algorithms. Examples can be found in Ref. [60], where the shape of trees is determined using triangulation, and in Ref. [61], in which the obtained improvement of the resolution is illustrated by the shape measurement of a car tyre.

As objects are normally resting on a base, the bottom side is not accessible to the triangulation system. For regular objects having a flat bottom this is not a problem. Recovering the full shape of *irregular objects*, like agricultural products, is not possible in this way. An elegant optical sensing system allowing the determination of the full shape is presented in Refs [62,63]. The sensing system consists of a ring with a large number of transmitters and receivers located alternately around the ring and pointing inwards. The area inside the ring is scanned by successively activating the transmitters; the light from each transmitter is detected by all receivers simultaneously. When an object resides in the ring, some of the receivers are in the 'shadow' of the object and will not receive light whilst the others do. By cycling along the ring, a set of cords is obtained approximating the shape of the object (actually only the cross section with the plane of the ring). When the object moves through the ring (just by falling) a 3D image of the total shape is obtained. The method is cheap and simple and appears to be suitable for characterizing the shape of potatoes and similar products.

## 7.4.5  Navigation

Movement is change in position; hence, it can be measured by any position sensor, provided such sensor has an adequate range and response time. When the movement is beyond the range of a position sensor, for instance as may occur with moving robots and AGVs, other sensing techniques to capture the position are required. A simple optical navigation method is tracking a marked path. The path may consist simply of a contrasting line on the floor. An LED or other light emitter casts a light beam down to the floor and two detectors on either side pick up the reflected light. The vehicle is controlled in such a way that both detectors receive the same light intensity, that is, when they are equal distance from the reflecting (white) line. When the vehicle starts to deviate from the track, either the left or the right detector receives more light, from which a control signal is derived.

When the position of the robot or vehicle along the path must also be determined, the path can be coded along the track, for instance by a continuous pattern of light and dark fields or particular optical codes [64]. Similarly the movement of a free mobile robot can be tracked by some contrasting 2D pattern on the floor, using sensors detecting dark−light transitions or even an onboard camera viewing the floor [65]. Another possibility to measure speed makes use of the 'natural' irregularities of the floor surface. At two positions of the vehicle (front and rear) the random intensity variations are measured; the speed follows from correlation of the two random signals [38]. Absolute position is then obtained by dead-reckoning techniques, starting from a known position, and possibly updated from time to time with reference positions achieved by special markers or beacons.

*Beacons* for the determination of absolute position can be either passive or active. Passive beacons are for instance reflectors located at precisely known positions. The transmitters and receivers are located on the vehicle. Active beacons transmit signals that are received by the vehicle. Important design considerations are the directivity (or the scanning area) of the transmitter/receiver, the location of the beacons and the identification of beacons (the vehicle must be able to recognize the active beacons).

As an illustration of optical sensing for navigation, Figure 7.35 shows a robot with a multi-sensor navigation system [66]. This robot has been designed for automatic assembly of industrial and household products. Coarse navigation of this robot is performed by fixed cameras looking to optical markers on the robot. For fine positioning of the robot (to perform proper assembling tasks) a special optical sensing system has been developed using three light-emitting beacons (Figure 7.36).

The sensing system on the robot measures the angles between two beacon pairs: $\alpha$ between $B_1$ and $B_2$, and $\beta$ between $B_2$ and $B_3$. Then the robot must be located at a point $P$ lying on the crossings of the two circles $c_1$ through $(B_1, B_2, P)$ and $c_2$ through $(B_2, B_3, P)$. From the known beacon positions and the two angles, point $P$ can be calculated. The sensing system on the robot consists of a rotating photo diode (four turns/s), connected to an encoder that provides information on the angle of the optical axis with respect to the robot co-ordinates. During one complete turn,

(A)                                                   (B)

**Figure 7.35** MART optical navigation system: (A) mobile assembly robot and (B) details of optical navigation system (University of Twente, The Netherlands).

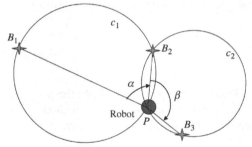

**Figure 7.36** High precision positioning of the mobile robot MART.

the detector receives three consecutive signals from the beacons, from which the angles $\alpha$ and $\beta$ are derived. The light from the beacons contains a unique code to distinguish them, which is decoded by the sensing system. The resolution of this angle measurement is better than $0.005°$ over a distance range from 1 to 10 m between robot and beacon [66].

Obviously the system just described requires a direct sight of view between beacons and vehicle. In many applications this is not the case, for instance when AGVs have to navigate in a complex area with many large obstacles or in corridors. Placing beacons in the plane of the AGV could be complicated, expensive or even impossible. A solution could be a higher position of the beacons (for instance on the ceiling of a hall), thus increasing the working range of the beacons. Since the beacons and sensors are not in one plane anymore, however, this complicates the optical configuration and the triangulation algorithm. de Cecco [67] gives a solution to this problem, based on a special optical scanning system able to accurately measure angles as well, resulting in a position accuracy better than 2 mm in a range of 10 m.

Navigation can also be performed by a simple 2D-PSD camera. When an extra d.o.f. is allowed, the 3D position can be derived with two such cameras. From the output of both 2D-PSD cameras the three co-ordinates $x_p$, $y_p$ and $z_p$ of the light spot can be calculated, as illustrated in Figure 7.37.

The focal point of the left camera coincides with the origin of the reference frame, so the light spot has co-ordinates $(x_{d1}, y_{d1}, -F_1)$, which satisfy:

$$\frac{x_p}{x_{d1}} = \frac{y_p}{y_{d1}} = \frac{z_p}{F_1} \qquad (7.24)$$

The second PSD camera is rotated over angles $\phi$ and $\theta$ around the $z$- and $y$-axis, respectively. The co-ordinates of the light spot with respect to a rotated PSD are related to that of the reference frame according to the transformation formula:

$$\begin{pmatrix} x \\ y \\ z \end{pmatrix} = \begin{pmatrix} \cos\phi\cos\vartheta & -\sin\phi & -\cos\phi\sin\vartheta \\ \sin\phi\cos\vartheta & \cos\phi & -\sin\phi\sin\vartheta \\ \sin\vartheta & 0 & \cos\vartheta \end{pmatrix} \begin{pmatrix} x_{d2} \\ y_{d2} \\ z_{d2} \end{pmatrix} \qquad (7.25)$$

When the camera is further translated over a vector $(x_T, y_T, z_T)$, the co-ordinates of the light spot satisfy the expression

$$\frac{x_p}{x_{d2} - x_T} = \frac{y_p}{y_{d2} - y_T} = \frac{z_p}{F_2 - z_T} \qquad (7.26)$$

which is similar to Eq. (7.24) for the left PSD. With Eqs (7.24)–(7.26) the position of the light spot can be expressed in terms of the output co-ordinates of both PSD cameras, their focal lengths and the geometrical parameters of the system.

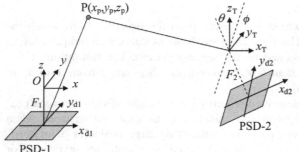

**Figure 7.37** 3D optical localization system with two 2D-PSD cameras.

If only speed information is required, inertial sensing techniques can be applied. Integrating the output of an accelerometer provides linear velocity. A gyro gives the angular rate. Most optical gyros, for instance the one discussed in Section 7.2.5, are based on interferometry. An extensive overview of commercially available optical and other gyros can be found in Ref. [68].

### 7.4.6   Force, Torque and Strain Sensing

Measurement of force and torque is mostly based on measuring the resultant deformation (compressive or tensile strain). In this sense in principle the displacement sensing methods discussed in previous sections of this chapter are candidates in the selection process of a force measurement method. The typical application, however, justifies a separate section on this kind of measurement.

An applied torque results in a twist of the object (for instance a shaft) over an angle $\phi$. This angle is about proportional to the torque and can be measured in various ways. To obtain a high sensitivity the distance between the end points should be large. However this may impede accurate measurement of the relative rotation. A simple solution is to transfer the rotation at the one end by a mechanical extension to near the other end (Figure 7.38A), simplifying the angle measurement. When the object angle is still too small an elastic element with known compliance can be inserted, as drawn in Figure 7.38B. The torque is reconstructed from the relative angle between two end points of the rotational spring element.

A special challenge is to measure torque on rotating shafts. In such an application contact-free sensing is required, which can be realized by optical means. In Ref. [69] this method is used for torque measurements on a wood cutting machine. The discs in Figure 7.38 are provided with optical markers on the circumference, viewed from the side by a pair of photo diodes. Even at a cutting speed of 24,000 rpm a proper torque signal could be obtained.

Another particular application of an optical force measuring system is the online measurement of 2D force on a pen for the purpose of signature verification [70]. The force on the pen tip is transferred to the other end of the ink channel, which can move relative to the pen holder. A small mirror, mounted on the end

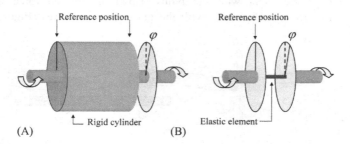

(A)                                    (B)

**Figure 7.38** Torque measurement of a shaft using (A) an extension part and (B) an elastic torque element.

side of the ink channel, modulates the direction of a laser beam, falling on a quad photo diode. From the four photo currents the 2D horizontal force on the pen tip can be reconstructed. Since the transfer from force via mirror displacement to the photo currents is rather complex, the device needs be calibrated. Experiments proved the feasibility of the optical method.

Very small deformations are best measured by interferometric methods, since this principle enables displacement measurements with very high resolution. An example is the study on electrostrictive properties of particular materials. Electrostriction is electric field-induced strain: deformations are in the sub-Ångström to sub-micron range. Yimnirun [71] presents a laser interferometric system that is able to measure such small displacement. The paper reports a displacement resolution of less than $10^{-5}$ nm, and operates in a frequency band from 3 to 20 kHz.

Many more examples of optical (interferometric) force and strain measurement systems have been reported in literature. A particular kind of force sensor is a *tactile sensor*, mainly encountered in robotics applications. Tactile sensors of the cutaneous type consist of an array or matrix of elementary pressure or displacement sensors. Many optical techniques have been developed to obtain information on the displacement of the individual elements (taxels). Four major methods are: variable transmission, variable contact area, total internal reflection and pressure sensitive optical fibres. A common problem is the resolution: in order to realize high resolution, the taxels should have small dimensions and be densely packed.

The principle of variable transmission is illustrated in Figure 7.39A. As with almost all tactile sensors, the applied force is transferred into a displacement by an elastic layer. In the variable transmission type, the displacement is measured by mechanical modulation of the intensity of a light beam, a method already published in 1983 [72]. The (commercial) device has a resolution of $16 \times 10$ tactile elements, and a pitch of about 0.3 cm. Despite this limited resolution the sensor is shown to be suitable for object recognition by touch [73].

The principle of variable contact surface is already addressed in Section 3.6.2, where the contact area is measured by resistive means, but optical readout is also an option. The variable contact area is achieved by protrusions at the back side of an elastic sheet, mounted on a flat transparent plate that serves as a waveguide to illuminate the elastic sheet with the bulbs. Upon an applied force the bulbs are impressed and the contact area with the glass plate increases (Figure 7.39B).

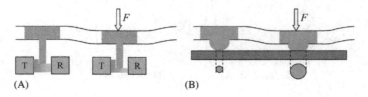

(A)                                                                   (B)

**Figure 7.39** Optical tactile sensing: (A) variable light transmission and (B) variable contact area.

This change in area is measured optically by a camera viewing the back side of the plate [74,75]. The resolution of this tactile sensor is restricted by the number of elastic bulbs; the camera is not limitative with respect to the resolution. However it makes the sensor rather bulky and unsuitable for incorporating in a robot gripper. An advantage of this setup is that scanning is performed by the camera electronics, hence the speed is high.

A more flat construction can be achieved by crossing optical fibres. The structure is similar to that of the carbon fibre tactile sensors. The crossing points act as taxels, and the fibres perform the scanning. When a force is applied to such a crossing point, the transmission of both fibres alters due to micro-bending. These fibre-optic tactile sensors have a potentially high resolution but more research is required to fully profit of this property. For less demanding applications, for instance wearable sensing, they offer good prospects [76]. Details on the construction and performance of fibre-optic tactile sensors are given in Refs [77−79].

# References and Literature

### References to Cited Literature

[1] D.J.W. Noorlag: Lateral-photoeffect position-sensitive detectors, PhD thesis, Delft University of Technology, Delft, The Netherlands, 1982.

[2] M. Tartagni, P. Perona: Progress in VLSI optical position-sensitive circuits, Sens. Actuators A, 67 (1998), 109−114.

[3] E. Fortunato, R. Martins: New materials for large-area position-sensitive sensors, Sens. Actuators A, 68 (1998), 244−248.

[4] T. Fujita, M. Idesawa: A new type of position sensitive device and its detecting characteristics, ISMCR'99, Tokyo, Japan, 10−11 June 1999; in: Proc. IMEKO-XV World Congress, Osaka, Japan, 13−18 June 1999, Vol. X, pp. 1−6.

[5] S. Iqbal, M.M.S. Gualini, A. Asundi: Measurement accuracy of lateral-effect position-sensitive devices in presence of stray illumination noise, Sens. Actuators A, 143 (2008), 286−292.

[6] D. Qian, W. Wang, I.J. Busch-Vishniac, A.B Buckman: A method for measurement of multiple light spot positions on one position-sensitive sensor (PSD), IEEE Trans. Instrum. Meas., 42(1) (Feb. 1993), 14−18.

[7] W.H. Ko, K.M. Chang, G.J. Hwang: A fibre-optic reflective displacement micrometer, Sens. Actuators A, 49 (1995), 51−55.

[8] M. de Bakker: The PSD chip: high speed acquisition of range images; PhD thesis, Delft University of Technology, The Netherlands, 2000, p. 123.

[9] D. Hüser, H. Rothe: Robust averaging of signals for triangulation sensors, Meas. Sci. Technol., 9 (1998), 1017−1023.

[10] D. Maschera, A. Simoni, M. Gootardi, L. Gonzo, S. Gregori, V. Liberali, G. Torelli: An automatically compensated readout channel for rotary encoder systems, IEEE Trans. Instrum. Meas., 50(6) (2001), 1801−1807.

[11] L.M. Sanchez-Brea, T. Morlanes: Metrological errors in optical encoders, Meas. Sci. Technol. 19 (2008) 115104 8 doi:10.1088/0957-0233/19/11/115104.

[12] K.K. Tan, H.X. Zhou, T.H Lee: New interpolation method for quadrature encoder signals, IEEE Trans. Instrum. Meas., 51(5) (2002), 1073−1079.

[13] J. Rozman, A. Pleteršek: Linear optical encoder system with sinusoidal signal distortion below −60 dB, IEEE Trans. Instrum. Meas., 59(6) (2010), 1544−1549.

[14] S. Wekhande, V. Agarwal: High-resolution absolute position Vernier shaft encoder suitable for high-performance PMSM servo drives, IEEE Trans. Instrum. Meas., 55(1) (2006), 357−364.

[15] E.M. Yeatman, P.J. Kushner, D.A. Roberts: Use of scanned detection in optical position encoders, IEEE Trans. Instrum. Meas., 53(1) (2004), 37−44.

[16] W. Zhou, L. Cai: An angular displacement interferometer based on total internal reflection, Meas. Sci. Technol., 9 (1998), 1647−1652.

[17] J.H. Zhang, C.H. Menq: A linear angular interferometer capable of measuring large angular motion, Meas. Sci. Technol., 9 (1998), 1247−1253.

[18] G.C. Hill, R. Melamud, F.E. Declercq, A.A. Davenport, I.H. Chan, P.G. Hartwell, B.L. Pruitt: SU-8 MEMS Fabry−Perot pressure sensor, Sens. Actuators A, 138 (2007), 52−62.

[19] P.M. Nieval, N.E. McGruer, G.G. Adams: Design and characterization of a micromachined Fabry−Perot vibration sensor for high-temperature applications, J. Micromech. Microeng., 16 (2006), 2618−2631.

[20] J. Yang, S. Jia, Y. Du: Novel optical accelerometer based on Fresnel diffractive micro lens, Sens. Actuators A, 151 (2009), 133−140.

[21] Heidenhain GmbH, Digitale Längen und Winkelmesstechnik, 1989, ISBN 3-478-93191-6.

[22] L. Zeng, K. Seta, H Matsumoto, S Iwashaki: Length measurement by a two-colour interferometer using two close wavelengths to reduce errors caused by air turbulence, Meas. Sci. Technol., 10 (1999), 587−591.

[23] D.J. Balek, R.B. Kelley: Using gripper mounted infrared proximity sensors for robot feedback control, IEEE Int. Conf. Rob. Autom., (1985), 282−287.

[24] B. Espiau, J.Y. Catros: Use of optical reflectance sensors in robotics applications, IEEE Trans. Syst. Man Cybern., SMC-10(12) (1980), 903−912.

[25] P.P.L. Regtien: Accurate optical proximity detector; IEEE Instrumentation and Measurement Technology Conference, San Jose, USA, 13−15 February 1990, Proceedings pp. 141−143.

[26] A. Bonen, R.E. Saad, K.C. Smith, B. Benhabib: A novel electrooptical proximity sensor for robotics: calibration and active sensing, IEEE Trans. Rob. Autom., 13(3) (June 1997), 377−386.

[27] C. Yüzbaşιoğlu, B. Barshan: Improved range estimation using simple infrared sensors without prior knowledge of surface characteristics, Meas. Sci. Technol., 16 (2005), pp. 1395−1409.

[28] S.R. Lang, D.J. Ryan, J.P. Bobis: Position sensing using an optical potentiometer, IEEE Trans. Instrum. Meas., 41(6) (1992), 902−905.

[29] Y. Alayli, S. Topçu, D. Wang, R. Dib, L. Chassagne: Applications of a high accuracy optical fibre displacement sensor to vibrometry and profilimetry, Sens. Actuators A, 116 (2004), 85−90.

[30] X.P Liu, R.C Spooncer, B.E Jones: An optical fibre displacement sensor with extended range using two-wavelength referencing, Sens. Actuators A, 25−27 (1991), 197−200.

[31] T. Oka, H. Nakajima, M. Tsugai, U. Hollenbach, U. Wallrabe, J. Mohr: Development of a micro-optical distance sensor, Sens. Actuators A, 102 (2003), 261−267.

[32] M. Norgia, S. Donati: A displacement-measuring instrument utilizing self-mixing interferometry, IEEE Trans. Instrum. Meas., 52(6) (2003), 1765−1770.

[33] S.H. Wang, C.J. Tay, C. Quan, H.M. Shang, Z.F. Zhou: Laser integrated measurement of surface roughness and microdisplacement, Meas. Sci. Technol., 11 (2000), 454−458.

[34] D.F. Garcia, M. Garcia, F. Obeso, V. Fernandez: Flatness measurement system based on a nonlinear optical triangulation technique, IEEE Trans. Instrum. Meas., 51(2) (2002), 188−195.

[35] T.W. Ng: The optical mouse as a two-dimensional displacement sensor, Sens. Actuators A, 107 (2003), 21−25.

[36] C.L. Chu, C.H. Lin, K.C. Fan: Two-dimensional optical accelerometer based on commercial DVD pick-up head, Meas. Sci. Technol., 18 (2007), 265−274.

[37] C.L. Chu, C.Y. Chiu: Development of a low-cost nanoscale touch trigger probe based on two commercial DVD pick-up heads, Meas. Sci. Technol., 18 (2007), 1831−1842.

[38] I. Gogoasa, M. Murphy, J. Szajman: An extrinsic optical fibre speed sensor based on cross correlation, Meas. Sci. Technol., 7 (1996), 1148−1152.

[39] K.C. Michel, O.F. Fiedler, A. Richter, K. Christofori, S. Bergeler: A novel spatial filtering velocimeter based on a photodetector array, IEEE Trans. Instrum. Meas., 47(1) (1998), 299−303.

[40] G. Plantier, N. Servagent, A. Sourice, T. Bosch: Real-time parametric estimation of velocity using optical feedback interferometry, IEEE Trans. Instrum. Meas., 50 (2001), 915−919.

[41] X. Raoul, T. Bosch, G. Plantier, N. Servagent: A double-laser diode onboard sensor for velocity measurements, IEEE Trans. Instrum. Meas., 53(1) (2004), 95−101.

[42] M.Q. Feng, D.-H. Kim: Novel fibre optic accelerometer system using geometric Moiré fringe, Sens. Actuators A, 128 (2006), 37−42.

[43] A. Khiat, F. Lamarque, C. Prelle, N. Bencheikh, E. Dupont: High-resolution fibre-optic sensor for angular displacement measurements, Meas. Sci. Technol., 21 (2010) 025306 10, doi:10.1088/0957-0233/21/2/025306.

[44] H. Kato, M. Kojima, M. Gattoh, Y. Okumura, S. Morinaga: Photoelectric inclination sensor and its application to the measurement of the shapes of 3-D objects, IEEE Trans. Instrum. Meas., 40(6) (1991), 1021−1026.

[45] I. Godler, A. Akahane, K. Ohnishi, T Yamashita: A novel rotary acceleration sensor, IEEE Control Syst., (1995), 56−60.

[46] J. Yuan, S.L. Yu: End-effector position-orientation measurement, IEEE Trans. Rob. Autom., 15(3) (1999), 592−595.

[47] T. Takatsuji, M. Goto, T. Kurosawa, Y. Tanimura, Y. Koseki: The first measurement of a three-dimensional co-ordinate by use of a laser tracking interferometer system based on trilateration, Meas. Sci. Technol., 9 (1998), 38−41.

[48] A. Pesatori, M. Norgia, C. Svelto: Optical sensor for online turbine edge measurement, Meas. Sci. Technol., 20 (2009) 104006 7, doi:10.1088/0957-0233/20/10/104006.

[49] G. de Anda-Rodríguez, E. Castillo-Castañeda, S. Guel-Sandoval, J. Hurtado-Ramos, M. Eugenia Navarrete-Sánchez: On-line wear detection of milling tools using a displacement fibre optic sensor; XVIII IMEKO WORLD CONGRESS, 17−22 September 2006, Rio de Janeiro, Brazil.

[50] T. Oiwa, H. Nishitani: Three-dimensional touch probe using three fibre optic displacement sensors, Meas. Sci. Technol., 15 (2004), 84−90.

[51] T. Oiwa, T. Tanaka: Miniaturized three-dimensional touch trigger probe using optical fibre bundle, Meas. Sci. Technol., 16 (2005), 1574−1581.

[52] L. Zeng, H. Matsumoto, K. Kawachi: Two-directional scanning method for reducing the shadow effects in laser triangulation, Meas. Sci. Technol., 8 (1997), 262−266.

[53] J. Clark, A.M. Wallace, G.L. Pronzato: Measuring range using a triangulation sensor with variable geometry, IEEE Trans. Rob. Autom., 14(1) (1998), 60—68.

[54] M. Rioux: Laser range finder based on synchronized scanners, Appl. Opt., 23(21) (1984), 3837—3844.

[55] E. Wu, Y. Ke, B. Du: Noncontact laser inspection based on a PSD for the inner surface of minidiameter pipes, IEEE Trans. Instrum. Meas., 58(7) (2009), 2169—2173.

[56] M. Baba, T. Konishi: A new fast rangefinder based on the continuous electric scanning mechanism, ISMCR'99, Tokyo, Japan, 10—11 June 1999; in: Proc. IMEKO-XV World Congress, Osaka, Japan, 13—18 June 1999, Vol. X, pp. 7—14.

[57] A.J. Mäkynen, J.T. Kostamovaara, R.A. Myllylä: Tracking laser radar for 3-D shape measurements of large industrial objects based on time-of-flight laser rangefinding and position-sensitive techniques, IEEE Trans. Instrum. Meas., 43(1) (1994), 40—49.

[58] E.R. van Dop: Multi-sensor object recognition: the case of electronics recycling, Ph.D. thesis, University of Twente, The Netherlands, 1999; ISBN 90-36512689.

[59] E.R. van Dop, P.P.L. Regtien: Fitting undeformed superquadrics to range data: improving model recovery and classification; Proceedings of the 1998 IEEE Computer Society Conference on Computer Vision and Pattern Recognition, Santa Barbara, California, USA, 23 June 1998, pp. 396—402; ISBN 0-8186-8497-6.

[60] M. Demeyere, Noncontact dimensional metrology by triangulation under laser plane lighting — development of new ambulatory instruments, PhD thesis, Université catholique de Louvain, Louvain-la-Neuve, Belgium, 2006.

[61] Z. Wei, F. Zhou, G. Zhang: 3D coordinates measurement based on structured light sensor, Sens. Actuators A, 120 (2005), 527—535.

[62] H. Gall: A ring sensor system using a modified polar co-ordinate system to describe the shape of irregular objects, Meas. Sci. Technol., 8 (1997), 1228—1235.

[63] H. Gall, A. Muir, J. Fleming, R. Pohlmann, L. Göcke, W. Hossack: A ring sensor system for the determination of volume and axis measurements of irregular objects, Meas. Sci. Technol., 9 (1998), 1809—1820.

[64] E.M. Petriu, J.S. Basran: On the position measurement of automated guided vehicles using pseudorandom encoding, IEEE Trans. Instrum. Meas., 38(3) (1989), 799—803.

[65] E.M. Petriu, W.S. McMath, S.K. Yeung, N. Trif, T. Bieseman: Two-dimensional position recovery for a free-ranging automated guided vehicle, IEEE Trans. Instrum. Meas., 42(3) (1993), 701—706.

[66] A.J. de Graaf: On-line measuring systems for a mobile vehicle and a manipulator gripper, PhD thesis, University of Twente, The Netherlands, 1994; ISBN 90-9007766-9.

[67] M. de Cecco: A new concept for triangulation measurement of AGV attitude and position, Meas. Sci. Technol., 11 (2000), N105—N110.

[68] H.R. Everett: Sensors for mobile robots; 1995; ISBN 1-56881-048-2; Chapter 13.

[69] A. Pantaleo, A. Pellerano, S. Pellerano: An optical torque transducer for high-speed cutting, Meas. Sci. Technol., 17 (2006), 331—339.

[70] H. Shimizu, S. Kiyono, T. Motoki, W. Gao: An electrical pen for signature verification using a two-dimensional optical angle sensor, Sens. Actuators A, 111 (2004), 216—221.

[71] R. Yimnirun, P.J. Moses, R.J. Meyer Jr, R.E. Newnham: A single-beam interferometer with sub-Ångström displacement resolution for electrostriction measurements, Meas. Sci. Technol., 14 (2003), 766—772.

[72] J. Rebman, K.A. Morris: A tactile sensor with electrooptical transduction; Proc. 3rd Int. Conf. on Robot Vision and Sensory Controls, 1983, pp. 210—216.

[73] R.C. Luo, H.H. Loh: Tactile array sensor for object identification using complex moments, J. Rob. Syst., 5(1) (1988), 1−12.

[74] R.M. White, A.A. King: Tactile array for robotics employing a rubbery skin and a solid-state optical sensor, Proc. 3rd Int. Conf. on Solid-State Sensors and Actuators Philadelphia, Penn., (USA), June 1985, pp. 18−21.

[75] S. Begej: Planar and finger-shaped optical tactile sensors for robotic applications, IEEE J. Rob. Autom., 4(5) (1988), 472−484.

[76] M. Rothmaier, M.P. Luong, F. Clemens: Textile pressure sensor made of flexible plastic optical fibers, Sensors, 8 (2008), 4318−4329.

[77] D.T. Jenstrom, C.-L. Chen: A fibre optic microbend tactile sensor array, Sens. Actuators, 20 (1989), 239−248.

[78] N. Fürstenau: Investigations in a tactile sensor based on fibre-optic interferometric strain gauges, SPIE, 1169, Fibre optic and laser sensors VII, 1989, pp. 531−538.

[79] S.R. Emge, C.-L. Chen: Two-dimensional contour imaging with a fibre optic microbend tactile sensor array, Sens. Actuators B, 3 (1991), 31−42.

# Literature for Further Reading

### Recommended Books on Optoelectronics

[1] R. Hui, M.S. O'Sullivan: Fiber optic measurement techniques; Elsevier/Academic Press, Amsterdam, Lausanne, New York, Oxford, Shannon, Singapore, Tokyo, 2009, ISBN 978-0-12-373865-3J.

[2] J. Wilson, J.F.B. Hawkes: Optoelectronics: an introduction; Prentice Hall Int., New York, London, Toronto, Sydney, Tokyo, Singapore, 1989; ISBN 0-13-638495-1 PBK.

[3] J. Dakin, B. Culshaw: Optical fibre sensors IV: applications, analysis, and future trends; Artech House, Boston, 1997; ISBN 0-89006-940-9.

[4] W. Stadler: Analytical robotics and mechatronics, Chapter 6: From photon to image identification; McGraw-Hill, New York, 1995; ISBN 0-07-113792-0.

[5] G. Dudek, M. Jenkin: Computational principles of mobile robotics; Cambridge University Press, Cambridge, 2000; ISBN 0-521-56876-5, Part 2: Sensing.

# 8 Piezoelectric Sensors

In this chapter we discuss principles, properties and applications of piezoelectric sensors. A phenomenological description of the piezoelectric effect is followed by an overview of the most common piezoelectric parameters. Piezoelectric materials are used in force sensors and accelerometers, and for that reason they are found in many applications. Piezoelectricity can also be applied to construct acoustic transducers; this is the subject of Chapter 9.

## 8.1   Piezoelectricity

### 8.1.1   Piezoelectric Materials

Piezoelectricity is encountered in some classes of crystalline materials. Deformation of such materials results in a change in electric polarization (the direct piezoelectric effect in Table 2.3): the positions of the positive and negative charges in the crystal are displaced relative to each other, causing a net polarization or change in intrinsic polarization. Figure 8.1 shows a simplified 2D representation of this effect.

In Figure 8.1A the structure is fully symmetric: the centre of gravity of all positive charges coincides with that of the negative charges (the whole crystal is electrically neutral). When compressed in the horizontal direction, (Figure 8.1B), the centre of positive charges has shifted downwards, resulting in a non-zero polarization. In case of a vertical compression, (Figure 8.1C), the centre of positive charges has shifted upwards, resulting in a non-zero polarization in the other direction. This shift of positive charges relative to the negative charges produces opposite charges on the opposing surfaces of the crystal.

Evidently Figure 8.1 is a strongly simplified representation of the piezoelectric phenomenon. Crystal structure, addition of dopants and other treatments of the material substantially determine the piezoelectric properties; for instance Ref. [1] shows the charge displacements for various compositions of PZT (a particular piezoelectric material).

The surface charge per unit area ($C/m^2$) is proportional to the applied stress ($N/m^2$), the charge (C) is proportional to the applied force (N). Generally the material is shaped as small rectangular blocks, cylinders, plates or even sheets, with two parallel faces that are provided with a conducting layer. The construction behaves as a flat-plate capacitor, for which $Q = C \cdot V$; so the output signal is available as a voltage. The sensitivity of such piezoelectric sensors is characterized either by the charge sensitivity $S_q = Q/F$ (C/N) or by the voltage sensitivity $S_u = V/F = S_q/C$ (V/N). The piezoelectric

**Sensors for Mechatronics.** DOI: 10.1016/B978-0-12-391497-2.00008-X
© 2012 Elsevier Inc. All rights reserved.

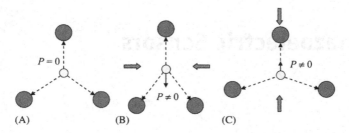

**Figure 8.1** Piezoelectricity; larger circles represent positive charges, the smaller negative charges.

effect is reversible (see Chapter 2): a voltage applied to the piezoelectric material results in a (small) deformation (the converse piezoelectric effect in Table 2.3); this property is used in piezoelectric actuators.

The following groups of materials exhibit piezoelectricity:

* natural piezoelectric materials: a well-known example is quartz (crystalline $SiO_2$)
* ceramic materials (polycrystalline), for instance barium titanate ($BaTiO_3$)
* polymers, for instance PVDF (or $PVF_2$).

Natural piezoelectricity was discovered already in 1880 by the brothers Curie, in several crystalline materials [2]. The most known is quartz (crystalline silicon dioxide). It has a rather low but stable piezoelectricity: about 2 pC/N.

Materials of the second group contain electrical dipoles but are not piezoelectric, because the random orientation of the dipoles averages out, resulting in a zero net polarization. However they can be made piezoelectric by poling: the material is heated up to the so-called Curie temperature, where the dipoles obtain high mobility. In this state a strong electric field is applied during a certain time period and the dipoles are oriented in the direction of the field. After slowly cooling down the dipoles maintain their orientation, giving the material its piezoelectric properties.

Compared to quartz poled ceramic materials have a much higher piezoelectricity, ranging from 100 up to over 1000 pC/N. The orientation of the dipoles slowly tends to a disordered state, resulting in a negative exponential decay in sensitivity over time:

$$\frac{S(t) - S(t_0)}{S(t_0)} = c \cdot \log \frac{t}{t_0} \tag{8.1}$$

where $S(t)$ and $S(t_0)$ are the piezoelectric sensitivities at time $t$ and the end of the polarization treatment time $t_0$. The factor $c$ in this expression lies in the range $-5 \times 10^{-2}$ to $-5 \times 10^{-3}$ per decade at room temperature. For this reason the poled material is aged artificially, before being applied as a transducer. Mechanical and thermal shocks, however, may initiate new aging effects. Clearly the decay goes faster at higher temperatures. Note that even quartz looses its piezoelectricity above the Curie temperature, which is 573°C.

In 1969 Kawai discovered that some polymers can be made piezoelectric by poling under particular conditions [3,4]. The most popular piezoelectric polymer is polyvinylidene fluoride (PVDF, a chain of repeating units of $CH_2-CF_2$). The material is available in sheets with various thicknesses ($6-100\ \mu m$), with conducting material deposited on its surface for connection purposes. PVDF has the highest piezoelectricity of all known polymers, about 25 pC/N. Poling is performed during stretching (in one or both directions), to obtain reasonable piezoelectricity. Piezoelectric PVDF has a limited temperature range, because its melting point is about 150°C, whereas the Curie temperature lies around 125°C. Since shortly after the material became commercially available, the material has been successfully applied as a transducer in the thermal, mechanical and acoustical domains [5,6].

An additional advantage of PVDF over ceramics is the compatibility with silicon technology. The thin sheets can be glued on top of a silicon wafer that contains part of the interface electronics. More recently piezoelectric copolymers of PVDF, for instance VDF/TrFE, have been developed. This material is available as a solution. It can be deposited on wafers by spinning [7]. Polymer-on-silicon technology opens up the possibility for the creation of smart (integrated) sensing devices, for the detection of pressure, force, sound and thermal energy (the latter due to the fact that PVDF is pyroelectric as well) [8,9]. The major advantage of polymers over ceramic materials is their flexibility and small size (sheet thickness of several $\mu m$).

Another development concerns the deposition of thin film ceramics. This technology, too, can benefit from compatibility with integrated circuits and, using the reversibility of piezoelectricity, for micro-actuators [10].

The primary output quantity of a piezoelectric sensor is charge. In Section 8.3 we will show two basic schemes to convert that charge into a voltage: with a very high load impedance (open terminals) and with an almost zero load impedance (short-circuited terminals).

## 8.1.2 Piezoelectric Parameters

In literature piezoelectric relations and properties are described in several ways. In this book we use the two-port description, which is analogous to the description of electrical two ports (Figure 8.2). Two pairs of variables are involved: two input variables and two output variables. Each pair consists of one extensive and one intensive variable (see Section 2.2).

The relation between the four quantities can be described in various ways. For a linear electric network we can express the two voltages $V_1$ and $V_2$ as a linear function of the two currents $I_1$ and $I_2$:

$$V_1 = Z_{11}I_1 + Z_{12}I_2$$
$$V_2 = Z_{21}I_1 + Z_{22}I_2$$

(8.2)

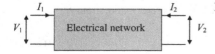

**Figure 8.2** Electrical two-port system.

or, in matrix notation:

$$\begin{pmatrix} V_1 \\ V_2 \end{pmatrix} = \begin{pmatrix} Z_{11} & Z_{12} \\ Z_{21} & Z_{22} \end{pmatrix} \cdot \begin{pmatrix} I_1 \\ I_2 \end{pmatrix} \qquad (8.3)$$

In this description the matrix elements have a clear physical meaning: they represent electrical impedances. It is easy to find experimentally the input and output impedances of the system. For systems with no internal energy sources $Z_{12} = Z_{21}$. We assume this condition fulfilled in the remaining part of this chapter.

The input impedance is the ratio between input voltage and input current; likewise, the output impedance is the ratio between output voltage and output current. Note that the input impedance depends on the circuit connected to the output, and vice versa. We distinguish two cases: open- and short-circuited terminals. The input impedance of the system in Figure 8.2 is:

$$Z_{io} = Z_{11} \quad \text{(at open output: } I_2 = 0\text{)} \qquad (8.4)$$

$$Z_{is} = Z_{11}\left(1 - \frac{Z_{12}^2}{Z_{11}Z_{22}}\right) \quad \text{(at short-circuited output: } V_2 = 0\text{)} \qquad (8.5)$$

For the output impedance we find:

$$Z_{oo} = Z_{22} \quad \text{(at open input: } I_1 = 0\text{)} \qquad (8.6)$$

$$Z_{os} = Z_{22}\left(1 - \frac{Z_{12}^2}{Z_{11}Z_{22}}\right) \quad \text{(at short-circuited input: } V_1 = 0\text{)} \qquad (8.7)$$

Apparently input characteristics depend on what is connected at the output and vice versa. The degree of this mutual influence is expressed by the *coupling factor* $\kappa$, defined as:

$$\kappa^2 = \frac{Z_{12}^2}{Z_{11}Z_{22}} \qquad (8.8)$$

The coupling factor is also related to the power transfer of the system. The maximum power transfer occurs for a load impedance equal to

$$Z_L = Z_{22}\sqrt{1 - \kappa^2} \qquad (8.9)$$

Now we consider a piezoelectric sensor as a two-port network, analogous to the electrical system in Figure 8.2, but this time with a mechanical input port and an electrical output port. In Chapter 2 we defined the relation between stress $T$ and strain $S$:

$$T = cS \quad \text{(N/m}^2\text{)} \quad \text{or} \quad S = sT(-) \qquad (8.10)$$

Similarly the relation between dielectric displacement $D$ and electrical field strength $E$ is given by

$$D = \varepsilon \cdot E \quad (\text{C/m}^2) \quad \text{or} \quad E = \left(\frac{1}{\varepsilon}\right) \cdot D \quad (\text{V/m}) \tag{8.11}$$

where $\varepsilon$ is the dielectric constant or permittivity (F/m). In piezoelectric materials mechanical and electrical quantities are coupled. A two-port model of a piezoelectric system is shown in Figure 8.3.

The two-port equations of this system are, in general

$$\begin{aligned} S &= f_1(T, D) \\ E &= f_2(T, D) \end{aligned} \tag{8.12}$$

and for linear systems in particular

$$\begin{aligned} S &= s^D \cdot T + g \cdot D \\ E &= -g \cdot T + \left(\frac{1}{\varepsilon^T}\right) \cdot D \end{aligned} \tag{8.13}$$

that is, Eqs (8.10) and (8.11) extended by the parameter $g$ (m²/C or V m/N). Equation (8.13) is the *constitutive equation* of the piezoelectric system. Mathematically $s^D$ is the partial derivative of $S$ to $T$ at constant $D$, and $\varepsilon^T$ the partial derivative of $D$ to $E$ at constant mechanical tension $T$. From a physical point of view $s^D$ is the compliance at open electrical terminals ($\Delta D = 0$), and $\varepsilon^T$ is the permittivity at open mechanical terminals ($\Delta T = 0$). The latter means that the material can freely deform: it is not clamped. At open electrical terminals the voltage generated by an applied force is $E = -gT$, which explains the name piezoelectric voltage constant for $g$.

Further as we have seen before, the impedance of the connecting circuits (both mechanical and electrical) influence the properties of the material and hence the piezoelectric behaviour. For instance the material's stiffness is higher at short-circuited electrical terminals than when these terminals are open.

Another set of constitutive equations is:

$$\begin{aligned} S &= f_3(T, E) \\ D &= f_4(T, E) \end{aligned} \tag{8.14}$$

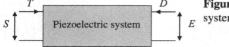

**Figure 8.3** Two-port model of a piezoelectric system.

**Table 8.1** The Four Piezoelectric Parameters

| Parameter | Definition | Unit |
|-----------|-----------|------|
| $d$ | $D/T\|_E$ or $S/E\|_T$ | C/N = m/V |
| $e$ | $D/S\|_E$ or $T/E\|_S$ | C/m$^2$ = N/V m |
| $g$ | $E/T\|_D$ or $S/D\|_T$ | V m/N = m$^2$/C |
| $h$ | $E/S\|_D$ or $T/D\|_S$ | V/m = N/C |

which, for linear systems, becomes:

$$S = s^E \cdot T + d \cdot E$$
$$D = d \cdot T + \varepsilon^T \cdot E \tag{8.15}$$

where $s^E$ is the compliance at short-circuited electrical terminals ($\Delta E = 0$) and $d$ (C/N or m/V) is the *piezoelectric charge constant*. At short-circuited terminals the electrical displacement, generated by an applied force, is found to be $D = dT$, which explains the name piezoelectric charge constant for the parameter $d$.

Other possible combinations of the electrical and mechanical quantities result in other material parameters, for instance $\varepsilon^S$: the permittivity at clamped state ($\Delta S = 0$). Further in literature the parameters $e$ and $h$ can be found. Table 8.1 shows an overview of all these piezoelectric parameters. The definitions are given as $A/B\|_C$, with $A$ the quantity affected by $B$ at constant $C$. We will use only $d$ and $g$ in the remaining of this chapter.

Relations between these four piezoelectric parameters follow from the respective constituent relations that define them (see for instance Refs [11,12]). Irrespective of which set of constitutive equations is taken, they all describe the same material. Hence there is a relation between the various piezoelectric parameters. For instance the relation between $d$ and $g$ is:

$$d = \varepsilon^T \cdot g \tag{8.16}$$

Both $g$ and $d$ describe the strength of the piezoelectric effect of the material only; they do not depend on dimensions. The relations between $s^D$ and $s^E$, respectively between $\varepsilon^S$ and $\varepsilon^T$ can be derived from the constitutive equations as follows:

$$s^D = s^E \cdot (1 - \kappa^2) \tag{8.17}$$

$$\varepsilon^S = \varepsilon^T \cdot (1 - \kappa^2) \tag{8.18}$$

where

$$\kappa^2 = g^2 \frac{\varepsilon^T}{s^D} \tag{8.19}$$

is the piezoelectric coupling constant of the material (compare the coupling constant for an electrical two port in Eq. (8.8)). This coupling constant is an important parameter for the strength of the piezoelectric effect. Just as in the case of the electrical system, the piezoelectric constant $\kappa^2$ denotes the maximal fraction of (mechanical) energy stored in the crystal that can be converted into electrical energy (see for instance in Ref. [13]). Note that $\kappa^2$ differs from the overall efficiency.

The equations given earlier only apply for isotropic materials, that is materials whose properties do not depend on the orientation within the material. Many piezoelectric materials are anisotropic. So similar to the parameters $s$ and $c$, the parameter $\varepsilon$ should also be given as a matrix:

$$\varepsilon = \begin{pmatrix} \varepsilon_{11} & \varepsilon_{12} & \varepsilon_{13} \\ \varepsilon_{21} & \varepsilon_{22} & \varepsilon_{23} \\ \varepsilon_{31} & \varepsilon_{32} & \varepsilon_{33} \end{pmatrix} \tag{8.20}$$

Depending on symmetry in the crystal, some of the matrix elements are zero and others have equal numerical values. As an example Eq. (8.21) shows the permittivity matrices for quartz and ferroelectric Rochelle salt, respectively:

$$\begin{pmatrix} \varepsilon_{11} & 0 & 0 \\ 0 & \varepsilon_{11} & 0 \\ 0 & 0 & \varepsilon_{33} \end{pmatrix} \quad \text{and} \quad T \begin{pmatrix} \varepsilon_{11} & 0 & \varepsilon_{13} \\ 0 & \varepsilon_{22} & 0 \\ \varepsilon_{13} & 0 & \varepsilon_{33} \end{pmatrix} \tag{8.21}$$

Hence the dielectric properties of quartz can be described by only two parameters: $\varepsilon_{11}$ and $\varepsilon_{33}$ (apparently $\varepsilon_{11} = \varepsilon_{22}$), whereas for Rochelle salt ($NaKC_4H_4O_6 \cdot H_2O$) four parameters suffice for the characterization of its piezoelectric properties.

Now the relationship between $E$ and $D$ is described with three linear equations:

$$\begin{aligned} D_x &= \varepsilon_{xx}E_x + \varepsilon_{xy}E_y + \varepsilon_{xz}E_z \\ D_y &= \varepsilon_{yx}E_x + \varepsilon_{yy}E_y + \varepsilon_{yz}E_z \\ D_z &= \varepsilon_{zx}E_x + \varepsilon_{zy}E_y + \varepsilon_{zz}E_z \end{aligned} \tag{8.22}$$

or in matrix notation:

$$D_i = \varepsilon_{ij} \cdot E_j \tag{8.23}$$

where $i, j$ are the directions $x$, $y$ and $z$ in a given co-ordinate system. The parameters $d$ and $g$, too, depend on the orientation of the mechanical parameters. Each of the six force components may result in a dielectric displacement in three directions. So for a piezoelectric material

$$D_i = \varepsilon_{ij}E_j + d_{ipq}T_q \tag{8.24}$$

or using the notation as presented in Chapter 2:

$$D_i = \varepsilon_{ij}E_j + d_{ik}T_k \tag{8.25}$$

where $i, j = 1 \ldots 3$; $k = 1 \ldots 6$. Applied, for instance, to the displacement in $x$-direction (1-direction), this results in

$$D_1 = \varepsilon_{11}E_1 + \varepsilon_{12}E_2 + \varepsilon_{13}E_3 + d_{11}T_1 + d_{12}T_2 + d_{13}T_3 + d_{14}T_4 + d_{15}T_5 + d_{16}T_6 \tag{8.26}$$

Similar equations can be derived for $D$ in the $y$- and $z$-directions, $D_2$ and $D_3$. In total the piezoelectric matrix contains 18 elements. Many of them are zero, due to the particular crystal structure. Moreover many of them are mutually dependent. As an example, for quartz, $d_{12} = -d_{11}$, $d_{25} = -d_{14}$ and $d_{26} = -2d_{11}$, due to the specific crystal structure of this material. So the $3 \times 6$ $d$-matrix in the equation $D_i = d_{ik} \cdot T_k$ is [2, p. 18]:

$$\begin{pmatrix} d_{11} & -d_{11} & 0 & d_{14} & 0 & 0 \\ 0 & 0 & 0 & 0 & -d_{14} & -2d_{11} \\ 0 & 0 & 0 & 0 & 0 & 0 \end{pmatrix}$$

with the numerical values $d_{11} = 2.3 \times 10^{-12}$ C/N and $d_{14} = -0.7 \times 10^{-12}$ C/N. Hence the equations for $D$ for quartz can be written as

$$\begin{aligned} D_1 &= \varepsilon_{11}E_1 + d_{11}T_1 - d_{11}T_2 + d_{14}T_4 \\ D_2 &= \varepsilon_{11}E_2 - d_{14}T_{25} - 2d_{11}T_6 \\ D_3 &= \varepsilon_{33}E_3 \end{aligned} \tag{8.27}$$

As an illustration we give here the piezoelectric $d$-matrices for two other piezoelectric materials: LiTaO$_3$ (poled ceramic) and PVDF (poled after uni-axial stretching), respectively

$$\begin{pmatrix} 0 & 0 & 0 & 0 & d_{15} & -d_{22} \\ -d_{22} & d_{22} & 0 & d_{15} & 0 & 0 \\ d_{31} & d_{31} & d_{33} & 0 & 0 & 0 \end{pmatrix} \quad \text{and} \quad \begin{pmatrix} 0 & 0 & 0 & 0 & d_{15} & 0 \\ 0 & 0 & 0 & d_{24} & 0 & 0 \\ d_{31} & d_{32} & d_{33} & 0 & 0 & 0 \end{pmatrix} \tag{8.28}$$

For a unique notation with respect to the orientation of polarized crystals, an agreement has been made saying that the $z$-axis (or 3-axis) is the direction of polarization. Figure 8.4 visualizes various components of the $d$-matrix. In all these examples the electrical output (voltage or current) is measured in the direction of the polarization vector (3-direction), so the components of the matrix are all of the form $d_{3i}$. In Figure 8.4A the applied force is in the polarization direction. Many crystals have high piezoelectric sensitivity in this direction, which explains the often encountered parameter $d_{33}$ in specification sheets. This mode of operation is called longitudinal, since the mechanical and electrical directions are the same.

In Figure 8.4B a force is applied perpendicular to the direction of polarization: this is the transversal mode of operation. Transversal operation is used with crystals having a high value of $d_{32}$. Figure 8.4C shows a shear force in transversal direction (here the 4-direction), yielding the parameter $d_{34}$. Finally in Figure 8.4D the shear force is applied in the 6-direction, so the parameter $d_{36}$ shows up.

Table 8.2 presents, for comparison purposes, various properties of a ceramic material based on PZT (lead−zirconium−titanate), of quartz and of PVDF. Note that the numerical values of all piezoelectric parameters depend on temperature, on

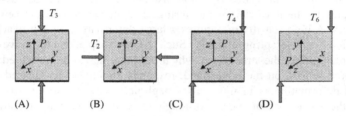

**Figure 8.4** Major piezoelectric parameters for various directions of the applied force: (A) $d_{33}$, (B) $d_{32}$, (C) $d_{34}$ and (D) $d_{36}$. In all these cases the electrical output is taken over the $z$-direction (thick lines in (A) to (C) represent the electrodes).

**Table 8.2** Comparison of Piezoelectric Properties PZT, Quartz and PVDF

| Parameter | Symbol | PZT[a] | Quartz | PVDF |
|---|---|---|---|---|
| Density (kg/m$^3$) | $\rho$ | $7.6 \times 10^3$ | $2.65 \times 10^3$ | $1.78 \times 10^3$ |
| Curie temperature (°C) | $\Theta_C$ | 285 | 550 | 100−150 |
| Compliances (m$^2$/N) | $s^E_{33}$ | $19 \times 10^{-12}$ | $9.7 \times 10^{-12}$ | $4 \times 10^{-10}$ |
| | $s^E_{11}$ | $16 \times 10^{-12}$ | $12.8 \times 10^{-12}$ | |
| Permittivities | $\varepsilon^T_{33}/\varepsilon_0$ | 1700 | 4.6 | 10−13 |
| | $\varepsilon^T_{11}/\varepsilon_0$ | 1730 | 4.5 | |
| p.e. voltage constants (V m/N) | $g_{33}$ | $27 \times 10^{-3}$ | | |
| | $g_{31}$ | $-11 \times 10^{-3}$ | | |
| | $g_{11}$ | | 0.06 | |
| | $g_{14}$ | | 0.02 | |
| | $g_{15}$ | $33 \times 10^{-3}$ | | |
| p.e. charge constants (C/N) | $d_{33}$ | $425 \times 10^{-12}$ | | $-18 \times 10^{-12}$ |
| | $d_{31}$ | $-170 \times 10^{-12}$ | | |
| | $d_{11}$ | | $2.3 \times 10^{-12}$ | |
| | $d_{14}$ | | $0.7 \times 10^{-12}$ | |
| | $d_{15}$ | $515 \times 10^{-12}$ | | |

From data sheets of various manufacturers.
[a]Data for a modified form of lead−zirconate−titanate; small modifications have a large effect on most piezoelectric properties.

composition and on the manufacturing process. Detailed data are found in for instance Ref. [2]. Piezoelectric coefficients of PVDF and a measurement method for obtaining numerical values can be found in Ref. [14], for instance, where all piezoelectric constants according to Eq. (8.28) are presented.

## 8.2 Force, Pressure and Acceleration Sensors

### 8.2.1 Construction

Most force sensors are based on an elastic or spring element: the force generates a deformation which is measured by some displacement sensor (as discussed in previous chapters). The elasticity of the spring element determines the sensitivity of such sensors. A high sensitivity requires a large deformation, which is achieved by a low stiffness of the spring element. Such a large displacement, however, could unintentionally affect the structure in which the force has to be measured. A piezoelectric force sensor, on the other hand, responds directly to an applied force: the associated deformation is in most cases negligibly small, assuring small loading errors in the force measurement. Although force is the primary quantity that is measured by a piezoelectric sensor, other quantities such as pressure, strain and acceleration can easily be measured as well, using a proper construction.

Piezoelectric sensors have disadvantages, too. The surface charge produced by an applied force might be neutralized easily by charges from the environment (airborne charges), by current leakage (due to a non-zero conductivity of the dielectric) or just by the input resistance of the connected electronics (discussed further in Section 8.3). This makes the sensor behave as a high-pass filter for input signals, impeding pure static measurements.

A further point of attention is the temperature sensitivity. There are various causes for this temperature effect. First piezoelectric materials are pyroelectric, so they also respond to temperature changes. Secondly a temperature change may induce crystal deformation and hence an electrical output as well. Further when materials connected to the piezoelectric crystals have different thermal expansion coefficients (for instance clamping parts and electrodes), the crystal experiences unwanted forces. Fortunately as long as these changes are slow, they would not limit the applicability because of the previously discussed intrinsic high-pass character.

Usually the elements of the transducer are kept together by clamping rather than by gluing. A consequence is that the crystals become preloaded. Figure 8.5 shows in a very schematic way some basic constructions.

In a force sensor (Figure 8.5A) the force to be measured is directly transferred to the piezoelectric crystal. Obviously the construction provides a means to connect the electrodes of the crystal to an external connector (not shown in the figure). Usually the case acts as the ground terminal.

In a piezoelectric pressure sensor the pressure to be measured is applied to a thin metal membrane. The total force on the membrane — that is the pressure times

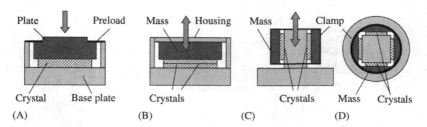

**Figure 8.5** Principle of piezoelectric sensors (A) force sensor (longitudinal), (B) compression type accelerometer, (C) shear-type accelerometer, (D) top view of (C); the arrows show the main axis of sensitivity; electrical terminals are not shown.

the active area of the membrane − is mechanically transferred to the crystal. Piezoelectric pressure sensors are also sensitive to acceleration, because the mass of the housing produces an inertial force on the crystal when accelerated. For applications where pressure has to be measured in a vibrating or otherwise moving environment, special pressure sensors are designed with a compensating crystal, to minimize the acceleration sensitivity.

An accelerometer consists basically of one or more piezoelectric crystals and a *proof mass* (or *seismic mass*). Here, also, the mass is fixed by a preload onto the crystal. Figure 8.5B shows a compression mode accelerometer. This example comprises two crystals, mounted back to back between the seismic mass and the base plate. One electrode is connected to the common surface of the crystals, the other to the housing. In this configuration the crystals are electrically in parallel and mechanically in series, resulting in a doubled sensitivity. Moreover some compensation for common interferences is achieved. The shear-type accelerometer shown in Figure 8.5C and D contains four crystals mounted on a rectangular post. The masses and crystals are fixed between a centre post and a clamping ring (as preload).

The absence of moving parts allows a piezoelectric sensor to be mounted in a robust package and hermetically sealed. Figure 8.6 gives an impression of the real construction of the two accelerometer types shown in Figure 8.5.

In both cases the main axis is in the vertical direction, where the sensitivity has a maximum value. Ideally the sensitivity in the orthogonal direction is zero. Commercial sensors have a non-zero sensitivity in directions perpendicular to the main axis, due to construction tolerances (not well-aligned crystals) or the cross sensitivity of the crystal itself. The crystal type and orientation should be chosen such that the highest value of $d$ is in line with the main axis and zero for all other directions. For instance when $d_{15}$ in the shear type is responsible for maximum output (in the *1*-direction), the values of $d_{12}$ and $d_{13}$ should be zero, since these forces normal to the crystal may not generate an electrical output in that direction. Commercial accelerometers have a cross sensitivity that is just a few percent of the main sensitivity.

When two or all three components of a force or acceleration vector have to be measured, multi-axis sensors should be applied. The earlier three-axes versions

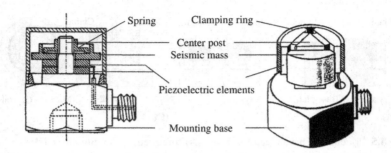

**Figure 8.6** Schematic drawing of piezoelectric accelerometers; left: compression type; right: three-crystal shear type (Brüel & Kjaer).

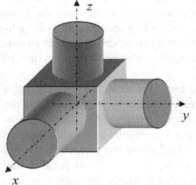

**Figure 8.7** Three-axis piezoelectric accelerometer consisting of three stand-alone single axis sensors.

usually consist of three separate sensors, mounted orthogonally in a single housing, as shown schematically in Figure 8.7. The smallest dimensions of such sensors are comparable with a die and have a weight down to 20 g, and sensitivities of around 10 mV/g, where $g$ is the gravitational acceleration in m/s$^2$. More recently MEMS tri-axial accelerometers have flooded the market, mainly for consumer products (e.g. cameras and games). They are also available with integrated interface circuits. Typical sensitivities are 10−100 mV/g, and are 10−50 g in weight.

## 8.2.2 Characteristics of Accelerometers

The frequency band of a piezoelectric accelerometer is limited at the lower side of the spectrum by the shunt resistance of the crystal (Section 8.2.1). At the upper side of the spectrum it is limited by the resonance frequency. Figure 8.8 shows the useful frequency range of a standard type piezoelectric accelerometer. Here $S(\omega)$ is the sensitivity, that is the ratio between applied acceleration and electrical output, and $S_{nom}$ is the nominal sensitivity.

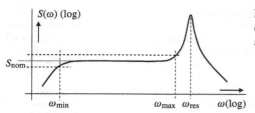

**Figure 8.8** Typical frequency characteristic of a piezoelectric accelerometer.

**Table 8.3** Specifications of Various Accelerometers

| Application Area | Range (FS) | Sensitivity (mV/g) | Resolution (g) | Frequency Range (Hz) |
|---|---|---|---|---|
| General | $\pm 10$ to $\pm 500$ | $10-100$ | $0.001$ | $1-10^4$ |
| Robust | $\pm 50$ | $100$ | $10^{-3}$ | $1-1000$ |
| Seismic | $\pm 5$ | $0.05$ | $10^{-4}$ | $1-3000$ |
| Ballistic | $\pm 10^5$ | $1000$ | $1$ | $1-8000$ |

Most types cover a temperature range from $-70°$ to $+120°C$.

A piezoelectric accelerometer consists of a base plate, a seismic mass and a (preloaded) crystal. The mechanical characteristics of this spring-mass system follow from the equation of motion:

$$m\frac{d^2x}{dt^2} + \alpha\frac{dx}{dt} + kx = F(t) \tag{8.29}$$

where $m$ is the seismic mass, $\alpha$ the damping constant and $k$ the spring constant (between seismic mass and base). The natural frequency of the system (i.e. the resonance frequency for $\alpha = 0$) equals:

$$\omega_o = \sqrt{\frac{k}{m}} \tag{8.30}$$

The damping constant of a piezoelectric accelerometer is usually very small; in that case the (damped) resonance frequency $\omega_{res}$ in Figure 8.8 is close to $\omega_o$. A high resonance frequency requires a small mass $m$ and a large stiffness $k$. The mass determines the sensitivity in the nominal frequency range and is chosen in accordance to the application. The resonance frequency of commercial accelerometers range from 1 to 250 kHz: the smaller the size, the higher the resonance frequency.

Many different types of piezoelectric sensors exist, for various applications. The smallest commercial accelerometers have a weight of 0.2 g and a resonance frequency at 50 kHz.

Table 8.3 lists the major properties of several groups of accelerometers and Table 8.4 shows some maximum ratings.

**Table 8.4** Maximum Ratings of Piezoelectric Sensors

| Type | Range (FS) | Sensitivity | $T_{max}$ (°C) |
|---|---|---|---|
| Acceleration | $10^3 - 10^6$ m/s$^2$ | 0.1 − 50 pC/ms$^{-2}$ | 500 |
| Force | $10^2 - 10^6$ N | 2 − 4 pC/N | 300 |
| Pressure | $10^7 - 10^8$ Pa | 20 − 800 pC/MPa | 200 |

When applying piezoelectric sensors the following points should be considered:

- Material properties depend on the electric load (open or closed electrical terminals); hence all sensor properties (sensitivity $S$, resonance frequency $\omega_a$) depend on the load too.
- When mounting a sensor on an object, the latter is loaded mechanically, resulting in a change of acceleration $a$ and resonance frequency $\omega_a$.

When $M$ is the mass of the object and $m$ the mass of the accelerometer, the acceleration is lowered by a factor $1 + m/M$, and the resonance frequency by a factor $\sqrt{(1 + m/M)}$, according to Eq. (8.30):

$$a_L = a_o \frac{M}{M + m}$$

$$\omega_{aL} = \omega_{ao} \sqrt{\frac{M}{M + m}} \tag{8.31}$$

where subscripts o and L denote the unloaded and loaded situation, respectively. These equations allow quick assessment of the loading effects when applying an accelerometer.

## 8.3 Interfacing

The primary signal of a piezoelectric sensor is charge $Q$. The sensor impedance behaves as a capacitance $C_e$, corresponding to the electrical capacity of a flat-plate capacitor with (piezoelectric) dielectric. This simple impedance model can be extended by a resistance $R_s$, modelling leakage currents (Figure 8.9).

The charge signal can be measured in two different ways. The first method is based on the relation between charge and voltage over a capacitor: $Q = C \cdot V$. Hence the interface circuit consists of a voltage amplifier. In the second method the charge is allowed to flow through an impedance, preferably a capacitor. The voltage across this capacitor is proportional to the charge. This type of interface is commonly called a charge amplifier; a better name would be charge−voltage converter.

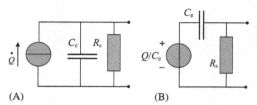

**Figure 8.9** Two equivalent models of a piezoelectric sensor: (A) current source and (B) voltage source.

(A)                        (B)

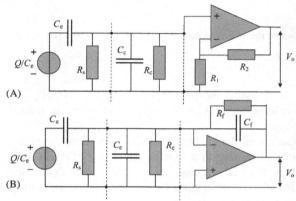

**Figure 8.10** Interfacing a piezoelectric sensor: (A) voltage amplifier and (B) charge amplifier.

(A)

(B)

Figure 8.10 shows examples of these interface circuits, together with models for the sensor and the connecting cable (the part between the dotted lines). It will be shown that the cable capacitance can have a substantial influence on the signal transfer of the system, due to the capacitive character of the sensor. The sensor is modelled by the voltage source model of Figure 8.9B.

We calculate, for both interface circuits, the total signal transfer, using the formula for operational amplifier configurations in Appendix C. First we consider the voltage readout (Figure 8.10A). The transfer of this circuit is given by

$$H_v = \frac{V_o}{Q/C_e} = A \cdot \frac{j\omega R_p C_e}{1 + j\omega R_p(C_e + C_c)} = A \cdot \frac{C_e}{C_e + C_c} \cdot \frac{j\omega R_p(C_e + C_c)}{1 + j\omega R_p(C_e + C_c)}$$

(8.32)

where $R_p = R_s // R_c$ (parallel connection) and $A = 1 + R_2/R_1$, the gain of the non-inverting amplifier. Obviously the transfer shows a high-pass character (Figure 8.11A): for frequencies satisfying $\omega R_p(C_e + C_c) \gg 1$ the transfer equals $C_e/(C_e + C_c)$, so it is frequency independent. The cable capacitance $C_c$ causes signal attenuation. Hence the total signal transfer depends on the length of the cable. This requires recalibration each time the connection cable is replaced.

The interface circuit of Figure 8.10B is better, in this respect. Due to the virtual ground of the operational amplifier the voltage across the sensor and the cable is kept at zero: neither the cable impedance nor the input impedance of the amplifier influences the transfer, and hence can be ignored. Assuming an ideal operational

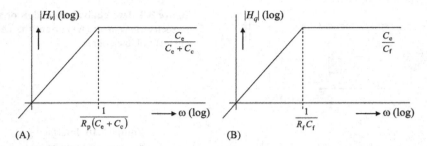

**Figure 8.11** Transfer characteristics of the interface circuits in (A) Figure 8.10A (B) Figure 8.10B.

amplifier, its output voltage is $-Z_2/Z_1$. Without resistor $R_f$, $Z_1 = 1/j\omega C_e$ and $Z_2 = 1/j\omega C_f$, hence the output voltage equals $V_o = -(C_e/C_f)(Q/C_e)$ which, indeed, does not depend on cable properties.

Ideally the circuit with a capacitor in the feedback path behaves as an integrator: the input current flows through $C_f$ resulting in a voltage equal to $(1/C_f)\int I dt$. Here the input current is generated by the piezoelectric sensor and equals $dQ/dt$, that is the time derivative of the piezoelectrically induced charge $Q$. Hence the voltage across $C_f$ is just $Q/C_f$ explaining the name charge amplifier.

Unfortunately the integrator does not only integrate the input charge, but also the unavoidable offset voltage and bias current of the operational amplifier. To prevent the amplifier from overloading, the feedback resistor $R_f$ cannot be left out. This results in a transfer given by

$$H_q = \frac{V_o}{Q/C_e} = \frac{-j\omega R_f C_e}{1 + j\omega R_f C_f} = -\frac{C_e}{C_f} \cdot \frac{j\omega R_f C_f}{1 + j\omega R_f C_f} \tag{8.33}$$

Just as in Eq. (8.32), this transfer has a high-pass characteristic (Figure 8.11B): the cut-off frequency is set by the components of the amplifier only: for $\omega R_f C_f \gg 1$ the voltage transfer is $-(C_e/C_f)$. When using a high quality operational amplifier (low offset, low bias current), the value of the cut-off frequency can be chosen down to 0.01 Hz. A true static measurement of acceleration or force, however, is not possible.

## 8.4 Applications

Piezoelectric force sensors and accelerometers are available for a wide range of applications. They can be mounted on proper places of the mechatronic construction, taking into account mechanical loading and direction of sensitivity. If only little space is available for mounting a commercial device, single piezoelectric crystals could be used, taking note of the polarization direction, the orientation of the electrical contact surfaces and a proper interfacing. Charge amplifiers are preferred in most cases.

In this section we present some examples of special designs of piezoelectric sensors for various applications.

### 8.4.1 Stress and Pressure

Axially stressed and poled PVDF shows a rather strong piezoelectric effect (Table 8.2). A straightforward design of a pressure sensor using this property is a sheet of PVDF on top of a rigid backing, making use of lateral deformation. However the flexibility of PVDF allows the material to bend easily, so it can serve as a membrane, resulting in a much higher pressure sensitivity. Two pressure sensors designed according to this concept are presented in Ref. [15]. A circularly shaped sheet of piezoelectric PVDF 5 mm in diameter and about 25 μm thickness, with metallization layers on both sides, is clamped between two parts of a likewise circular housing made up of PVDF. The sensor is applied in pneumatic and hydraulic systems, for pressures up to 200 kPa and temperatures up to 125°C and is resistant to a wide range of chemicals. Many properties of the prototype have been determined experimentally: sensitivity to temperature and humidity, response time (less than 100 μs), frequency range and stability over time (i.e. aging and creep). As outlined in preceding sections, the sensor element is also pyroelectric, so it responds to temperature changes as well, and shows a high-pass characteristic that depends on the temperature-dependent loss factor (resistance) of the membrane.

Ice deposition, for instance on overhead power transmission lines and power pylons, may cause much damage, and it is useful to study the adhesion at different conditions. In Ref. [16] PVDF is used to measure interfacial stress between an ice layer deposited on an aluminium substrate. Clearly the small thickness of PVDF as well as its flexibility makes this sensing material an excellent candidate for this task. It is shown that an embedded PVDF layer provides useful information about ice de-bonding stress and propagation.

Many examples can be found in literature illustrating the versatility of PVDF as a means to measure pressure and obtain pressure images. For instance Ref. [17] describes a system for measuring the dynamic (normal) pressure between a car's tyre and the ground, and in Refs [18,19] the material is used for automatic verification of hand-written signatures.

### 8.4.2 Acceleration

Most piezoelectric accelerometers for industrial application are designed according to Figure 8.5. The present market offers a vast variety of types for almost any application. However proposals for alternative designs aiming at even better performances or for particular applications appear on a regular basis in scientific literature. For instance the advent of piezoelectric PVDF and the fast development of microtechnology motivate researchers for looking to new solutions for both old and new measurement problems.

Soon after its discovery piezoelectric PVDF was recognized as being a suitable sensing material. Moreover the light-weight film-shaped material allows the construction of small and cheap accelerometers. An early attempt to create a PVDF-based accelerometer is found in Ref. [20], which shows a sensor for accelerations up to 50,000 g. An angular acceleration sensor can be created as well, as shown in Ref. [21]. By a special geometric arrangement of four piezoelectric elements (PVDF) and two seismic masses, an angular sensitivity of about 0.1 pC/rad/s$^2$ has been achieved.

By suspending a seismic mass on a sheet of PVDF, a very sensitive accelerometer has been achieved, to be used for very low frequencies as a replacement for a geophone [22]. The resonance frequency (Figure 8.8) amounts to 265 Hz for a particular choice of the parameters; the sensitivity is almost constant up to 100 Hz.

Another way to realize small, sensitive accelerometers is the use of thin films of PZT (typically 1−5 μm). In this case the piezoelectric material is one of a sandwich of thin layers deposited on a silicon substrate [23]. Other layers serve as electrodes and isolation. The seismic mass is formed by selective etching of the substrate, resulting in a mass-spring system similar to the one shown in Figure 4.18. The structure is sensitive in three directions, 22 pC/g for transversal and 8 pC/g for in-plane acceleration. An extensive FEM analysis of such thin-film piezoelectric accelerometers is given in Ref. [24], providing guidelines for the design of devices with specified properties.

A different approach is followed in Ref. [25], in which the PZT is deposited in a thick layer (typically 50 μm), on an alumina substrate. This device is composed of two sensing elements, to compensate for temperature effects. They also serve as seismic mass. The sensitivity can be set by an additional amplifier and is constant between 1 Hz and 10 kHz.

The reversibility of the piezoelectric effect provides solutions with a combination of actuation and sensing. In Ref. [26] such a combined piezoelectric sensor−actuator system is used to reduce vibrations that may occur in precision mechatronic systems, in this example a waver stepper. A piezoelectric ceramic sensor (1 mm thickness) measures vibrations in a particular part of the construction, and a stack of piezoelectric actuators (4 mm thickness) generates a counter-movement in the same part of the construction. With a proper design of the control loop, the amplitude of the dominant vibration mode is reduced by a factor of 7.

### 8.4.3 Tactile Sensors

Piezoelectricity is one of the many physical effects that have been explored for the realization of tactile sensors. In particular the piezoelectric polymer PVDF has attracted much attention. The small dimensions in the direction of the force, as well as the flexibility of the material, simplify integration of PVDF tactile sensors in a robot gripper, even on the fingers of a dextrous gripper [27]. One drawback of the material is the charge leakage (preventing the measurement of static images). Moreover the material is also pyroelectric, which makes the sensor sensitive to temperature changes and gradients as well; hence this property could be put to use in robotics when thermal data about objects (i.e. temperature and thermal conductivity) add useful information to the control system, somewhat similar to the human skin [28–30].

Like other tactile sensors discussed in preceding chapters, piezoelectric tactile sensors too are used for two important tasks: contour recognition and improving gripper control [31,32]. In Ref. [32] a study on a thick (1 mm) piezoelectric tactile sensor is presented. The sensor is modelled as a second-order system and shows a frequency characteristic similar to the one given in Figure 8.8, with a resonance frequency of 6.1 kHz. In applications where dynamic performance matters (for instance as an active rate of force sensor described in Ref. [33]), PVDF sensors appear to be suitable devices.

A common problem in tactile sensing for contour measurement is the low spatial resolution. Piezoelectric sensors suffer from this problem too, although sensors have been realized that are able to determine the orientation of objects on a test bed equipped with a piezoelectric polymer [30].

Piezoelectric ceramics and polymers are also applied in the acoustic domain. Application examples, including tactile sensors, will be discussed in the next chapter on acoustic sensors.

The integration of polymer technology and silicon technology offers interesting prospects for future developments of compact, high-resolution tactile sensors with integrated electronics.

# References and Literature

### References to Cited Literature

[1] J.F. Tressler, S. Alkoy, R.E. Newnham: Piezoelectric sensors and sensor materials, J. Electroceram., 2(4) (1998), 257–272.

[2] G. Gautschi: Piezoelectric sensorics; Springer-Verlag, Berlin, 2002; ISBN 3540 422 595; Chapter 2.

[3] H. Kawai: The piezoelectricity of PVDF, Jpn. J. Appl. Phys., 8 (1969), 975–976.

[4] E. Fukada, T. Furukawa: Piezoelectricity and ferroelectricity in polyvinylidene fluoride, Ultrasonics, (1981), 31–39.

[5] H.R. Gallantree, R.M Quilliam: Polarized poly(venylidene fluoride) – its application to pyroelectric and piezoelectric devices, Marconi Rev., 39 (1976), 189–200.

[6] H.R Gallantree: Review of transducer applications of polyvinylidene fluoride, IEE Proc., 130(5) (1983), 219–222.

[7] D. Setiadi: Integrated VDF/TrFE copolymer-on-silicon pyroelectric sensors; PhD thesis, University of Twente, Enschede, The Netherlands, 1995; ISBN 90-900-8925-X.

[8] A. Lee, A.S. Fiorello, J. van der Spiegel, P.E. Bloomfield, J. Dao, P. Dario: Design and fabrication of a silicon-P(VDF-TrFE) piezoelectric sensor, Thin Solid Films, 181 (1989), 245–250.

[9] P.C.A. Hammes: Infrared matrix sensor using PVDF on silicon, PhD thesis, Delft University of Technology, The Netherlands, 1994.

[10] P. Schiller, D.L. Polla, Integrated piezoelectric microactuators based on PZT thin films, Proc. 7th Int. Conf. Solid-State Sensors and Actuators (Transducers '93), Yokohama, Japan, June 7-10, pp. 154–157.

[11] R. Holland, E.P. Eernisse: Design of resonant piezoelectric devices; Res. Monograph 56, M.I.T.Press, 1969, ISBN 0-262-08033-8.

[12] W. Wersing: Applications of piezoelectric materials: an introductory review; in: N. Setter (ed.): Piezoelectric materials in devices, N. Setter (Ceramics Laboratory EPFL, Lausanne, Switzerland, 2002); ISBN 2-9700346-0-3.

[13] A. Preumont: Mechatronics − dynamics of electromechanical and piezoelectric systems; Springer-Verlag, 2006; ISBN 1-4020-4695-2; Chapter 4.

[14] E.L. Nix, I.M. Ward: The measurement of the shear piezoelectric coefficients of polyvinylidene fluoride, Ferroelectrics, 67 (1986), 137–141.

[15] A.V. Shirinov, W.K. Schomburg: Pressure sensor from a PVDF film, Sens. Actuators, A, 142 (2008), 48–55.

[16] C. Akitegetse, C. Volat, M. Farzaneh: Measuring bending stress on an ice/aluminium composite beam interface using an embedded piezoelectric PVDF (polyvinylidenefluoride) film sensor, Meas. Sci. Technol., 19 (2008), 9pp (065703).

[17] R. Marsili: Measurement of the dynamic normal pressure between tire and ground using PVDF piezoelectric films, IEEE Instrum. Meas., 49(4) (2000), 736–740.

[18] P. de Bruyne: Piezo-electric film as a sensor element in signature verification, Proceedings 6th International Symposium on Electrets (ISE 6), Oxford, UK, 1–3 September 1988, pp. 229–233.

[19] D. Wang, Y. Zhang, C. Yao, J. Wu, H. Jiao, M. Liu: Toward force-based signature verification: a pen-type sensor and preliminary validation, IEEE Instrum. Meas., 59(4) (2010), 752–762.

[20] B. André, J. Clot, E. Partouche, J.J. Simonne: Thin film PVDF sensors applied to high acceleration measurements, Sens. Actuators, A, 33 (1992), 111–114.

[21] R. Marat-Mendes, C.J. Dias, J.N. Marat-Mendes: Measurement of the angular acceleration using a PVDF and a piezo-composite, Sens. Actuators, A, 76 (1999), 310–313.

[22] B.L.F. Daku, E.M.A. Mohamed, A.F. Prugger: A PVDF transducer for low-frequency acceleration measurements, ISA Trans., 43 (2004), 319–328.

[23] K. Kunz, P. Enoksson, R. Wright, G. Stemme: Highly sensitive triaxial silicon accelerometer with integrated PZT thin film detectors, Sens. Actuators, A, 92 (2001), 156–160.

[24] Q.M. Wang, Z.C. Yang, F. Li, P. Smolinski: Analysis of thin film piezoelectric microaccelerometer using analytical and finite element modeling, Sens. Actuators, A, 113 (2004), 1–11.

[25] D. Crescini, D. Marioli, A. Taroni: Large bandwidth and thermal compensated piezoelectric thick-film acceleration transducer, Sens. Actuators, A, 87 (2001), 131–138.

[26] J. Holterman, Th.J.A. de Vries: Active damping within an advanced microlithography system using piezoelectric smart disks, Mechatron., 14 (2004), 15–34.

[27] P. Dario, A. Bicchi, F. Vivaldi, P.C. Pinotti: Tendon actuated exploratory finger with polymeric, skin-like tactile sensor, Proceeding of International Conference Robotics and Automation, St. Louis, Miss., March 1985, pp. 701–706.

[28] R. Bardelli, P. Dario, D. de Rossi, P.C. Pinotti: PM- and pyroelectric polymers skinlike tactile sensors for robots and prostheses, 13th International Symposium on Industrial Robots, Chicago, IL, 1983, pp. 1845–1856.

[29] G.J. Monkman, P.M. Taylor: Thermal tactile sensing, IEEE Trans. Rob. Autom., 9(3) (1993), 313–318.

[30] P. Dario, R. Bardelli, D. de Rossi, L.R. Wang, P.C. Pinotti: Touch-sensitive polymer skin uses piezoelectric properties to recognise orientation of objects, Sens. Rev., (1982), 194–198.

[31] A.S. Fiorello, P. Dario, M. Bergamasco: A sensorized robot gripper, Robotics, 4 (1988), 49–55.

[32] H.S. Tzou, S. Pandita: A multi-purpose dynamic and tactile sensor for robot manipulators, J. Rob. Syst., 4(6) (1987), 719–741.

[33] M.F. Barsky, D.K. Lindner, R.O. Claus: Robot gripper control system using PVDF piezoelectric sensors, IEEE Trans. Ultrason. Ferroelectr. Freq. Control, 36 (1989), 129–134.

# Literature for Further Reading

## Some Books and Articles on Piezoelectric Materials, Sensors, and Applications

[1] A. Arnau (ed.): Piezoelectric transducers and applications; Springer, Berlin, 2004, ISBN 3-540-20998-0.

[2] N. Setter (ed.): Piezoelectric materials and devices; N. Setter, Lausanne, 2002, ISBN 2-9700346-0-3.

[3] G. Gautschi: Piezoelectric sensorics; Springer-Verlag, Berlin, 2002, ISBN 3540 422 595.

[4] J.F. Tressler, S. Alkoy, R.E. Newnham: Piezoelectric sensors and sensor materials, J. Electroceram., 2(4) (1998), 257–272.

[5] G. Caliano, N. Lamberti, A. Iula, M. Pappalardo: A piezoelectric bimorph static pressure sensor, Sens. Actuators, A46–47 (1995), 176–178.

[6] J.T. Broch, J. Courrech: Mechanical vibration and shock measurement; Bruel & Kjaer Corporation, Denmark, 1994, ISBN 87-87355-34-5.

[7] G. Asch: Les capteurs en instrumentation industrielle. Chapitre 10: Capteurs de force, pesage, couple; Bordas, Paris, 1983.

[8] I. Tichý, G. Gautschi: Piëzoelektrische Meßtechnik; Springer-Verlag, Berlin, 1980.

[9] J. van Randeraat, R.E. Setterington (eds.): Piezoelectric ceramics; Mullard, London, 1974.

[10] B. Jaffe, W.R. Cook, H. Jaffe: Piezoelectric ceramics; Academic Press, London, 1971.

# 9 Acoustic Sensors

The primary measurand of an acoustic sensor is sound intensity. In combination with a controllable sound source, acoustic sensors can be used for measuring various other physical parameters, for instance distance, velocity of fluids, material properties, chemical composition and many more. They are also suitable for navigation (of mobile robots), imaging (e.g. rangefinding and object recognition) and quality testing. In mechatronics they mainly fulfill the task of non-contact measurement of distances and derived quantities. Similar to optical sensing systems, such acoustic sensing systems also comprise basically three parts: a source, a receiver and a modulating medium (in mechatronics the medium is mostly air). Many acoustic sensing systems are based on the measurement of the travel time or ToF, that is the time it takes for an acoustic wave or signal to travel a certain path

$$x = v \cdot t \tag{9.1}$$

This path is made dependent on the physical parameter to be determined, for instance a distance. Such a time-based measurement approach is preferred over intensity-based measurements, because the latter are less accurate and the relation between distance and intensity is often affected by unpredictable sound reflections and scattering.

Acoustic sensing has some advantages over optical sensing:

- the output is much less sensitive to smoke, dust, vapour, etc.;
- no (artificial) illumination is required, allowing operation in the dark;
- cheap transducers (transmitters and receivers) will do in most cases.

We start the chapter with an overview of acoustic parameters and properties. Next the operation and performance of some acoustic (ultrasonic) sensors are discussed. In the ensuing section various methods for contact-free distance measurements are given. Finally, in the application section, we discuss examples of ranging and imaging.

## 9.1 Properties of the Acoustic Medium

Unlike light (electromagnetic waves) that can propagate in a vacuum, sound waves need an elastic medium to travel. The most common waves are longitudinal waves and shear waves. In longitudinal waves the particle motion is in the same direction as the propagation; in shear waves the motion is perpendicular to the propagation direction. Some sensors utilize surface acoustic waves (SAW). An absorbing

**Sensors for Mechatronics. DOI: 10.1016/B978-0-12-391497-2.00009-1**
© 2012 Elsevier Inc. All rights reserved.

substance is deposited on the surface of a device on which the waves are travelling: the SAW's velocity is influenced by the concentration of the absorbed matter. Such SAW devices are suitable for chemical sensing. The selectivity is greatly determined by the chemical interface.

In this chapter we only discuss acoustic sensors for the measurement of mechanical quantities in mechatronics, where the propagation medium is generally air. Since the quality of an acoustic measurement strongly depends on the acoustic properties of the medium, we present here an overview of the major acoustic characteristics of air. But first we give some definitions of terms commonly in use in acoustics.

### 9.1.1  Sound Intensity and Pressure

Any piece of vibrating material radiates acoustic energy. The rate at which this energy is radiating is called the acoustical or sound power (W). Sound *intensity* is the rate of energy flow through a unit surface area; hence intensity is expressed as $W/m^2$ (compare the optical quantity *flux*). A sound wave is characterized by two parameters: the *sound pressure* (a scalar, the local pressure changes with respect to ambient) and the *particle velocity* (a vector). The intensity is the time-averaged product of these two parameters. It may vary from zero (when the two signals are 90° out of phase) to a maximum (at in-phase signals). The relation between intensity and pressure in the free field (no reflections) is simply given by the equation

$$I = \frac{p^2}{\rho v} \qquad (9.2)$$

where $p$ is the pressure (in RMS value), $\rho$ the density and $v$ the speed of sound. This equation is only valid in a free field.

Often sound power, pressure and intensity are expressed in dB, which means relative to a reference value. Common reference values are: $P_{ref} = 1$ pW, $I_{ref} = 1$ pW/$m^2$ and $p_{ref}$ (threshold of hearing) about 20 µPa.

For example for the intensity at 2 m away from a sound source of $P = 1$ mW at the ground, the surface area of the hemisphere is about 25 $m^2$ (half of $4\pi r^2$), so the intensity in dB amounts $10 \log(I/I_{ref}) = 76$ dB.

In the free field pressure and intensity levels are numerically the same (when using the same references), but in practice there is a difference between these two levels. Most sensors do not measure sound intensity, but sound pressure only. Note that for a sine wave (a pure sound tone), the RMS pressure is $\frac{1}{2}\sqrt{2}$ times the pressure amplitude, similar to the case of electrical signals.

### 9.1.2  Sound Propagation Speed

In general the propagation speed of sound in a material is given by the equation

$$v_a = \sqrt{\frac{c}{\rho}} \qquad (9.3)$$

where $c$ is the stiffness (modulus of elasticity) and $\rho$ the specific mass of that material. For ideal gases the expression for the speed of sound is:

$$v_a = \sqrt{\frac{c_p}{c_v} \cdot \frac{p}{\rho}} = \sqrt{\frac{c_p R}{c_v M} \Theta} \tag{9.4}$$

where $\Theta$ is the absolute temperature, $R$ the gas constant and $M$ the molecular mass. Substitution of numerical values for air yields:

$$v_a = 331.4(1 + 1.83 \cdot 10^{-3}\vartheta) \text{ m/s} \tag{9.5}$$

with $\vartheta$ the temperature in °C. Hence, as a rule of thumb, the temperature coefficient of $v_a$ is about 2% per 10 K. The influence of air humidity is relatively small and is only relevant when high accuracy is required.

The speed of sound is roughly $10^6$ times lower than the speed of light. Hence ToF is easier to measure, but consequently the measurement time is longer. Moreover the much larger wavelength of sound waves compared to light waves makes it more difficult to manipulate with sound beams (for instance focussing or the creation of a narrow beam).

Acoustic velocity differs from particle velocity $u$. The latter follows from Euler's equation for the acceleration of a fluid:

$$a = -\frac{1}{\rho}\nabla p \rightarrow \frac{du}{dt} = -\frac{1}{\rho}\frac{dp}{dx} \rightarrow u = -\frac{1}{\rho}\int \frac{dp}{dx} \, dt \tag{9.6}$$

for the particle velocity in the $x$-direction.

### 9.1.3 Acoustic Damping

An acoustic wave is attenuated by molecular absorption of sound energy and by scattering: the wave gradually looses energy when propagating. For a plane wave Beer's law applies:

$$P_o = P_i e^{-\alpha \Delta x} \tag{9.7}$$

$P_i$ is the acoustic power at a certain place $x$ in space, $P_o$ is the power at a place $\Delta x$ farther in the direction of propagation. The attenuation or damping coefficient $\alpha$ comprises two effects: absorption loss and scattering loss. In solids the first effect is proportional to frequency, and the second proportional to the square of the frequency. In gases the second term dominates, hence $\alpha = a \cdot f^2$: the attenuation of the wave increases with squared frequency. Acoustic damping also depends on gas composition. For air it means that acoustic attenuation is affected by the air humidity. Table 9.1 shows some values of damping in dry and humid air.

**Table 9.1** Attenuation Coefficients for Air [1]

| Medium | Frequency (Hz) | $\alpha$ (dB/m) |
|--------|----------------|-----------------|
| Dry air | $10^6$ | 160 |
| 10% RH | $10^5$ | 18 |
| 90% RH | $10^5$ | 42 |

**Table 9.2** Acoustic Properties of Some Materials

| Medium | Density (kg/m³) | Velocity (m/s) | Impedance (kg m$^{-2}$ s$^{-1}$) |
|--------|-----------------|----------------|-----------------------------------|
| Air | 1.3 | 330 | $0.4 \times 10^3$ |
| Water | $10^3$ | $1.5 \times 10^3$ | $1.5 \times 10^6$ |
| PVDF | $1.8 \times 10^3$ | $2.2 \times 10^3$ | $2.5 \times 10^6$ |
| Perspex | $1.2 \times 10^3$ | $2.7 \times 10^3$ | $3.2 \times 10^6$ |
| Quartz | $2.6 \times 10^3$ | $5.7 \times 10^3$ | $1.5 \times 10^7$ |
| PZT (ceramic) | $7.5 \times 10^3$ | $3.8 \times 10^3$ | $3 \times 10^7$ |
| Stainless steel | $5.8 \times 10^3$ | $3.3 \times 10^3$ | $4.7 \times 10^7$ |

## 9.1.4 Acoustic Impedance

The acoustic impedance is defined as the ratio between the (sinusoidal) acoustic pressure wave $p$ and the particle velocity $u$ in that wave. For a sound wave that propagates only in one direction, the acoustic impedance $Z$ is found to be:

$$Z = \frac{p}{u} = \sqrt{\rho \cdot c} = v_a \cdot \rho \tag{9.8}$$

Table 9.2 displays typical values for various materials. Clearly air has a very low acoustic impedance; liquids and solids are much 'harder' materials.

The acoustic impedance is an important parameter with respect to the transfer of acoustic energy between two media. Similar to light waves which show reflection and refraction on the interface of two media with different optical properties (i.e. refractive index, dielectric constant and conductivity), sound waves are also reflected at the interface of two media with different acoustic properties. When a sound wave arrives at such a boundary, part of the acoustic energy is reflected back and the remaining part enters the other medium. The amount of reflected or transmitted energy can be expressed in terms of the acoustic impedances. For sound waves Snell's law applies to calculate the direction and amplitude of the refracted sound. In the special case of a wave perpendicular to the boundary plane, the ratio between reflected and incident power (the reflection ratio) is given by

$$R = \left(\frac{Z_1 - Z_2}{Z_1 + Z_2}\right)^2 \tag{9.9}$$

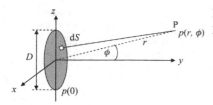

**Figure 9.1** Calculation of the acoustic field for a piston type emitter.

$Z_1$ is the acoustic impedance of the medium of the incident wave, $Z_2$ that of the medium behind the boundary plane. Obviously there is no reflection (hence total transmission) if both media have equal acoustic impedance. On the other hand when the impedances show a strong difference, the power transmission is small.

This has two important consequences for the S/N ratio in ToF measurements where emitter and receiver are both facing the reflecting surface. First an airborne sound wave that arrives at a solid object reflects in the same way as light on a mirror. Waves with an oblique angle of incidence reflect away from the direction of the emitter, so the reflected sound does only partially arrive at the receiver which is usually placed very near the emitter. Secondly transmission from an acoustic 'soft' material to an acoustic 'hard' material is poor. This applies in particular for acoustic sensors in air applications because of the wide gap between impedances of air and most sensor materials. Since many systems consist of a source and a detector, the power loss occurs twice, resulting in a low S/N ratio in acoustic sensing systems.

# 9.2  Acoustic Sensors

## 9.2.1  General Properties

Commonly used acoustic sensors belong to one of the following types:

* piezoelectric
* electrostatic
* electromagnetic and
* magnetostrictive.

The first two types enjoy great popularity in various ultrasonic applications (frequencies above 20 kHz). The transduction effects are reversible, which means that an acoustic transducer might be used as a source (transmitter) as well as a sensor (receiver). ToF systems may have both a transmitter and a receiver, or only one transducer alternately in use as a transmitter and a receiver.

We will now derive an approximated expression for the spatial distribution of the acoustic power emitted by an ultrasonic transducer. A common model for a transducer, used for this purpose, is the plane piston model: a circular disc vibrating in thickness mode (Figure 9.1).

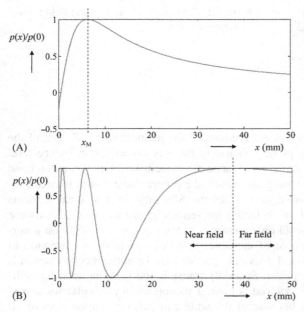

**Figure 9.2** Axial pressure distribution for $D = 15$ mm and wavelength (A) 8 and (B) 1.5 mm, respectively.

Each element d$S$ of the disc surface acts as a point source for acoustic waves. The sound intensity in any point $P$ in the hemisphere around the piston can be calculated by summing all contributions to the pressure variations in that point from each of the point sources d$S$. At places where the sound waves are (partly) out of phase, the sound intensity is low; at places where the waves are in phase, the sound intensity is high. This results in an interference pattern of acoustic waves in front of the transducer: the acoustic pressure appears to have peaks and lows in particular directions and at particular distances from the transmitter.

For simplicity we will consider two special cases: the pressure along the main axis of the transducer ($\phi = 0$), and the pressure at a distance $r$ far from the centre, for arbitrary angle $\phi$.

The acoustic pressure along the main or acoustic axis $p(x)$ is given by [2]

$$p(x) = 2p(0)\sin\left\{\frac{\pi}{\lambda}\left[\sqrt{\frac{D^2}{4} + x^2} - x\right]\right\} \tag{9.10}$$

where $p(0)$ is the amplitude of the acoustic pressure on the plate surface ($x = 0$) and $x$ the distance from the plate on the main axis. This is just the result of integrating the contributions to the pressure from all surface patches d$S$. Figure 9.2 shows a picture of the relative pressure along the acoustic axis for two cases: $D = 15$ mm, $\lambda = 8$ mm and $D = 15$ mm, $\lambda = 1.5$ mm. Close to the emitter the pressure shows strong pressure variations with distance. The place of the last maximum, at distance

$x_M$ from the plate, is taken as the boundary between the *near field* (or Fresnel zone) and the *far field* (or Fraunhofer zone) of the transducer.

Extreme values of $p(x)$ in Eq. (9.10) occur for

$$x_{\text{extr}} = \frac{1}{2}\left[\frac{D^2}{4\lambda[k+(1/2)]} - \lambda[k+(1/2)]\right] \quad k = 0, \pm 1, \pm 2, \cdots \tag{9.11}$$

The maximum for the largest value of $x$ is found for $k = 0$ and amounts:

$$x_M = \frac{D^2 - \lambda^2}{4\lambda} \tag{9.12}$$

At frequencies for which the wavelength is small compared with the dimension of the transmitter, the near field covers a range which is roughly $D^2/4\lambda$. The typical application example of Figure 9.2A with $D = 15$ mm, $\lambda = 8$ mm (40 kHz in air) shows that the near field ranges about 6 mm only. It means that in most mechatronic applications (in particular ToF measurements in air) we only need to consider the propagation in the far field.

An important sensor characteristic is the *directivity diagram*, representing the sound pattern in the far field. For a circular piston this directivity diagram is rotationally symmetric and can be represented in a plane. Imagine a sphere (or circle) with radius $r$ around the centre of the transmitter. The pressure distribution over this sphere can be described by the expression [2, p. 78]

$$p(r, \phi) = 2p(r)J_1\left(\frac{\pi D}{\lambda}\sin\phi\right)\left(\frac{\pi D}{\lambda}\sin\phi\right)^{-1} \tag{9.13}$$

where $p(r)$ is the pressure at distance $r$, $\lambda$ the wavelength of the sound signal, $D$ the plate diameter and $J_1(\pi D \sin \phi/\phi)$ is an order $-1$ Bessel function. The diagram consists of several 'lobes', a main lobe and side lobes. The width of the main lobe follows from Eq. (9.13). An approximated value for the *half angle* $\phi_h$ (or *angle of divergence* or the *cone angle*) of the main lobe is given by [1, p. 73]

$$\sin \phi \approx 1.22 \cdot \frac{\lambda}{D} \tag{9.14}$$

Apparently the ratio $D/\lambda$ determines the direction selectivity of the transducer: a narrow beam is achieved by a large ratio ($D \gg \lambda$). The theoretical directivity diagram derived by Eq. (9.13) is just a very rough approximation, since practical transducers do not behave as a piston. Figure 9.3 gives the directivity diagram of two real transducers, a piezoelectric and an electrostatic transducer (see Section 9.2).

The smaller the size of the transducer (characterized by the diameter $D$), or the larger the wavelength $\lambda$, the more it behaves as a point source. For a narrow beam

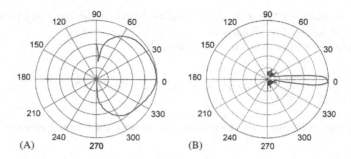

**Figure 9.3** Directivity plots of (A) a wide beam and (B) a narrow beam transducer.

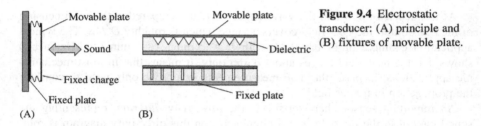

**Figure 9.4** Electrostatic transducer: (A) principle and (B) fixtures of movable plate.

(required for optimal S/N ratio), the transducer should be large compared to the wavelength. This means, to obtain narrow beams, we have to choose either a large transducer (which might be impractical in many applications) or a high frequency (resulting in increased damping and therefore a reduced range).

Since actual transducers may have shapes that differ from the circular piston model, the pressure field patterns have much more complicated shapes; see for instance Ref. [3]. By special design the beam angle can be made smaller, resulting in so-called focussed probes [2, chapter 5].

We will now resume the operation principle and properties of two types of acoustic sensors most encountered in mechatronics: the electrostatic and the piezoelectric transducers.

### 9.2.2 Electrostatic Transducers

An electrostatic (or capacitive) ultrasonic transducer consists of two flat conductive plates, one fixed to the housing, the other movable relative to the fixed plate, together constituting a flat-plate capacitor (Figure 9.4A). For this construction the equations $V = Q/C$ and $C = \varepsilon_0 \varepsilon_r A/d$ apply (Chapter 5).

In the *receiver mode* the moving plate is charged with a more or less constant charge, which is realized by connecting the plate via a resistor to a rather high DC voltage of some hundred volts. Sound waves (moving air particles) bring the plate into motion. Due to the constant charge and the changing plate distance, the voltage across the plates varies in accordance with the sound wave. To obtain a high sensitivity the plate is made very thin and flexible, resulting in an enhanced acoustic coupling to the air (see Section 9.1.4 on acoustic impedance). Obviously the sensitivity is also proportional to the DC voltage on the movable plate. The necessity of an external high voltage is the major drawback of the electrostatic transducer.

An alternative to such an external voltage is the electret version in which a charge is permanently stored in a sheet of a suitable piezoelectric material, making up the fixed plate of the capacitor. Such electret (or electret condenser) microphones for air-coupled applications operate mainly in the audible range (20 Hz−20 kHz).

In the *transmitter mode* an AC voltage is put on the movable plate of the capacitor. A voltage difference $V$ between the plates results in a force equal to

$$F = \varepsilon S \frac{V^2}{d^2} \tag{9.15}$$

with $S$ the surface area of the plates and $d$ the gap distance. Since this electrostatic force (explaining the name of this type of transducer) is proportional to the square of the voltage, it is always contractive, driving the movable plate towards the fixed plate until it is in equilibrium with the counterforce of the elastic bond. Should the plate move linearly proportional to an applied AC voltage (to generate the proper sound waves), the capacitor should be biased with a DC voltage higher than the amplitude of the AC voltage. Equation (9.15) also shows that the sensitivity of the emitter increases with decreasing initial plate distance.

To make a robust transducer with small initial gap and a thin movable plate (membrane), the latter is sustained not only at its edges, but over a larger part of the surface (Figure 9.4B), accomplished by, for instance, a grooved plate or other isolating structures like a grid or a net [4]. Obviously the membrane only vibrates at places where it is free to move. This multi-mode vibration explains the large deviations in the directivity diagrams of real electrostatic transducers as compared to those given in Figure 9.3. Moreover the layout of the structure strongly determines the frequency characteristics of the transducer [5]. On the other hand it allows designers of electrostatic transducers to optimize transducer characteristics for particular applications.

Like many other sensors ultrasound transducers can also be manufactured using silicon technology. The membrane in such devices is made from silicon by anisotropic etching, similar to the piezoresistive pressure sensors discussed in Chapter 4. It acts as the movable (vibrating) electrode of a capacitor-like structure [6,7]. An important advantage of this technology is the ability to make large numbers of membranes on a single chip. Silicon membrane sensors have a smaller bandwidth, due to (multi-mode) resonance frequencies of the mechanical structure. The output power and sensitivity are substantially less than the macro-devices, because of the much smaller dimensions and deflections of the micro-membranes.

### 9.2.3  Piezoelectric Transducers

A piezoelectric acoustic transducer consists of a piece of piezoelectric material (Chapter 8), configured as a flat-plate capacitor (Figure 9.5). An AC voltage

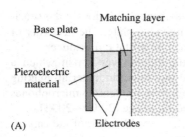
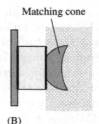

**Figure 9.5** Piezoelectric transducer for ultrasound applications (A) in liquid and (B) in air.

applied to the crystal causes the material to vibrate, resulting in the generation of acoustic energy. Conversely when the crystal is deformed by incident sound waves, a piezoelectric voltage is induced. Hence the transducer might be used in transmitter mode and in receiver mode, similar to the electrostatic transducer but without the need of a biasing DC voltage.

Irrespective of the application the transfer of sound energy from transducer to medium or vice versa is crucial for the S/N ratio of the measurement signal. As explained in Section 9.1.4 the transfer is governed by the acoustic impedances of the materials involved. Obviously since a piezoelectric ceramic material has a high acoustic impedance compared to air, the efficiency of power transfer to air is extremely poor. Also for a liquid or solid medium, power loss can be considerable too. In case of a solid, air voids between the transducer and the solid significantly lowers the transmission efficiency, a problem often encountered in acoustic NDT of objects. In such applications a coupling gel between the transducer and the (solid) object under test is essential.

There are many ways to improve the transmission of sound from the transducer to its acoustic environment. However efficiency is not the only characteristic of interest: bandwidth and beam shape are equally important parameters. Optimization is often a compromise between these parameters. Advanced models of the acoustic arrangement, based on transmission lines, are required to find an optimal solution for a particular application.

Basically there are two major approaches: matching layers and shaping. The first is usually applied for liquid and solid contacts (Figure 9.5A). Two essential parameters of such a layer are the acoustic impedance and the thickness. The acoustic impedance of the matching layer should have a value between that of both materials at either side of this layer. According to very simple models, the optimum value is the harmonic mean of these two values, but in practice other values appear to be more effective [2, chapter 4]. The thickness of the plate should be equal to a quarter of the sound wavelength, to minimize the reflection at the front side of the matching layer, in favour of the transmitted waves. Having found the optimum impedance value, the next problem is finding a material having that particular impedance. Matching layer design is still the subject of continuous research, including the search for suitable materials, the use of multiple matching layers and backing layers, extension of the frequency range and manufacturing issues – for example the influence of the glue between the layers [8] and the use of silicon-compatible technologies.

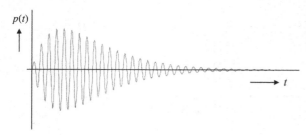

**Figure 9.6** Typical response of a PE transducer to an input burst signal.

To overcome the enormous mismatch in airborne applications, the transmission efficiency is improved by placing a precisely shaped piece of an acoustically soft material on the front side of the ceramic element, for instance a tiny horn as shown in Figure 9.5B. A horn or cone also affects the frequency range as well as the radiation field of the transducer [1, p. 102]. Most commercial low-cost piezoelectric transducers are provided with such a horn.

Poled PVDF has a relatively low acoustic impedance (see Table 9.2), making it an attractive material for ultrasound applications. The flexibility of the material allows operation in bending mode, whereas the ceramic sensors usually operate in thickness mode [9,10]. Other piezoelectric materials are being studied for their suitability as acoustic transducer, for instance porous films with artificially electric dipoles [11].

Piezoelectric transducers operate best at resonance: the high stiffness of ceramic materials results in a narrow bandwidth (in other words, a high mechanical Q-factor). The resonance frequency is determined by the dimensions of the crystal and, to a lesser extent, by the matching elements. Popular frequencies are 40 and 200 kHz. Many acoustic detection systems use tone burst signals. A typical response of a piezoelectric transducer to a burst signal at the resonance frequency is shown in Figure 9.6. Apparently the output burst shows strong distortion at the edges, due the narrow bandwidth of the device. This has important consequences for the determination of the ToF measurement, as will be explained in Section 9.3.

### 9.2.4 Arrays

As shown in Section 9.2.1 the beam width of a sound emitter is determined by its lateral dimensions and the frequency of the emitted sound signal. With a properly designed (linear) array of transducers a much narrower beam can be obtained, using spatial interference of the individual acoustic signals (Figure 9.7).

For points in the hemisphere where the travelled distances differ a multiple of the wavelength, the waves add up (for instance in $P$). In other points, as in $Q$, the waves (partly) cancel. Moreover the direction of this beam can be controlled over a limited range ($\pm30°$) with the phase difference of the applied electrical signals. Similarly the sensitivity angle of an acoustic transducer operating in receiver mode can be narrowed using an array of such receivers. At the receiving side the individual outputs are added after a properly chosen time delay. By controlling the delay

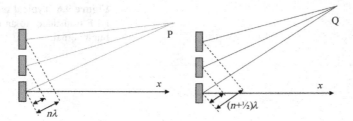

**Figure 9.7** Beam narrowing and steering by phased arrays.

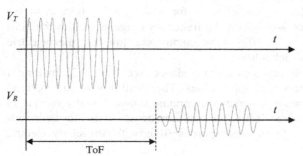

**Figure 9.8** Distance determination based on ToF.

times the main sensitivity axis can be varied over a limited angle. The application of this general principle to acoustic transducers has already been described in earlier literature [12−15]. Due to improved technology phased arrays are gaining interest for acoustic ranging and imaging [16−19]. They eliminate the need for a mechanical (hence slow) scanning mechanism but have a limited angular scanning range. In general at an increasing deflection angle the amplitude of the side lobes (Figure 9.3) increases as well, deteriorating the beam quality.

## 9.3   Measurement Methods

The majority of acoustic distance measurement systems are based on the ToF method. A measurement based on distance-dependent attenuation (as with light) is rather susceptible to environmental influences and the absorption properties of the objects involved. In most acoustic ToF systems the elapsed time between the excitation of a sound pulse and the arrival of its echo is measured (Figure 9.8). Since the transduction effects are reversible, the transducer can be switched from transmitter to receiver mode, within one measurement. The travelled distance $x$ follows directly from the ToF $t$, using the relation $v_a = x/t$ (for direct travel) or $v_a = 2x/t$ for echo systems (where the sound wave travels twice the distance).

The sound signal can be of any shape. Most popular are the burst (a number of periods of a sine wave), a continuous wave with constant frequency (CW) and an

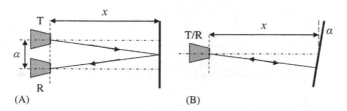

**Figure 9.9** ToF inaccuracies due to (A) separate transmitter and receiver and (B) tilted reflecting surface.

FM-modulated sine wave (FMCW or 'chirp'). We discuss the major characteristics of these three types. Pulses and bandwidth limited Gaussian noise are other possibilities.

### 9.3.1 Burst

The tone burst is the most common signal type used for acoustic distance measurements in air, in particular for applications where high accuracy is not required (e.g. ranging, navigation and obstacle avoidance). A transmitter emits a short burst, a few up to about 10 periods of a sine wave, in the direction of an object. The sound reflects (in accordance with the reflection properties of the object's surface and its orientation, as discussed in Section 9.1), travels back and is received by the same or a second transducer, which detects the moment of arrival.

Due to various causes the accuracy of the ToF measurement — and so the distance measurement — is limited. First there is noise of electrical and acoustic origin, which becomes important at larger distances, when the echo signal is low due to divergence of the sound waves. Next, spurious reflections can also mask the arrival time of the main echo. A further cause of inaccuracy is the (unknown) orientation of the reflecting surface relative to the main axis of the transducers. Finally the narrow bandwidth of a piezoelectric transducer makes the edges of the burst less pronounced, obstructing an accurate detection of the starting point of the pulse. Some but not all of these causes are typical for the burst method.

In case of an individual transmitter and receiver with spacing $a$ (Figure 9.9A) the measured distance $x'$ is:

$$x' = \sqrt{\frac{a^2}{4} + x^2} \tag{9.16}$$

from which follows a relative error equal to

$$\frac{\Delta x}{x} \approx \frac{1}{8}\frac{a^2}{x^2} \tag{9.17}$$

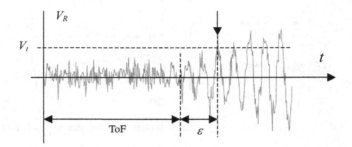

**Figure 9.10** Noise may introduce large errors in the measured ToF.

So, for a distance $x$ 3.5 times the space between T and R, the relative error is 1% only, which is acceptable in most practical situations.

A tilted reflecting surface (Figure 9.9B) introduces a typical cosine error of $-\alpha^2/2$, with $\alpha$ in radians: only waves normal to the surface will return to the receiver. As an example, a 1% error occurs when the tilt angle is about 8°. Note, however, that the sound intensity is reduced according to the directivity. When in this numerical example the half angles of both R and T are also 8°, the intensity of the echo has dropped by a factor of 4.

When a single transducer is used, switching from transmitter to receiver mode is only possible after the complete burst has been transmitted. This limits the minimum detectable distance: there is a dead band or dead zone, equal to half the number of wavelengths (because sound is travelling twice the distance). Shortening the pulse duration reduces this dead band, but also the sound energy, and hence the S/N ratio is reduced.

The response of a piezoelectric transducer on a burst (Figure 9.6) shows a rather large rise time due to the high quality factor of this transducer type. Actually this effect occurs twice: first at the transmitter and a second time at the receiver. The starting point of the echo is then easily obscured by noise.

The most common detection method for the echo burst is a threshold applied to the receiver signal. The ToF is set by the first time the echo exceeds the threshold. The threshold must be well above the noise level, to prevent noise being interpreted as the incoming echo. On the other hand when the threshold is too high, one or more periods of the received signal can be missed, as shown in Figure 9.10. At 40 kHz this results in an error of (a multiple of) about 8 mm, which is the penalty for this simple signal processing scheme.

Many methods have been published over the past years, all aiming at a higher accuracy of the burst ToF measurement, despite noise and other interfering effects such as temperature changes and air turbulence. One such approach is to use the complete echo signal and not just the starting point to determine the ToF, for instance by autocorrelation or cross correlation [20,21]. These techniques yield typical errors below 1 mm, depending on the S/N ratio and the travelled distance but need longer processing time. Another approach is based on knowledge about the

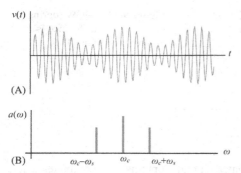

**Figure 9.11** AM-signal: (A) time domain and (B) frequency spectrum.

shape of the echo's envelope, which is determined by the transducer's transfer function [22]. Alternatively instead of the starting point of the echo signal, where the S/N ratio is low, the arrival time of a point around the maximum amplitude could be chosen as a measure for the ToF. With this method the typical distance error can be less than 0.1 mm [23]. Other researchers use the phase information comprised in the echo signal, the typical shape of the envelope, or just more advanced signal processing techniques [24–28]. Noise as the input signal is studied in Ref. [29]. A comparison of various detection methods is presented in Ref. [30].

### 9.3.2 Continuous Sine Wave (CW)

The average output power of a burst is relatively small, and may result in a poor S/N ratio, in particular at increased distances. In this respect the emission of a continuous acoustic signal is better. Distance information is obtained from the phase difference between the transmitted and the received waves. This method has two main drawbacks. First emission and detection cannot share the same transducer, hence two transducers are required, making the system larger and more expensive. It also introduces crosstalk because sound waves propagate through the construction directly from the transmitter to the receiver giving signals that may be larger than the echo itself. Secondly the unambiguous measurement range is only one period of the sound wave. The resolution, however, can be very high: for instance one degree phase resolution and 7 mm wavelength corresponds with a distance resolution of 19.4 $\mu$m. Further there is continuous information about the measured distance. So for distance control over a short range, the CW method is preferred over the burst method [31].

The unambiguous range can easily be enlarged by amplitude modulation (Figure 9.11A). Suppose the transmitted signal is an AM signal described by the general expression:

$$V_T = A(1 + m \cos \omega_s t)\cos \omega_c t \tag{9.18}$$

where $\omega_c$ is the carrier frequency, $\omega_s$ the modulation frequency, and $\omega_s \ll \omega_c$. When using piezoelectric transducers the carrier frequency should equal the resonance

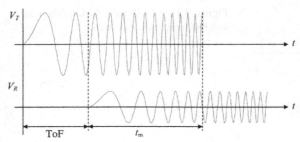

**Figure 9.12** FMCW; during $t_m$ the frequency difference is proportional to the ToF.

frequency. The spectrum of this modulated signal contains the components $\omega_c$, $\omega_c + \omega_s$ and $\omega_c - \omega_s$ (Figure 9.11B).

All components should fall within the bandwidth of the transducer, even when using a piezoelectric type. It means that the period time of the 'envelope' is much larger than the wavelength of the carrier. The received wave, delayed over the ToF $\tau$, is written as

$$V_R = A\{(1 + m \cos \omega_s(t - \tau)) \cos \omega_c(t - \tau)\} \tag{9.19}$$

The phase difference of the carrier is $\phi_c = \omega_c\tau$, that of the envelope $\phi_s = \omega_s\tau$. So for a ToF corresponding to one period $(2\pi)$ of the envelope, the carrier has shifted over $\omega_c/\omega_s$ as many periods. Obviously the unambiguous range is increased by a factor of $\omega_c/\omega_s$ when using the phase of the envelope signal (to be reconstructed by for instance a rectifying circuit). The method has the advantage of the continuous mode, but with enlarged range. Here, also, the accuracy in the ToF measurement is limited by the resolution of the phase measurement and the uncertainty in sound velocity, according to Eq. (9.1). In Ref. [32] the carrier and modulation frequencies are 40 kHz and 150 Hz, respectively, corresponding to a low-frequency period of about 2 m. Using a digital phase detection circuit, an accuracy of 2 mm over a 1.5 m range was obtained.

### 9.3.3 Frequency-Modulated Continuous Waves (FMCW)

The unambiguous range of the continuous mode can be increased by frequency modulation of the sound wave. Figure 9.12 shows the transmitted and the received signals (sometimes called a 'chirp', referring to the whistling of some birds).

Assume the frequency varies linearly with time, starting from a lowest value $f_L$, and increasing with a rate $k$ (Hz/s): $f(t) = f_L + k \cdot t$. Due to the time delay of the reflected wave, its frequency in the same time frame is $f(t) = f_L + k \cdot (t - \tau)$. At any moment within the time period $t_m$ where the signals occur simultaneously, the frequency difference between the transmitted and the received wave equals:

$$\Delta f = f_L + k \cdot t - \{f_L + k(t - \tau)\} = k \cdot \tau \tag{9.20}$$

with $\tau$ the ToF. The distance to the reflecting surface follows from

$$x = \frac{1}{2}v_a\tau = \frac{v_a}{2k} \cdot \Delta f \tag{9.21}$$

Apparently this distance is directly proportional to the frequency difference and unambiguous over a time $t_m$ as illustrated in Figure 9.12. The CWFM method combines the advantages of the first two methods: the transmitted signal is continuous (favourable for the S/N ratio, uninterrupted distance information) and the range is larger (determined by the frequency sweep). Disadvantages over the burst method are the more complex interfacing and the fact that only wide-band transducers can be applied since the frequency varies. Further since transmission is essentially continuous, two transducers are required for simultaneous transmitting and receiving.

Important parameters of a CWFM system are the sweep length, the start and stop frequencies and the sweep rate $k$ (in Hz/s). The sweep length should correspond to the largest distance to be measured, to guarantee sufficient time overlap between the transmitted and reflected waves. The start and stop frequencies are set by the characteristics of the transducer. Electrostatic transducers for low-cost applications have a rather limited frequency range, typically from 50 to 100 kHz, so just one octave (a factor of 2). Usually the size of an electrostatic transducer in this frequency range is larger than that of a piezoelectric transducer. This may be a disadvantage from the viewpoint of construction; however, as has been shown in Section 9.2.1, the half angle is much smaller, contributing to a better S/N ratio of the echo signal.

According to Eq. (9.21) the distance measurement requires the determination of the frequency difference between transmitted and received signals. Usually this is done by some correlation procedure, performed in the digital signal domain (see for instance Refs [33,34]). This has the advantage of noise reduction (depending on the type of filtering), but the disadvantage of the need to digitize and sample both signals. However the procedure is essentially based on the multiplication of both signals, followed by some filtering. The basic goniometric relation of Eq. (9.22) is used:

$$\cos \alpha \cdot \cos \beta = \frac{1}{2}\{\cos(\alpha + \beta) + \cos(\alpha - \beta)\} \tag{9.22}$$

So multiplying two sine waves results in a signal containing sum and difference frequencies. Now let the transmitted FM signal be:

$$x_T(t) = A \cos[2\pi(f_L + k \cdot t)t] \tag{9.23}$$

The reflected wave signal delayed over a time $\tau$ becomes:

$$x_R(t) = B \cos[2\pi\{f_L + k(t - \tau)\}(t - \tau)] \tag{9.24}$$

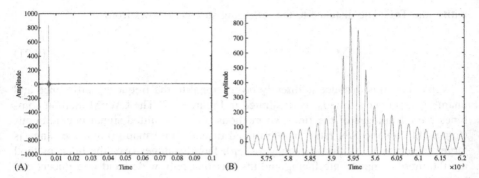

**Figure 9.13** Example of a CWFM received signal, after being processed with a matched filter: (A) time scale equal to the sweep period and (B) zoomed in at the peak value.

The product of these two signals contains components with sum and difference frequencies. Since we are only interested in the frequency difference, the much higher frequencies are filtered out and what remains is of the form

$$
\begin{aligned}
AB \cos 2\pi[(f_L + k\cdot t)t - \{f_L + k(t-\tau)\}(t-\tau)] \\
= C \cos 2\pi\{2k\cdot t\cdot\tau + f_L\tau - k\cdot\tau^2\} \\
= C \cos 2\pi\{2\Delta f\cdot t + \phi(\tau)\}
\end{aligned}
\tag{9.25}
$$

This clearly shows that multiplication produces a signal containing the frequency difference. Obviously for an accurate determination of this frequency all other components due to the multiplication procedure as well as noise should be sufficiently suppressed by some method of filtering. A treatment of the various signal processing algorithms goes beyond the scope of this book. To get an impression of a possible output Figure 9.13 presents the result of a ToF-measurement based on CWFM with matched filtering, applied to the signal reflected from a flat surface at 1 m distance. The horizontal time scale corresponds to the delay time (or ToF). In Figure 9.13A the full scale is just the cycle period (100 ms). At around 0.006 s a pronounced maximum is noticeable. In the zoomed version Figure 9.13B this peak appears to occur at 5.95 ms, corresponding with a distance of 1.02 m (at a measured sound frequency of 343 m/s). This figure also shows that the resolution of the measurement is limited: due to noise it may happen that an adjacent peak has a larger amplitude, introducing an error in the ToF determination.

Whether burst mode or continuous mode should be applied strongly depends on the requirements and the application. In general the burst method is simple to realize but is discontinuous and noise sensitive. CWFM is less sensitive to noise but requires more complicated interfacing and signal processing [35,36]. Generally speaking the CWFM method outperforms the burst method in terms of range and noise immunity.

### 9.3.4   Other Signal Types

Common to all methods considered thus far is the accurate determination of the time delay of an acoustic wave. Apart from the three signals discussed, many other signal types could be used to measure the ToF. Assuming the total transfer from transmitter, acoustic medium and receiver is linear (which is the case in the majority of the applications considered here), any signal containing some time-discriminating parameter will do.

A short, needle-like pulse is one of these. However the bandwidth of most transducers is too small for handling such signals and, moreover, for air applications the energy content is too low. Gaussian noise is another one. The autocorrelation of wide-band Gaussian noise is a narrow peak. When such a noise signal is applied to the transmitter, cross correlation with the received signal results in a narrow peak in the correlogram at a position equal to the ToF. Limitations of this method are mainly of a practical nature: the noise signal and the transducers are always band limited, while sampling, AD conversion, and the finite measurement time further limit the resolution and accuracy of the digitally processed signals.

Another approach is phase shift or frequency shift keying. The signal is a (sinusoidal) carrier modulated in phase or frequency by a binary coded signal. Such signals are now widely used in telecommunication, but may also play a role in acoustic distance sensing. An example of an application of binary phase shift keying signals for acoustic distance measurement in air is given in Ref. [37].

## 9.4   Applications

The versatility of ultrasound sensing is widely recognized. Ultrasonic sensors are extensively used in the processing industry for a variety of parameters [38], but they can also be found in numerous other fields, such as transportation, medical diagnostics, material research and so on. The focus of this chapter is on the application in mechatronics, for a variety of tasks, with distance as the basic measurement quantity. These tasks comprise movement control in production machines, obstacle avoidance in traffic, navigation of AGVs, object recognition and inspection and many more. Acoustic sensors are cheap, small and easily mountable on or in a mechatronic construction. When moderate distance accuracy is required, the interfacing can be kept simple. All measurements are based on the ToF principle as outlined in the previous section. We review two main tasks: navigation and inspection.

### 9.4.1   Navigation

#### General

Acoustic transducers are suitable for short-range navigation in structured or unstructured (unknown) environments. Application examples are mobile robots (e.g. path finding and collision avoidance), aids for visually impaired people and

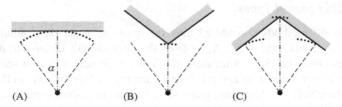

**Figure 9.14** Sonar maps from a scanning echo location system.

localization of tools in a work space. Common to these applications is the extraction of 2D or 3D information about the relative position of persons and objects in a mostly unconditioned environment. The easiest task is *collision avoidance*, because the demands with respect to accuracy and speed are not severe. Depending on further requirements various options can be considered. A simple anti-collision system for robots and other vehicles like cars and wheelchairs consists of a number of fixed transducers at the front or around the circumference of the vehicle. The method works well if the reflecting obstacle has sufficiently large dimensions (at least a few wavelengths of the sound signal) and is oriented perpendicularly to the main axis of the transducer and in the absence of multiple echoes.

When the reflecting surface is not perpendicular to the main axis of the transducer, an error is introduced (compare Figure 9.9). As a consequence the range data obtained by an acoustic scanning system show the typical behaviour as seen in the sonar maps in Figure 9.14. Simulations of such errors are extensively discussed in literature [39].

In Figure 9.14 the transmitter/receiver combination rotates over an angle $\alpha$ with respect to the surface normal, and the dots represent measurement points. Range data are only available when the echo signal exceeds the detection threshold. Assume that this happens only when the reflected beam falls within the half angle of the receiver (an arbitrary choice because the threshold level may be adjusted to any other value). In Figure 9.14A range data lie on an arc spanning twice the half angle of the transducers. In case of a rather sharp edge as in Figure 9.14B, the acoustic wave reflects away from the receiver, so no echo is received. In practice, however, such an edge gives an echo regardless, but only over a small scanning angle. In Figure 9.14C the received echo of a wave transmitted into the direction of the edge is mainly the result of multiple reflections in that edge. As can be seen in Figure 9.14B and C even a small rotation angle results in a rather large distance error. Obviously in these cases the detectivity depends greatly on the angle between the two surfaces.

Erroneous distance data can be avoided by narrowing down the beam angle of the transmitter and receiver, but this is at the expense of lost data points (dead angles).

Similar to the optical system the ultrasonic counterpart may be disturbed too by spurious signals from other sources or unwanted reflections. In a natural, unconditioned environment, spurious echoes are unavoidable.

When such echoes interfere with the main echo, phase information is easily corrupted, even when the amplitude of the unwanted echo is small. Systems using ToF

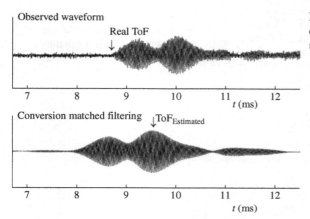

**Figure 9.15** Interfering echo, causing a peak time shift error [40].

determination by thresholding detection are rather insensitive to this phenomenon: time windowing may suppress most of the unwanted echoes since the direct wave arrives first in time. In systems based on envelope detection such spurious echoes may introduce large ToF errors, as shown in Figure 9.15. More advanced detection methods are required to overcome this problem [40,41].

The performance of acoustic echo location in air is further determined by the power and frequency of the emitted pulse. Typical working distances of low-cost anti-collision systems range from tens of cm (limited by the dead zone) to a few m.

## Navigation for Mobile Robots

Ultrasonic rangefinding is an excellent technique for the exploration of an unknown environment, to find the position and orientation of a robot in a known environment and to navigate. There is an overwhelming number of papers on this topic, dealing with many kinds of system concepts and signal processing algorithms. Most of these concepts are applied for autonomous navigation without beacons. Navigation with acoustic beacons is a low-cost alternative for the optical beacon systems discussed in Chapter 7.

Important design consideration is that the whole environment (e.g. walls, obstacles and persons) fall in the field-of-view of the sensing system, so not only in the direction of movement but also sideways and backwards. Three approaches can be followed:

1. many transducers fixed all around the robot
2. a rotatable T/R combination on top of the robot
3. fixed transducers while using the robot itself as scanner.

An example of the first approach is presented in Ref. [42], in which an angular accuracy of less than 1° using eight transducers is reported. An obvious advantage of a (mechanical) scanning system is the need for just one or a very few pair of transducers; the scanning, however, is slow. A further point of consideration is the

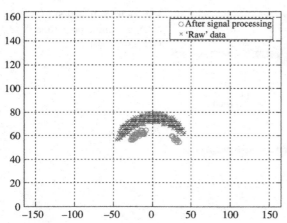

**Figure 9.16** Resolution improved by convolution; scales in cm (unpublished from laboratory experiments by the author and students).

directivity of the transducers: a high lateral resolution over a wide viewing angle (possibly up to 360°) requires a larger number of transducers with high directionality; the same coverage can be achieved with fewer transducers but with reduced lateral resolution.

Due to the relatively large wavelength of ultrasonic sound in air, the lateral resolution is rather low. This complicates the detection of small-sized objects or to distinguish between closely spaced objects, which is important in applications like a floor-cleaning robot [43] or navigation systems for blind persons. Enhancement of the lateral resolution without greatly increasing the number of transducers or their directivity (which is limited anyway) can be achieved by more advanced signal processing, for instance potential fields [44] (applied for wheel chair navigation) and a one-transmitter three-receiver configuration [45]. A well-known method is the convolution of the echo signals with the transmitted signals. This may potentially improve both the lateral and the radial resolutions of an echo-location system, as is shown in Figure 9.16. This figure shows the outcome of an experiment with two coffee cups placed on a flat table; the rotating piezoelectric transducers are positioned in the origin. Due to the wide beam angle the two cups are seen as one object. By deconvolution (using the known pulse response of the transducers) they can be distinguished separately.

Another approach to handling multiple objects is presented in Ref. [46] where the information from the echo amplitude is also used to construct a sonar map.

Navigation of vehicles or robots can be accomplished by a totally different approach based on beacons. Such beacons can be transmitting, receiving or passive. In case of receiving beacons the robot generates continuous or intermittent signals all around. From the signals received by the beacons the distance to the robot can be derived. With known beacon positions the position of the robot is easily derived (Figure 9.17). In general a minimum of three beacons is required to determine unambiguously the robot's 2D position in an open space.

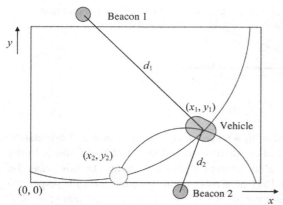

**Figure 9.17** Navigation with beacons; unambiguous position determination requires at least three beacons. In this example the vehicle's position is on one of the two possible intersections of two circles with radius equal to the measured distances from the two beacons.

The reverse also works: transmitting beacons send out unique signals, picked up by a robot equipped with an omnidirectional receiver. From these signals an on-board robot computer determines its distance from the beacons and subsequently its 2D position. A combination of these two possibilities is applied in an accurate docking system for Automated Guided Vehicles (AGVs) [47].

In a system with passive beacons the robot transmits and receives signals, reflected by the beacons. Concerning the beacons this is by far the most simple and flexible concept. They need no power supply and the only requirements are an accurately known position and sufficient reflectivity for the signals. A disadvantage is a higher demand on environmental conditions, to prevent spurious reflections.

## Navigation Tool for Visually Impaired Persons

This particular application of ultrasound sensing has attracted increased interest over the last few decades. Researchers have developed various concepts based on ultrasound ranging, like electronic dogs, canes, belts and hats, all comprising some acoustic ranging device [48–51]. Fixing the acoustic transducers somewhere on the body leaves the hands free; attaching the transducers on a person's head allows exploration of the environment in a more natural way. The detection area of the transducers may also be adapted to the immediate need and situation [52].

For this application it is important to not only avoid obstacles, but also to characterize or identify objects in the person's immediate environment. The FMCW method yields such information [53], but the number of classes is very limited.

The technical problems to overcome are similar to those encountered in robot navigation. A particular challenge is the demand for safe operation outdoors, or in a noisy, heavily unstructured and unconditioned environment. A completely different approach to many of these systems is still in the experimental phase, but there is noticeable progress in performance. With the availability of cheap cameras (webcams) nowadays, navigation tools are being developed based on a combination of acoustic ranging and vision.

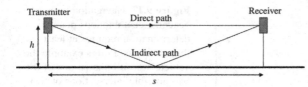

**Figure 9.18** Ground floor echoes.

## Localization of Tools

When a moving object is provided with a sound source, its position and orientation can be determined by using a set of acoustic sensors located at fixed points in the working space, similar to the beacon concept discussed earlier. In this way, the (3D) position of the *end-effector* [54] of a robot or its orientation with respect to a reference plane [55] can be determined. A different approach, presented in Ref. [56], combines a 2D vision system (fixed above the manipulator) and an acoustic distance sensor, together enabling 3D positioning of the robot gripper with respect to the object. Another example is the localization of surgical tools [57], where a position accuracy of 40 μm is reported.

In all these applications, special care should be taken to secondary echoes from floor, walls and other objects in the surrounding tools. They may cause phase errors and a shift in the envelope top of the received signal (see Figure 9.15), of which the consequences have already been discussed. Figure 9.18 illustrates a situation where reflections via the ground floor may cause interference errors.

Assume the sound frequency is 40 kHz, corresponding with a wavelength of 8 mm. When the path difference equals an odd number of half a wavelength (i.e. at 4 mm, 12 mm and so on) destructive interference of the received signals occurs. On the other hand at multiples of a full wavelength path difference (i.e. 8 mm, 16 mm and so on) constructive interference takes place. The differences are rather common in many practical situations. Since the amplitudes of the sound waves differ, interference is only partial, to an extent that depends on the angles with respect to the acoustical axis of the transducers and on the reflectivity of the surfaces. Serious ToF errors occur at smaller distances and in the vicinity of many reflecting objects.

### 9.4.2 Inspection

#### General

Compared to 3D vision the technique of acoustic ranging suffers from a much lower spatial resolution. Conversely an acoustic ToF measurement system can be implemented in a cost-effective way, provides depth information and is less susceptible to environmental conditions (except for temperature variations, see Section 9.1). This explains why ultrasound ranging techniques for object recognition have been studied since the first decade of industrial robotics; see for instance Refs [58], presenting a system based on ToF in the $z$-direction and spatial scanning using an $x-y$ table, and [59] (2D angular scanning). We discuss three particular approaches: acoustic imaging in air as an alternative for imaging by vision, NDT and inspection by touch using ultrasound-based tactile sensors. A series of examples for process monitoring using ultrasound concludes this chapter.

## Acoustic Imaging

The goal of any imaging system is the acquisition of an electronic picture from which information about an object's size and shape can be inferred, preferably including as many details as feasible. In an adverse environment in particular, acoustic imaging could be a good alternative for the well-known optical imaging with cameras. Since the spatial resolution of an acoustic distance measurement is relatively low, some method of scanning is required to obtain detailed shape information. The performance of an ultrasonic recognition system is limited by the nature of sound waves, the properties of the transducers, the quality of the scanning system, the number of transducers and the signal processing. In general the acquired image data are sparse and not accurate, which complicates recognition. However in many applications the resolution requirements are not severe.

Scanning can be performed in various ways. Either the objects move relative to fixed transducers, or the transducers are moving relative to a fixed object. In a robotics environment a robot can perform the scanning, thereby making use of the many d.o.f. of the manipulator to perform 3D scanning all around the object [60]. The encoders in the joints provide the orientation of the scanning range finder. A height map is obtained by Cartesian scanning in a plane above the scene. Sideways scanning enables the reconstruction of the circumference of an object, from the position and orientation of the flat or curved sides [61−63]. Several processing algorithms have been proposed to reconstruct the object shape from the obtained data or to classify the objects, including digital filtering, holographic algorithms and artificial neural networks [64−66].

In all these examples mechanical scanning is used. The recognition of objects from a fixed sensory system has been studied as well. Usually such systems comprise an array of transducers with relatively large spacing [67]. For an accurate characterization of large objects (3−10 m), a set of transmitters/receivers is positioned around the object in such a way that together they cover the whole area of interest. An accuracy analysis of such an approach can be found in Ref. [68]. One of the difficulties in such a system is to distinguish between multiple objects and between various shapes returning similar echo patterns. These problems are solved using a set of transmitters and/or receivers at known relative positions and proper signal processing.

For the classification of objects with known shape, the 'signature method' can be applied [62,69,70]. In this approach no attempt is made to reconstruct the object shape from the range data. Instead the system is taught to associate each object class with a characteristic echo pattern. A typical problem to be overcome in this approach is to obtain image data that are position and orientation invariant. On the other hand different but similarly shaped objects may result in echo patterns that are also similar. Correct classification on the basis of the echo patterns then requires special processing techniques, as described, for instance in Ref. [71]. For objects with clearly distinguishable echo patterns the method is quick, requires simple hardware and software and is cost effective.

## NDT and Material Properties

Sound waves reflect at the boundary of two materials with different acoustic impedances. This property allows the measurement of many geometrical parameters of an object, just by applying sound waves to the device under test and the analysis of the waves reflecting at existing boundaries. One of the applications is the detection and localization of defects (e.g. voids and flaws) in a solid, even when invisible at the outside. The detection of small defects requires waves with a correspondingly short wavelength and therefore a much higher frequency compared to applications in air, typically in the MHz range. To a certain extent the size and shape of defects can also be reconstructed from the echo pattern [72−74].

Another application area is the measurement of layer thickness; the acoustic technique is used in a wide spectrum of materials and products, ranging from (steel) plates [75] to layers of paint [76] and in a human tooth [77]. To achieve high accuracy advanced signal processing is required. As an example in Ref. [78] the thickness of a multilayered structure is measured with an accuracy less than 10 μm for the individual layers. With the touch sensor in Ref. [79], thickness and hardness are simultaneously measured by a combination of acoustic and capacitive sensing elements.

Where acoustic damping (see Section 9.1.3) is an unwelcome effect in most of the applications discussed so far, this property of sound can be used to identify, monitor or determine particular properties of a material. It is a suitable method in cases where the relation between the unknown parameter and the damping constant is known, for instance by an acknowledged model or by calibration. Again there are numerous examples to illustrate this technique. We give just one example: the measurement of bubbles in liquids [80]. Both the acoustic attenuation and the speed of sound are related to the volume ratio of the two phases. By measuring these parameters over a relevant frequency range, various liquid properties can be deduced or evaluated.

## Production and Process Control

The next list is an illustration of the versatility of acoustic sensing as an aid in production and process control; the references given are only examples of the many scientific papers published on these topics over the last decades.

- vibration measurements (example: a vibrating cantilever, phase detection of an 80 kHz sound wave [81]);
- seam tracking (example: short-range scanning with a MHz piezoelectric transducer with focussing surface [82]);
- respiratory air flow (example: two CW waves from piezoelectric transducers, with slightly different frequencies [83]);

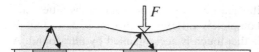

**Figure 9.19** Basic construction of an ultrasound ToF tactile sensor (after [93]).

- liquid level (examples: in bottles, 220 kHz piezoelectric sensors, burst mode [84]; as an acoustic resonator [85]);
- go/no-go decision on product quality (examples: quality of foundry pieces; signature imaging and Artificial Neural Network (ANN) for identification [86]; defect detection in food can production line: same method, according to Figure 1.4 [87]);
- air bubbles volume in liquids (example: PVDF-based T/R system, burst mode at 2 MHz bursts [88]);
- speed of sound measurement (example: acoustic resonance determination by a piezoelectric (ceramic) tube [89]);
- automotive (example: active suspension control; ToF measurement of backside vehicle to ground, 40 kHz burst mode [90]);
- body movement (example: head movement relative to a computer screen, using three transmitters on the head and three receivers near the screen [91]);
- paper roughness (using the roughness-dependent reflection of ultrasound [92]).

The number of applications where a ToF measurement can be used is almost unlimited, and the reader is referred to the vast amount of literature in books and journals on NDT for further examples and inspiration.

## Tactile Sensors

Tactile sensors are used either to extract shape information on the basis of touch or to provide haptic feedback to the operator of manipulators as for instance in surgery. A tactile sensor consists of a matrix of force-sensitive elements (taxels). The tactile force is transformed into a deformation of some elastic material, and this deformation is measured using some displacement measuring method. So a tactile sensor provides a pressure or deformation image of the object with which it is brought into contact. General design aspects of tactile sensors have been considered in Section 4.4.3 on resistive sensors, where the aspect of taxel selection was also discussed. In this section we focus on ultrasonic measurement of the deformation, which is mostly based on the ToF method. Evidently it requires an acoustic transmitter—receiver combination for each taxel and a means for selecting the individual taxels. Usually a piezoelectric material is used, allowing small dimensions of the taxels. PVDF has several advantages over ceramic materials: it is available in sheets, it is a flexible material (matching well with the shape of robot grippers and fingers) and it has a favourable acoustic impedance (Section 9.1.4).

One of the first sound-based tactile matrix sensors is described in Ref. [93]. Figure 9.19 shows the basic structure. The elastic top layer of this $3 \times 4$ matrix is responsible for the force-deformation transfer. Clearly the echo time changes proportionally to the compression of the layer and hence the applied force. The spatial resolution is about 1 mm. The ToF, determined by the layer thickness and the acoustic velocity, amounts to about 5 μs.

A similar construction, but with the elastic layer in between transmitter and receiver, is proposed in Ref. [94]. A completely different approach is described in Ref. [95]. The basic assumption here is that touch is accompanied by ultrasound. A quad PVDF sensor at the bottom of a sphere-shaped touch element enables the detection of this sound as well as the direction and hence the point of touch. The authors expect the sensor also be sensitive to slip. Indeed when an object is slipping this will generate vibrations, and hence sound; this information (as a binary sensor) can be used to control delicate gripping [96]. Another type of tactile sensor, presented in Ref. [97], is able to detect the friction coefficient, basically by measuring the resonance frequencies of an elastic cavity comprising a transmitter–receiver combination.

## Conclusion

This section on applications clearly demonstrates that the applicability of acoustic (ultrasound) sensors is almost unlimited. The major advantages of acoustic sensing over optical sensing are the lower sensitivity to dirt, smoke and environmental light; the instantaneously obtained depth information (in ranging); and the ability to sense inside an object (in NDT). Significant disadvantages are the low resolution and the temperature sensitivity (due to the temperature dependence of sound speed). The latter drawback can easily be eliminated by an additional temperature compensation. For very high accuracy one should also consider the effect of humidity on the speed of sound, and possible air turbulence, introducing noise. When designing a mechatronic system it is worthwhile to make an extensive comparison between the two basic sensing concepts: acoustic versus optical.

# References and Literature

## References to Cited Literature
[1] R.C. Asher: Ultrasonic sensors for chemical and process plant; IOP Publ., 1997; ISBN 0-7503-0361-1. p. 355.
[2] M.G. Silk: Ultrasonic transducers for nondestructive testing; Adam Hilger, Bristol, 1984; ISBN 0-85274-436-6. p. 76.
[3] A. Weyns: Radiation field calculations of pulsed ultrasonic transducers; Part 1: planar circular, square and annular transducers, Part 2: spherical disc- and ring-shaped transducers; Ultrasonics, July 1980, pp. 183–188 and September 1980, pp. 219–223.
[4] P. Mattila, J. Stor-Pellinen, J. Ignatius, J. Hietanen, M. Luukkala: Capacitive ultrasonic transducer with net backplate, Meas. Sci. Technol., 11 (2000), 1119–1125.
[5] P. Mattila, J. Hietanen: Bandwidth control of an electrostatic ultrasonic transducer, Sens. Actuators A, 45 (1994), 203–208.
[6] K. Inoue, Y. Suzuki, S. Ogawa: Fabrication of ultrasonic sensor using silicon membrane; Transducers '95 – Eurosensors IX, Stockholm, Sweden, 25–29 June, pp. 616–619.

[7] G. Caliano, A. Savoia, A. Caronti, V. Foglietti, E. Cianci, M. Pappalardo: Capacitive micromachined ultrasonic transducer with an open-cells structure, Sens. Actuators A, 121 (2005), 382−387.

[8] D. Callens, Chr. Bruneel, J. Assaad: Matching ultrasonic transducer using two matching layers where one of them is glue, NDT&E Int., 37 (2004), 591−596.

[9] F. Harnisch, N. Kroemer, W. Manthey: Ultrasonic transducers with piezoelectric polymer foil, Sens. Actuators A, 25−27 (1991), 549−552.

[10] L. Capineri, M. Calzolai, S. Rocchi, A.S. Fiorillo: A digital system for accurate ranging with airborne PVDF ultrasonic transducers; Proceedings of the ISMCR '95 Conference, Smolenice, Slovakia, 1995, pp. 71−74.

[11] A. Jiménez, Á. Hernández, J. Ureña, M.C. Pérez, F.J. Álvarez, C. De Marziani, J.J. García, J.M. Villadangos: EMFi-based ultrasonic transducer for robotics applications, Sens. Actuators A, 148 (2008), 342−349.

[12] W.L. Beaver: A method of three-dimensional electronic focussing and beam steering using electronic delay lines; Proceedings of Ultrasonics Symposium, 22−24 September 1975, Los Angeles, CA, pp. 88−90.

[13] R.A. Mucci: A comparison of efficient beamforming algorithms, IEEE Trans. Acoust. Speech Signal Process., ASSP−32(3) (1984), 548−558.

[14] J.T. Walker, J.D. Meindl: A digitally controlled CCD dynamically focussed phased array; Proceedings of Ultrasonics Symposium 22−24 September, 1975, Los Angeles, CA, pp. 80−83.

[15] K. Higuchi, K. Suzuki, H. Tanigawa: Ultrasonic phased array transducer for acoustic imaging in air; IEEE Ultrasonic Symposium, 1986, pp. 559−562.

[16] K. Yamashita, L. Chamsomphou, H. Nishimoto, M. Okuyama: A new method of position measurement using ultrasonic array sensor without angular scanning, Sens. Actuators A, 121 (2005), 1−5.

[17] R. Queirós, R.C. Martins, P.S. Girão, A.Cruz Serra: A new method for high resolution ultrasonic ranging in air; Proceedings XVIII IMEKO World Congress, Rio de Janeiro, Brazil, 17−22 September 2006.

[18] K. Yamashita, H. Katata, M. Okuyama: High-directivity array of ultrasonic micro sensor using PZT thin film on Si diaphragm; Transducers 01/Eurosensors XV, Munich, Germany, 10−14 June 2001.

[19] T. Yamaguchi, H. Kashiwagi, H. Harada: A multi-beam ultrasonic echo location system for multi-object environment; ISMCR'99, Tokyo, Japan, 10−11 June 1999; in: Proceedings of IMEKO-XV World Congress, Osaka, Japan, 13−18 June 1999, Vol. X, pp. 1−6.

[20] M. Parrella, J.J. Anaya, C. Fritsch: Digital signal processing techniques for high accuracy ultrasonic range measurements, IEEE Trans. Instrum. Meas., 40(4) (1991), 759−763.

[21] D. Marioli, C. Narduzzi, C. Offelli, D. Petri, E. Sardini, A. Taroni: Digital time-of-flight measurement for ultrasonic sensors, IEEE Trans. Instrum. Meas., 41(1) (1992), 93−97.

[22] J.M. Martín Abreu, R. Ceres, T. Freire: Ultrasonic ranging: envelope analysis gives improved accuracy, Sensor Rev., 12(1) (1992), 17−21.

[23] E. van Ginneken, P.P.L. Regtien: Accurate ultrasonic distance measurement using time-of-flight; AIM '92 (Australasian Instrumentation and Measurement Conference), Auckland, New Zealand, 24−27 November 1992, pp. 377−380.

[24] F.E. Gueuning, M. Varlan, C.E. Eugène, P. Dupuis: Accurate distance measurement by an autonomous ultrasonic system combining time-of-flight and phase-shift methods, IEEE Trans. Instrum. Meas., 46(6) (1997), 1236−1240.

[25] C.-C. Tong, J.F. Figueroa, E. Barbieri: A method for short or long range time-of-flight measurements using phase-detection with an analog circuit, IEEE Trans. Instrum. Meas., 50(5) (2002), 1324–1328.

[26] L. Angrisani, R. Schiano Lo Moriello: Estimating ultrasonic time-of-flight through quadrature demodulation, IEEE Trans. Instrum. Meas., 55(1) (2006).

[27] G. Andria, F. Attivissimo, N. Giaquinto: Digital signal processing techniques for accurate ultrasonic sensor measurement, Measurement, 30 (2001), 105–114.

[28] A. Egaña, F. Seco, R. Ceres: Processing of ultrasonic echo envelopes for object location with nearby receivers, IEEE Trans. Instrum. Meas., 57(12) (2008), 2751.

[29] P. Holmberg: Robust ultrasonic range finder – an FFT analysis, Meas. Sci. Technol., 3 (1992), 1025–1037.

[30] B. Barshan: Fast processing techniques for the accurate ultrasonic range measurements, Meas. Sci. Technol., 11 (2000), 45–50.

[31] A. Al-Sabbagh, P.A. Gaydecki: A non-contacting ultrasonic phase-sensitive displacement measurement system, Meas. Sci. Technol., 6 (1995), 1068–1071.

[32] H. Hua, Y. Wang, D. Yan: A low-cost dynamic range-finding device based on amplitude-modulated continuous ultrasonic wave, IEEE Trans. Instrum. Meas., 51(2) (2002), 362–367.

[33] K. Nakahira, T. Kodama, S. Morita, S. Okuma: Distance measurement by an ultrasonic system based on a digital polarity correlator, IEEE Trans. Instrum. Meas., 50(6) (2001), 1748–1752.

[34] K. Nakahira, T. Kodama, T. Furuhashi, S. Okuma: A self-adapting sonar ranging system based on digital polarity correlators, Meas. Sci. Technol., 15 (2004), 347–352.

[35] P.T. Gough, A. de Roos, M.J. Cusdin: Continuous transmission FM sonar with one octave bandwidth and no blind time, IEE Proc., 131(F) (1984), 270–274.

[36] G. Mauris, E. Benoit, L. Foulloy: Local measurement validation for an intelligent chirped-FM ultrasonic range sensor, IEEE Trans. Instrum. Meas., 49(4) (2000), 835–839.

[37] H. Piontek, M. Seyffer, J. Kaiser: Improving the accuracy of ultrasound-based localisation systems; in: Lecture Notes in Computer Science, Springer-Verlag; ISSN 0302-9743; pp. 132–143.

[38] P. Hauptmann, N. Hoppe, A. Püttmer: Application of ultrasonic sensors in the process industry, Meas. Sci. Technol., 13 (2002), R73–R83.

[39] R. Kuc, M.W. Siegel: Physically based simulation model for acoustic sensor robot navigation, IEEE Trans. Pattern Anal. Mach. Intell., PAMI-9(6) (1987), 766–778.

[40] F. van der Heijden, S. van Koningsveld, P.P.L. Regtien: Time-of-flight estimation using extended matched filtering; IEEE Sensors 2004 Conference, Vienna, 24–27 October 2004; ISBN 0-7803-8693-0, pp. 1461–1464.

[41] X. Li, R. Wu, S. Rasmi, J. Li, L.N. Cattafesta, M. Sheplak: Acoustic proximity ranging in the presence of secondary echoes, IEEE Trans. Instrum. Meas., 52(5) (2003), 1593–1605.

[42] E. Iriarte, J.M. Martin Abreu, L. Caldéron Estévez: Absolute-orientation sensor for mobile robots, Eurosensors XII, Southampton, UK, 13–16 September 1998, in: N.M. White (ed.), Eurosensors XII, Vol.II, pp. 917–920.

[43] J. Palacín, J.A. Salse, I. Valgañón, X. Clua: Building a mobile robot for a floor-cleaning operation in domestic environments, IEEE Trans. Instrum. Meas., 53(5) (2004), 1418–1424.

[44] P. Veelaert, W. Bogaerts: Ultrasonic potential field sensor for obstacle avoidance, IEEE Trans. Rob. Autom., 15 (1999), 774–779.

[45] H. Peremans, K. Audenaert, J.M. van Campenhout: A high-resolution sensor based on tri-aural perception, IEEE Trans. Rob. Autom., 9(1) (1993), 36–48.

[46] R. Kuc: Pseudoamplitude scan sonar map, Proc. IEEE Rob. Autom., 17(5) (2001), 767–770.

[47] F. Tong, S.K. Tso, T.Z. Xu: A high precision ultrasonic docking system used for automatic guided vehicle, Sens. Actuators A, 118 (2005), 183–189.

[48] J. Borenstein: The NavBelt – a computerized multi-sensor travel aid for active guidance of the blind; Proceedings of the Fifth Annual Conference on Technology and Persons with Disabilities, Los Angeles, CA, 21–24 March 1990, pp. 107–116.

[49] S. Shoval, J. Borenstein, Y. Koren: Mobile robot obstacle avoidance in a computerized travel aid for the blind; Proceedings of the IEEE Robotics and Automation Conference, San Diego, CA, 8–13 May 1994, pp. 2023–2029.

[50] P. Mihajlik, M. Guttermuth, K. Seres, P. Tatai: DSP-based ultrasonic navigation aid for the blind; IEEE Instrumentation and Measurement Technology Conference, Budapest, Hungary, 21–23 May 2001, pp. 1535–1540.

[51] I. Ulrich, J. Borenstein: The guidecane – applying mobile robot technologies to assist the visually impaired, IEEE Trans. Syst. Man Cybernet., 31(2) (2001), 131–136.

[52] J.M. Martín Abreu, L. Calderón, L.A. Pérez: Shaping the detection lobe of ultrasonic ranging devices, Meas. Sci. Technol., 8 (1997), 1279–1284.

[53] G. Kao, P. Probert, D. Lee: Object recognition with FM sonar: an assistive device for blind and visually impaired people; American Association for Artificial Intelligence, 1996.

[54] H.W. When, P.R. Bélanger: Ultrasound-based robot position estimation, IEEE Trans. Rob. Autom., 13(5) (1997), 682–692.

[55] D. Marioli, E. Sardini, A. Taroni: Ultrasonic distance measurement for linear and angular position control, IEEE Trans. Instrum. Meas., 37(4) (1988), 578–581.

[56] A. Nilsson, P. Holmberg: Combining a stable 2-D vision camera and an ultrasonic range detector for 3-D position estimation, IEEE Trans. Instrum.Meas., 43(2) (1994), 272–276.

[57] F. Tatar, J. Bastemeijer, J.R. Mollinger, A. Bossche: Two-frequency method for measuring the position of surgical tools with $\mu$m precision; Proceedings Eurosensors XIX, Barcelona, 11–14 September 2005, Vol. 1, p. TC12.

[58] R.C. Dorf, A. Nezamfar: A robot ultrasonic sensor for object recognition; Robots 8, Conference Proceedings of Future Considerations, Vol. 2, Society of Manufacturing Engineers 1984, pp. 21.44–21.55.

[59] J.S. Schoenwald, J.F. Martin, L.A. Ahlberg: Acoustic scanning for robotic range sensing and object pattern recognition; Ultrasonics Symposium 1982, pp. 945–949.

[60] T. Tsujimura, T. Yabuta, T. Morimitsu: Three-dimensional shape recognition method using ultrasonics for manipulator control system, J. Rob. Syst., 3(2) (1986), 205–216.

[61] M.K. Brown: Locating object surfaces with an ultrasonic range sensor; International Conference on Robotics and Automation, 1985, pp. 110–115.

[62] M. Lach, H. Ermert: An acoustic sensor system for object recognition, Sens. Actuators A, 25–27 (1991), 541–547.

[63] A. Lázaro, I. Serrano, J.P. Oria: Ultrasonic circular inspection for object recognition with sensor–robot integration, Sens. Actuators A, 77 (1999), 1–8.

[64] M. Brudka, A. Pacut: Intelligent robot control using ultrasonic measurements, IEEE Trans. Instrum. Meas., 51(3) (2002), 454–459.

[65] C. Barat, N. Ait Oufroukh: Classification of indoor environment using only one ultrasonic sensor; IEEE Instrumentation and Measurement Technology Conference, Budapest, Hungary, 21–23 May 2001, pp. 1750–1755.

[66] M.K. Brown: Feature extraction techniques for recognizing solid objects with an ultrasonic range sensor, IEEE J. Rob. Autom., RA-1(4) (1985), 191−205.

[67] P. Mattila, J. Siirtola, R. Suoranta: Two-dimensional object detection in air using ultrasonic transducer array and non-linear digital L-filter, Sens. Actuators A, 55 (1996), 107−113.

[68] F. Franceschini, D. Maisano, L. Mastrogiacomo, B. Pralio: Ultrasound transducers for large-scale metrology: a performance analysis for their use by the MScMS, IEEE Trans. Instrum. Meas., 59(1) (2010), 110−121.

[69] P.P.L. Regtien, H.C. Hakkesteegt: A low-cost sonar system for object identification; Proceedings of International Symposium on Industrial Robots, Lausanne, Switzerland, April 1988, pp. 201−210.

[70] J.M. Martín Abreu, T. Freire Bastos, L. Calderón: Ultrasonic echoes from complex surfaces: an application to object recognition, Sens. Actuators A, 31 (1992), 182−187.

[71] A.M. Sabatini: A digital-signal-processing technique for ultrasonic signal modeling and classification, IEEE Trans. Instrum. Meas., 50(1) (2001), 15−21.

[72] C.A. Chaloner, L.J. Bond: Ultrasonic signal processing using Born inversion, NDT Int., 19(3) (1986), 133−140.

[73] R. Ludwig, D. Roberti: A non-destructive imaging system for detection of flaws in metal blocks, IEEE Trans. Instrum. Meas., 38(1) (1989), 113−118.

[74] B.-C. Shin, J.-R Kwon: Ultrasonic transducers for continuous-cast billets, Sens. Actuators A, 51 (1996), 173−177.

[75] J.G. Cherng, X.F. Chen, V. Peng: Application of acoustic metrology for detection of plate thickness change, Measurement, 18(4) (1996), 207−214.

[76] B. Pant, S.R. Skinner, J.E. Steck: Paint thickness measurement using acoustic interference, IEEE Trans. Instrum. Meas., 55(5) (2006), 1720−1724.

[77] S. Toda, T. Fujita, H. Arakawa, K. Toda: An ultrasonic nondestructive technique for evaluating layer thickness in human teeth, Sens. Actuators A, 125 (2005), 1−9.

[78] R. Raišutis, R. Kažys, L. Mažeika: Ultrasonic thickness measurement of multilayered aluminum foam precursor material, IEEE Trans. Instrum. Meas., 57(12) (2008), 2846−2855.

[79] A. Kimoto, K. Shida: A new touch sensor for material discrimination and detection of thickness and hardness, Sens. Actuators A, 141 (2008), 238−244.

[80] A.H.G. Cents, D.W.F. Brilman, P. Wijnstra, P. Regtien: Measuring bubble, drop and particle sizes in multiphase systems with ultrasound, AICHE J., 50(11) (2004), 2750−2762.

[81] F. Figueroa, E. Barbieri: An ultrasonic ranging system for structural vibration measurements, IEEE Trans. Instrum. Meas., 40(4) (1991), 764−769.

[82] E.L. Estochen, C.P. Neuman, F.B. Prinz: Application of acoustic sensors to robotic seam tracking, IEEE Trans. Ind. Electron., IE-31(3) (1984), 219−224.

[83] B. Hök, A. Blückert, G. Sandberg: A non-contacting sensor system for respiratory air flow detection; Transducers '95 − Eurosensors IX, Stockholm, Sweden, 25−29 June 1995, pp. 424−427.

[84] E. Vargas, R. Ceres, J.M. Martín, L. Calderón: Ultrasonic sensor to inspect the liquid level in bottles for an industrial line; Eurosensors X, Leuven, Belgium, 8−11 September 1996, pp. 1385−1388.

[85] D. Donlagic, M. Zavrsnik, I. Sirotic: The use of one-dimensional acoustical gas resonator for fluid level measurements, IEEE Trans. Instrum. Meas., 49(5) (2000), 1095−1100.

[86] A. Lázaro, I. Serrano: Ultrasonic recognition technique for quality control in foundry pieces, Meas. Sci. Technol., 10 (1999), N113−N118.

[87] F. Seco, A. Ramón Jiménez, M. Dolores del Castillo: Air coupled ultrasonic detection of surface defects in food cans, Meas. Sci. Technol., 17 (2006), 1409–1416.
[88] P. Benech, E. Novakovu: Ultrasonic detection of air bubbles in ducts using PVDF, Meas. Sci. Technol., 10 (1999), 1032–1036.
[89] B.V. Antohe, D.B. Wallace: The determination of the speed of sound in liquids using acoustic resonance in piezoelectric tubes, Meas. Sci. Technol., 10 (1999), 994–998.
[90] A. Carullo, M. Parvis: An ultrasonic sensor for distance measurement in automotive applications, IEEE Sensor J., 1(2) (2001), 143–147.
[91] Y. Ebisawa: A pilot study on ultrasonic sensor-based measurement of head movement, IEEE Trans. Instrum. Meas., 51(5) (2002), 1109–1115.
[92] J. Stor-Pellinen, M. Luukkala: Paper roughness measurement using airborne ultrasound, Sens. Actuators A, 49 (1995), 37–40.
[93] A.R. Grahn, L. Astle: Robotic ultrasonic force sensor arrays; Robots 8 Conference and Proceedings of Society of Manufacturing Engineers, Dearborn, 1984, pp. 22-1–22-18.
[94] P. Dario, C. Domenici, R. Bardelli, D. de Rossi, P.C. Pinotti: Piezoelectric polymers: new sensor materials for robotic applications; 13th International Symposium on Industrial Robots, 17–21 April 1983, Chicago, IL, pp. 1434–1449.
[95] S. Ando, H. Shinoda: Ultrasonic emission tactile sensing, IEEE Control Systems, 1995, pp. 61–69.
[96] P.J. Kyberd, P.H. Chappell: Characterization of an optical and acoustic touch and slip sensor for autonomous manipulation, Meas. Sci. Technol., 3 (1992), 969–975.
[97] K. Nakamura, H. Shinoda: A tactile sensor instantaneously evaluating friction coefficients; Proceedings Transducers '01 /Eurosensors XV Conference on Solid State Sensors and Actuators, Munich, Germany, 10–14 June 2001, (4B1-20P) p. 1430.

# Literature for Further Reading

Some books and review articles on acoustics and ultrasonic measurement principles are listed below.

[1] H. Piontek, M. Seyffer, J. Kaiser: Improving the accuracy of ultrasound-based localisation systems; Springer-Verlag, Berlin, 2005; ISBN 978-3-540-25896-4.
[2] R.C. Asher: Ultrasonic sensors for chemical and process plant; IOP Publ., Bristol, Philadelphia, 1997; ISBN 0-7503-0361-1.
[3] Š. Kočiš, Z. Figura: Ultrasonic measurements and technologies, Chapman & Hall, London, Weinheim, New York, Tokyo, Melbourne, Madras, 1996; ISBN 0-412-63850-9.
[4] Q.X. Chen, P.A. Payne: Industrial applications of piezoelectric polymer transducers, Meas. Sci. Technol., 6 (1995), 249–267.
[5] J.J. Leonard, H.F. Durrant-Whyte: Directed sonar sensing for mobile robot navigation; Kluwer Academic Publishers, Norwell, MA, USA, 1992; ISBN 0-7923-9242-6.
[6] L.L. Beranek: Acoustics, American Institute of Physics, Cambridge, MA, 1986.
[7] M.G Silk: Ultrasonic transducers for nondestructive testing; Adam Hilger, Bristol, 1984; ISBN 0-85274-436-6.
[8] I. Grabec, M. Platte: A comparison of high-performance acoustic emission transducers, Sens. Actuators, 5 (1984), 275–284.
[9] D.I. Crecraft: Ultrasonic instrumentation: principles, methods, applications, J. Phys. E: Sci. Instrum., 16 (1983), 181–189.

[10] R.G. Swartz, J.D. Plummer: Monolithic silicon-PVF$_2$ piezoelectric arrays for ultrasonic imaging, in: A.F. Metherell (ed.), Acoustical imaging, Vol. 8, Plenum Press, New York, NY, 1980, pp. 69–95.

[11] C.S. Desilets, J.D Fraser, G.S Kino: The design of efficient broad-band piezoelectric transducers, IEEE Trans. Son. Ultrason., SU25(3) (1978), 115–125.

[12] J. Blitz: Ultrasonics, methods and applications; Butterworth, London, 1973 (Chapters 2–5).

[13] G.L. Gooberman: Ultrasonics, theory and application; The English University Press, London, 1968.

[14] G. Kossoff: The effects of backing and matching on the performance of piezoelectric ceramic transducers, IEEE Trans. Son. Ultrason., SU-13(1) (1966), 20–30.

# Appendix A

# Symbols and Notations

Sensors operate at the boundary of two physical domains. Despite international normalization of symbols for quantities and material properties, notations for physical quantities are not unambiguous when considering the various disciplines, which each have their own system of notation. This appendix offers a brief review of quantities in the electrical, thermal, mechanical and optical domains, together with their symbols and units, as used in this book. Further, where useful, relations between quantities are listed as well.

## A.1 The Electrical Domain

A sensor produces an electrical output. Table A.1 displays the most important energetic quantities used in the electrical domain. The table also shows magnetic quantities, since they are closely connected to electrical quantities. Relations between these variables are discussed in Chapter 6.

Table A.2 shows the major properties for the electrical domain.

Conductivity is the inverse of resistivity, and conductance the inverse of resistance. The electric permittivity $\varepsilon$ is the product of the permittivity of free space (vacuum) $\varepsilon_0$ and the relative permittivity (or dielectric constant) $\varepsilon_r$. Similarly the magnetic permeability $\mu$ is the product of the magnetic permeability of vacuum $\mu_0$ and the relative permeability $\mu_r$. Numerical values of $\varepsilon_0$ and $\mu_0$ are:

$$\varepsilon_0 = (8.85416 \pm 0.00003) \cdot 10^{-12} (\text{F/m})$$
$$\mu_0 = 4\pi \cdot 10^{-7} (\text{V s/A m})$$

The relative permittivity and permeability account for the dielectric and magnetic properties of a material.

In the electrical domain, some particular variables apply, associated to properties of (electrical) time-varying signals. They are listed in Table A.3.

The duty cycle is defined as the high—low ratio of one period in a periodic pulse signal. It varies from 0% (whole period low) to 100% (whole period high). A duty cycle of 50% refers to a symmetric square wave signal.

**Table A.1** Electrical and Magnetic Quantities

| Quantity | Symbol | Unit |
|---|---|---|
| Electric current | $I$ | A (ampère) |
| Current density | $J$ | A m$^{-2}$ |
| Electric charge | $Q$ | C (coulomb) = A s |
| Dielectric displacement | $D$ | C m$^{-2}$ |
| Electric field strength | $E$ | V m$^{-1}$ |
| Potential difference | $V$ | W A$^{-1}$ |
| Magnetic flux | $\Phi$ | Wb (weber) = J A$^{-1}$ |
| Magnetic induction | $B$ | T (tesla) = Wb m$^{-2}$ |
| Magnetic field strength | $H$ | A m$^{-1}$ |

**Table A.2** Electrical Properties

| Electrical Property | Symbol | Unit |
|---|---|---|
| Resistivity | $\rho$ | $\Omega$ m |
| Resistance | $R$ | $\Omega$ (ohm) |
| Conductivity | $\sigma$ | S m$^{-1}$ = $\Omega^{-1}$ m$^{-1}$ |
| Conductance | $G$ | S (siemens) |
| Capacitance | $C$ | F (farad) |
| Self-inductance | $L$ | H (henry) |
| Mutual inductance | $M$ | H |
| Permeability | $\mu = \mu_0 \cdot \mu_r$ | H m$^{-1}$ |
| Relative permeability | $\mu_r$ | – |
| Permittivity | $\varepsilon = \varepsilon_0 \cdot \varepsilon_r$ | F m$^{-1}$ |
| Dielectric constant | $\varepsilon_r$ | – |

**Table A.3** Signal Variables

| Signal Quantity | Symbol | Unit |
|---|---|---|
| Time | $t$ | s (second) |
| Frequency | $f$ | Hz (hertz) = s$^{-1}$ |
| Period | $T$ | s |
| Phase difference | $\phi$ | rad (radian) |
| Duty cycle | $\delta$ | – |
| Pulse width | $\tau$ | s |

## A.2 The Thermal Domain

Sensor systems suffer most from unwanted thermal influences. Changes in environmental or local temperature affect the mechanical and electrical performance of mechatronic systems. Therefore we will consider thermal effects when discussing the sensor properties. Tables A.4 and A.5 review the major thermal quantities.

The coefficient of heat transfer accounts for the heat flow across the boundary area of an object whose temperature is different from that of its (fluidic) environment. Its value strongly depends on the flow type along the object and the surface properties.

## A.3 The Mechanical Domain

Quantities in the mechanical domain describe state properties related to distance, force and motion. A possible categorization of these quantities is a division into:

- position or geometric quantities
- force or dynamometric quantities
- flow quantities.

We will briefly resume the first two groups. Flow quantities are not further discussed in this book. The basic position quantities and their units and symbols are listed in Table A.6.

Table A.7 shows the most common force quantities, symbols and units.

Some commonly used properties in the mechanical domain are listed in Table A.8.

**Table A.4** Thermal Variables

| Energetic Quantity | Symbol | Unit |
|---|---|---|
| (Thermodynamic) Temperature | $\Theta$ | K (kelvin) |
| Quantity of heat | $Q_{th}$ | J (joule) |
| Heat flow rate | $\Phi_{th}$ | W (watt) |

**Table A.5** Thermal Parameters

| Thermal Property | Symbol | Unit |
|---|---|---|
| Thermal expansion coefficient | $\alpha$ | $K^{-1}$ |
| Heat capacitance | $C_{th}$ | $J\,K^{-1}$ |
| Specific heat capacity | $c$ | $J\,kg^{-1}\,K^{-1}$ |
| Thermal resistance | $R_{th}$ | $K\,W^{-1}$ |
| Thermal conductivity | $g_{th}$ | $W\,m^{-1}\,K^{-1}$ |
| Coefficient of heat transfer | $h$ | $W\,m^{-1}\,K^{-1}$ |

**Table A.6** Geometric or Position Quantities

| Quantity | Symbol | Unit | Relation |
|---|---|---|---|
| Length | $l$ | m | |
| Velocity | $v$ | m s$^{-1}$ | $v = dx/dt$ |
| Acceleration | $a$ | m s$^{-2}$ | $a = dv/dt$ |
| Surface area | $A$ | m$^2$ | |
| Volume | $V$ | m$^3$ | |
| Angular displacement | $\phi$ | rad | |
| Angular velocity | $\omega$ | rad s$^{-1}$ | $\omega = d\phi/dt$ |

**Table A.7** Force Quantities

| Quantity | Symbol | Unit |
|---|---|---|
| Force | $F$ | N |
| Tension, stress | $T$ | Pa = N m$^{-2}$ |
| Shear stress | $\tau$ | Pa |
| Pressure | $p$ | Pa |
| Moment of force | $M$ | N m |
| Power | $P$ | W = J s$^{-1}$ |
| Energy | $E$ | J = N m |

**Table A.8** Mechanical Properties

| Property | Symbol | Unit | Relation |
|---|---|---|---|
| Mass density | $\rho$ | kg m$^{-3}$ | $M = \rho \cdot V$ |
| Mass | $M$ | kg | $F = M \cdot a$ |
| Weight | $G$ | kg | $F = G \cdot g$ |
| Elasticity | $c$ | N m$^{-2}$ | $T = c \cdot S$ |
| Compliance | $s$ | m$^2$ N$^{-1}$ | $s = 1/c$ |

The relation between tension $T$ and strain $S$ is given by Hooke's law:

$$T = c \cdot S \tag{A.1}$$

or

$$S = s \cdot T \tag{A.2}$$

Note that compliancy $s$ is the reciprocal of the elasticity $c$. A force (tension) results in a deformation (strain): longitudinal forces cause translational deformation; shear

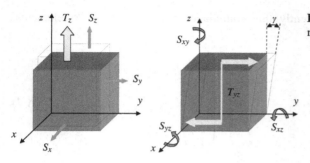

**Figure A.1** Six d.o.f. in the mechanical domain.

forces cause rotational deformation; the angle of rotation is called the shear angle. The mechanical domain has six degrees of freedom (d.o.f.): 3 for translation and 3 for rotation. As an example, in Figure A.1 a longitudinal force in $z$-direction (left) and a shear force around the $x$-direction (right) are shown. Note that in this example the force causes a positive strain in $z$-direction and negative strain in the other two directions. The shear force in this example causes only a shear strain in the $yz$ plane; the shear angle is denoted by $\gamma$.

In general a longitudinal force in the $z$-direction, $T_z$, causes the material to be deformed in three possible directions ($x$, $y$ and $z$), resulting in the strain components $S_z$ (usually the most important), $S_x$ and $S_y$. A shear force in the $z$-plane, $T_{xy}$, can cause the material to twist in three directions, resulting in the three torsion components $S_{xy}$ (usually the most important), $S_{xz}$ and $S_{yz}$. Sometimes a longitudinal force produces torsion. This can be described by, for instance, $S_{yz} = s_{yzz} \cdot T_z$. In general the relation between tension and strain is described by

$$S_{ij} = s_{ijkl} \cdot T_{kl} \tag{A.3}$$

or

$$T_{ij} = c_{ijkl} \cdot S_{kl} \tag{A.4}$$

with $i$, $j$, $k$, $l$ equal to $x$, $y$ and $z$. Note that for notation symmetry, longitudinal tension $T_i$ ($i = x,y,z$) is written as $T_{ii}$. To simplify the notation the tensor indices $ijkl$ are substituted by engineering indices 1−6, according to the next scheme:

$$
\begin{aligned}
xx &\rightarrow 1 & yz = zy &\rightarrow 4 \\
yy &\rightarrow 2 & xz = zx &\rightarrow 5 \\
zz &\rightarrow 3 & xy = yx &\rightarrow 6
\end{aligned}
$$

So, Hooke's law becomes:

$$S_\lambda = s_{\lambda\mu} \cdot T_\mu \tag{A.5}$$

or

$$T_\mu = c_{\mu\lambda} \cdot S_{\lambda\mu} \tag{A.6}$$

with $\lambda$, $\mu = 1 \ldots 6$. Apparently this compliance matrix contains 36 elements. Written in full

$$
\begin{pmatrix} S_1 \\ S_2 \\ \cdot \\ \cdot \\ \cdot \\ S_6 \end{pmatrix} = \begin{pmatrix} s_{11} & s_{12} & \cdot & \cdot & \cdot & s_{16} \\ s_{21} & s_{22} & \cdot & \cdot & \cdot & s_{26} \\ \cdot & \cdot & \cdot & \cdot & \cdot & \cdot \\ \cdot & \cdot & \cdot & \cdot & \cdot & \cdot \\ \cdot & \cdot & \cdot & \cdot & \cdot & \cdot \\ s_{61} & s_{62} & \cdot & \cdot & \cdot & s_{66} \end{pmatrix} \cdot \begin{pmatrix} T_1 \\ T_2 \\ \cdot \\ \cdot \\ \cdot \\ T_6 \end{pmatrix}
\tag{A.7}
$$

In most materials many of the matrix elements in Eq. (A.7) are zero, due to symmetry in the material crystal structure. To get an idea of the practical values Table A.9 shows numerical values of the elements of the compliance matrix of quartz (crystalline $SiO_2$) and of silicon (Si). Note that $s_{ij} = s_{ji}$. Silicon has an almost symmetrical crystal structure, leading to $s_{11} = s_{22} = s_{33}$, $s_{44} = s_{55} = s_{66}$ and $s_{12} = s_{13}$. The remaining elements in the matrix are zero. Obviously quartz has a less symmetrical structure, as can be deduced from the many different element values.

## A.4 The Optical Domain

### A.4.1 Optical Quantities

In the optical domain two groups of quantities are distinguished: radiometric and photometric quantities. The radiometric quantities are valid for the whole electromagnetic spectrum. Photometric quantities are only valid within the visible part of the spectrum ($0.35 < \lambda < 0.77$ μm). They are related to the mean, standardized sensitivity of the human eye and are applied in photometry. Table A.10 shows the major radiometric and corresponding photometric quantities. Some of them will be explained in more detail.

The *radiant intensity* $I$ (W/sr) is the emitted radiant power of a point source per unit of solid angle. For a source consisting of a flat radiant surface the emitted power is expressed in terms of $W/m^2$, called the *radiant emittance* $E_s$. It includes the total emitted energy in all directions. The quantity *radiance* accounts also for

**Table A.9** Matrix Elements of Compliancy for $SiO_2$ and Si; Unit $10^{-12}$ m$^2$/N

| SiO$_2^a$ | | | | Si | | | |
|---|---|---|---|---|---|---|---|
| $s_{11}$ | 12.77 | $s_{33}$ | 9.6 | $s_{11}$ | 7.68 | $s_{44}$ | 12.56 |
| $s_{12}$ | −1.79 | $s_{44}$ | 20.04 | $s_{12}$ | −2.14 | | |
| $s_{13}$ | −1.22 | $s_{66}$ | 29.12 | | | | |
| $s_{14}$ | 4.50 | | | | | | |

$^a$SiO$_2$ is piezoelectric: numerical values are at zero voltage ($s^E$); see Chapter 8.

**Table A.10** Radiometric and Photometric Quantities

| | **Radiometric Quantities** | | | **Photometric Quantities** | |
|---|---|---|---|---|---|
| | Quantity | Symbol | Unit | Quantity | Unit |
| Power | Radiant flux | $P = dU/dt$ | W | Luminous flux | lumen (lm) |
| Energy | Radiant energy | $U$ | J | Luminous energy | lm s |
| Emitting power per unit area | Radiant emittance | $E_s = dP_s/dS$ | W/m$^2$ | Luminous emittance | lm/m$^2$ = lux |
| Incident power per unit area | Irradiance | $E_d = dP_d/dS$ | W/m$^2$ | Illuminance | lm/m$^2$ = lux |
| Radiant power per solid angle | Radiant intensity | $I = dP/d\Omega$ | W/sr | Luminous intensity | lm/sr = candela (cd) |
| Power per solid angle per unit of projected surface area | Radiance | $L = dI/dS_p$ | W/m$^2$ sr | Luminance, brightness | lm/m$^2$ sr = cd/m$^2$ |

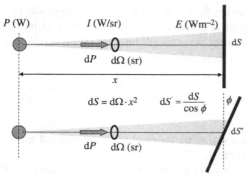

**Figure A.2** Relations between various optical quantities in Table A.10. d$S$ is the projected area.

the direction of the emitted light: it is the emitted power per unit area, per unit of solid angle (W/m$^2$ sr).

A flat surface can receive radiant energy from its environment, from all directions. The amount of incident power $E_d$ is expressed in terms of W/m$^2$ and is called the *irradiance*.

In radiometry the unit of (emitting or receiving) surface area is often expressed in terms of 'projected area'. A surface with (real) area $S$ that is irradiated from a direction perpendicular to its surface receives power equal to $E_d \cdot S$ (W). When that surface is rotated over an angle $\phi$, relative to the direction of radiation, it receives only an amount of radiant power equal to $E_d \cdot S \cdot \cos \phi$ (Figure A.2). To make these radiometric quantities independent of the viewing direction, some authors express them in terms of projected area instead of real area, thus avoiding the $\cos \phi$ in the expressions.

**Table A.11** Characteristic Parameters of Optical Sensors

| Parameter | Symbol | Unit |
|---|---|---|
| Output voltage, current | $S$ | V; A |
| Output power | $P_o$ | W |
| Input power | $P_i$ | W |
| Noise power | $N$ | V; A; W |
| Responsivity | $S/P_i$ | V/W; A/W |
| S/N ratio | $S/N$ | – |
| Detectivity | $D = (S/N)/P$ | 1/W |
| NEP | $NEP = 1/D$ | W |

In Table A.10 the *radiance L* of a surface is defined following this idea: it is the emitted radiant energy in a certain direction, per solid angle (to account for the directivity) and per unit of projected area, that is per unit area when the emitting surface is projected into that direction. These quantities are used for deriving formulas for sensors that are based on reflecting surfaces (Chapter 7).

All units have their counterpart in the photometric domain (right-hand side of the table). However we will not use them further in this book. The radiometric quantities in the table do not account for the frequency or wavelength dependence. All of them can be expressed also in spectral units, for example the spectral irradiance $E_d(\lambda)$, which is the irradiance per unit of wavelength interval (W/m$^2$ μm). The total irradiance over a spectral range from $\lambda_1$ to $\lambda_2$ then becomes:

$$E_d = \int_{\lambda_1}^{\lambda_2} E_d(\lambda)d\lambda \qquad (A.8)$$

To characterize optical sensors special signal parameters are used, related to the noise behaviour of these components (Table A.11).

Note that the signals can be expressed in voltage, current or power. Irrespective of this choice, the S/N ratio is a dimensionless quantity. The parameters $D$ and NEP represent the lowest perceivable optical signal. NEP is, in words, equal to the power of sinusoidally modulated monochromatic radiation that produces an RMS signal at the output of an ideal sensor, equal to the noise signal of the real sensor. This means that, when specifying NEP, the modulation frequency, the sensor bandwidth, the temperature and the sensitive surface area should also be specified, as all these parameters are included in the noise specification. To avoid wavelength and detector area dependency in the specifications, the parameters NEP* and $D^*$ are used: the spectral NEP and the spectral $D$, respectively. Since noise power of a sensor is usually proportional to the sensitive area $A$ and the bandwidth $\Delta f$, the noise current and noise voltage are proportional to $\sqrt{(A \Delta f)}$; hence, NEP* = NEP/$\sqrt{(A \Delta f)}$ and $D^* = \sqrt{(A \Delta f)}$/NEP.

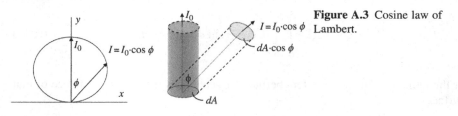

**Figure A.3** Cosine law of Lambert.

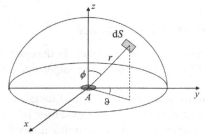

**Figure A.4** Derivation of the total radiant energy from a unit surface with Lambertian emission.

## A.4.2   Radiant Energy from a Unit Surface with Lambertian Emission

If a surface is a perfect diffuser (its radiance is not a function of angle), then $I = I_0 \cdot \cos \phi$ (Figure A.3).

This is known as the cosine law of Lambert. Many surfaces scatter the incident light in all directions equally. A surface that satisfies Lambert's law is called Lambertian.

All radiant energy emitted from a surface $A$ with radiant emittance $E_s$ has to pass a hemisphere around that surface (Figure A.4).

The area of surface element $dS$ on the hemisphere, corresponding with a solid angle $d\Omega$ equals:

$$dS = r^2 d\Omega = r^2 \sin \phi d\phi d\vartheta \tag{A.9}$$

where $r$ is the radius of the hemisphere. According to its definition in Table A.10 the radiant emittance from the emitting surface $A$ in the direction of $dS$ is:

$$\frac{dP_s}{dA} = L \cos \phi d\Omega \tag{A.10}$$

Since all radiant energy from $A$ passes the hemisphere, the total radiant emittance of $A$ equals the integral of $dP_s/dA$ over the full surface area of the hemisphere. Further since for a Lambertian surface the radiant emittance is independent of the direction, $L$ is constant. So

$$E_s = \iint\limits_{\text{hemisphere}} L \cos \phi d\Omega = \int_0^{2\pi} \int_0^{\pi/2} L \cos \phi \sin \phi d\phi d\vartheta = \pi L \qquad (A.11)$$

or the radiant emittance of a Lambertian surface equals $\pi$ times the radiance of that surface.

### A.4.3  Derivation of Relations Between Intensity and Distance

The transfer characteristic of an optical displacement sensor (Chapter 7) depends strongly on the geometry of the configuration. Figure A.5 depicts the major parameters involved in the direct mode.

Light energy radiates from the source with emitting surface area $S_1$ and radiance $L_e$ (W/sr m$^2$). The sensor surface area is $S_2$, the distance from source to sensor is $r$. Further the light beam as received by the sensor makes an angle $\vartheta_1$ with the normal on the emitting surface, and an angle $\vartheta_2$ with the normal on the receiving surface. We calculate the radiant power that arrives at the sensor. Assume the source dimensions being small compared to $r$. Then the light beam that arrives at the sensor has a solid angle equal to

$$d\Omega = \frac{S_2 \cos \vartheta_2}{r^2} \, (\text{sr}) \qquad (A.12)$$

where $S_2 \cos \vartheta_2$ is the projected surface area of the sensor. The emitter emits $L_e \cdot S_1$ W per steradian; hence, the part $\Delta P_e$ of the emitted light falling on the sensor is:

$$\Delta P_e = L_e S_1 \cos \vartheta_1 d\Omega \, (\text{W}) \qquad (A.13)$$

where $S_1 \cos \vartheta_1$ is the projected surface area of the source. By substituting Eq. (A.13) in Eq. (A.12) we find for the radiant power arriving at the sensor:

$$P_o = L_e \frac{S_1 S_2 \cos \vartheta_1 \cos \vartheta_2}{r^2} \, (\text{W}) \qquad (A.14)$$

from which follows that the output signal, indeed, is inversely proportional to the square of the distance.

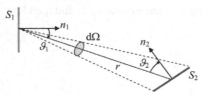

**Figure A.5** Parameters in the source–sensor system.

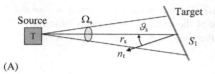

(A)

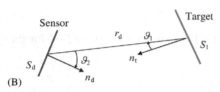

(B)

**Figure A.6** General configuration of an optical proximity sensor with a flat object (split representation).

Next we consider the indirect mode, where we distinguish between two different situations. In the first situation the reflecting target is only partly illuminated by the source: all the light from the source is intercepted by the object. In the second situation the whole object falls within the light cone of the source. Figure A.6 displays the first situation and defines the various parameters. For a better understanding the object (target) is presented twice, once as the receiving surface and once as the (secondary) emitting surface.

The target intercepts a light beam that makes an angle $\vartheta_s$ with the normal vector $n_t$ on the target's surface. The source can be characterized by an intensity $I_s$ (W/sr) and a beam solid angle $\Omega_s$. The surface area $S_1$ of the light spot on the object satisfies the equation:

$$S_1 \cos \vartheta_s = \Omega_s r_s^2 \ (\text{m}^2) \tag{A.15}$$

where $r_s$ is the distance between the source and the target. The irradiance of the object (actually the lighted part of it) is:

$$E_1 = \frac{I_s \Omega_s}{S_1} \ (\text{W/m}^2) \tag{A.16}$$

Since the object surface is assumed to be Lambertian it behaves as a radiant source with radiance $L_1 = E_1/\pi$ (see Eq. (A.11)). Part of the scattered light is received by the detector with sensitive area $S_d$ on a distance $r_d$ from the target. Similar to Eq. (A.12) this sensor receives light over a solid angle $\Omega_d$ equal to

$$\Omega_d = \frac{S_d \cos \vartheta_2}{r_d^2} \ (\text{sr}) \tag{A.17}$$

The radiant power received by the sensor is:

$$P_d = L_1 S_1 \cos \vartheta_1 \Omega_d = \frac{E_1 S_1 \cos \vartheta_1 \Omega_d}{\pi} \ (\text{W}) \tag{A.18}$$

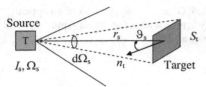

**Figure A.7** The target falls completely within the light cone of the source; light reflects from the target to the sensor according to Figure A.6B.

Finally when substituting Eqs (A.16) and (A.17) into Eq. (A.18) we find for the optical power received by the sensor:

$$P_{\mathrm{d}} = \frac{S_{\mathrm{d}} \cos \vartheta_1 \cos \vartheta_2}{\pi r_{\mathrm{d}}^2} \cdot I_{\mathrm{s}} \Omega_{\mathrm{s}} \text{ (W)} \tag{A.19}$$

So the sensor output is inversely proportional to the square of the distance to the sensor, as long as the object intercepts the whole light beam. Note that $P_{\mathrm{d}}$ does not depend on the angle $\vartheta_{\mathrm{s}}$, because all the light from the source falls on the object, whose surface is assumed to be Lambertian.

Next we consider the situation where the whole object is illuminated by the source, as can happen at large distances and with small objects (Figure A.7).

The solid angle of the beam falling on the object is:

$$d\Omega_{\mathrm{s}} = \frac{S_{\mathrm{t}} \cos \vartheta_{\mathrm{s}}}{r_{\mathrm{s}}^2} \text{ (sr)} \tag{A.20}$$

where $S_{\mathrm{t}}$ is the area of the illuminated part of the object. The radiant emittance now becomes:

$$E_1 = \frac{I_{\mathrm{s}} d\Omega_{\mathrm{s}}}{S_{\mathrm{t}}} = \frac{I_{\mathrm{s}} \cos \vartheta_{\mathrm{s}}}{r_{\mathrm{s}}^2} \text{ (W/m}^2) \tag{A.21}$$

Following the same steps as before, we find for the radiant power received by the sensor:

$$P_{\mathrm{d}} = L_1 S_{\mathrm{t}} \cos \vartheta_1 \Omega_{\mathrm{d}} = \frac{E_1}{\pi} \cdot S_{\mathrm{t}} \cos \vartheta_1 \cdot \frac{S_{\mathrm{d}} \cos \vartheta_2}{r_{\mathrm{d}}^2} = I_{\mathrm{s}} \cdot \frac{S_{\mathrm{t}} S_{\mathrm{d}} \cos \vartheta_1 \cos \vartheta_2 \cos \vartheta_{\mathrm{s}}}{\pi r_{\mathrm{s}}^2 r_{\mathrm{d}}^2} \tag{A.22}$$

Clearly the sensor output is inversely proportional to the fourth power of the distance between source/sensor and the target.

# Appendix B
# Relations Between Quantities

## B.1 Generalized Equations

In Chapter 2 a classification of quantities based on thermodynamics is presented. Here we review the basic equations extended with the magnetic domain.

The first law of thermodynamics is:

$$dQ = \Theta \, d\sigma \quad (J) \tag{B.1}$$

with $\Theta$ the absolute temperature and $d\sigma$ the increase in entropy of the system. According to the second law of thermodynamics the energy content of an infinitely small volume of an elastic dielectric material changes by adding or extracting heat $dQ$ (J) and by work $dW$ (J) exerted upon it

$$dU = dQ + dW \quad (J) \tag{B.2}$$

The work $dW$ is the sum of the different energy forms involved (here thermal, mechanical, electrical and magnetic energy), so

$$dU = \Theta \, d\sigma + F \, dx + V \, dQ + I \, d\Phi \quad (J) \tag{B.3}$$

When the heat is taken per unit of volume, this equation reads:

$$dU = \Theta \, d\sigma + T \, dS + E \, dD + H \, dB \quad (J/m^3) \tag{B.4}$$

that is the change in internal energy when thermal, mechanical, electrical and magnetic energy is supplied to the system. Note that for entropy we have used the same symbol but the dimension is now $J/Km^3$.

Apparently in this equation only through-variables affect the system. If, on the other hand, only across-variables affect the energy state of the system, the equation for the energy change per unit volume is:

$$dG = -\sigma \, d\Theta - S \, dT - D \, dE - B \, dH \quad (J/m^3) \tag{B.5}$$

where $G$ is the Gibbs-free energy of the system, which can be found from the free energy $U$ by a Legendre transformation.

We can generalize Eq. (B.4) as

$$dU = \sum_i A_i \, dB_i \quad (\mathrm{J/m^3}) \tag{B.6}$$

where $A_i$ and $B_i$ are conjugated pairs of variables. Note that the terms intensive and extensive have lost their original meaning in these equations. From Eq. (B.6) it follows for the parameters $A_i$

$$A_i = \frac{\partial U}{\partial B_i} \tag{B.7}$$

On the other hand the system configuration or the material itself couples the conjugate variables of each pair. For instance in the mechanical domain $T$ and $S$ are linked by Hooke's law:

$$T = c \cdot S \quad \text{or} \quad S = s \cdot T \tag{B.8}$$

where $c$ is the elasticity (or stiffness) and $s$ the compliancy.

In general the intensive variable $A_i$ and the extensive variable $B_i$ within one domain are linked according to

$$A_i = c_i \cdot B_i \quad \text{or} \quad B_i = s_i \cdot A_i \tag{B.9}$$

where $c_i$ is a (generalized) elasticity and $s_i$ is a (generalized) compliancy. These are the state equations of the system.

From Eqs (B.3) and (B.4) it follows that the system energy depends on all (generalized) parameters:

$$U = f(A_i, B_i), \quad i = 1 \cdots n \tag{B.10}$$

but since $A_i$ and $B_i$ are related by Eq. (B.9) we can also write:

$$U = f(B_i), \quad i = 1 \cdots n \tag{B.11}$$

So for small variations in the parameters $B_i$ the variation in energy (as given in Eq. (B.6)) becomes:

$$dU = \frac{\partial U}{\partial B_1} dB_1 + \cdots + \frac{\partial U}{\partial B_n} dB_n = \sum_i^n \frac{\partial U}{\partial B_i} dB_i = \sum_i^n A_i \, dB_i \tag{B.12}$$

Equation (B.9) links parameters within one domain. However variables in one domain are also linked to variables in another domain, resulting in cross effects, the basis for transducers. So

$$A_i = f(B_j), \quad i,j = 1 \cdots n \tag{B.13}$$

and consequently, assuming linearity (small variations only):

$$dA_i = \frac{\partial A_i}{\partial B_1} dB_i + \cdots + \frac{\partial A_i}{\partial B_n} dB_n = g_{i1} \, dB_1 + \cdots g_{in} \, dB_n = \sum_{j=1}^{n} g_{ij} \, dB_j \tag{B.14}$$

where $g_{ij}$ represents material properties linking domains $i$ and $j$. Inversely parameter $B_j$ depends on all parameters $A_i$. So the relations between small variations in parameters $A_i$ and $B_j$ can be described by

$$A_i = g_{ij} \cdot B_j \quad \text{or} \quad B_j = g_{ji} \cdot A_i \tag{B.15}$$

The coefficients $g_{ij}$ and $g_{ji}$ are equal. This can be proven by combining Eqs (B.14) and (B.7):

$$\frac{\partial A_i}{\partial B_j} = g_{ij} = \frac{\partial^2 U}{\partial B_j \partial B_i} = \frac{\partial^2 U}{\partial B_i \partial B_j} = g_{ji} \tag{B.16}$$

since the order of differentiation can be reversed.

## B.2 Application to Four Domains

The starting point is Eq. (B.5) for the mechanical, electrical, magnetic and thermal domains, which is repeated here in this order:

$$dG = -S \, dT - D \, dE - B \, dH - \sigma \, d\Theta \quad (\text{J/m}^3) \tag{B.17}$$

The through variables can be written as

$$S(T, E, H, \Theta) = -\left(\frac{\partial G}{\partial T}\right)_{E,H,\Theta}$$

$$D(T, E, H, \Theta) = -\left(\frac{\partial G}{\partial E}\right)_{T,H,\Theta}$$

$$B(T, E, H, \Theta) = -\left(\frac{\partial G}{\partial H}\right)_{T,E,\Theta} \tag{B.18}$$

$$\sigma(T, E, H, \Theta) = -\left(\frac{\partial G}{\partial \Theta}\right)_{T,E,H}$$

Since we consider only small variations, the variables $S$, $D$, $B$, and $\Delta\sigma$ are approximated by linear functions, so

$$
\begin{bmatrix}
dS(T,E,H,\Theta) \\
dD(T,E,H,\Theta) \\
dB(T,E,H,\Theta) \\
d\sigma(T,E,H,\Theta)
\end{bmatrix}
=
\begin{bmatrix}
\left(\dfrac{\partial S}{\partial T}\right)_{E,H,\Theta} & \left(\dfrac{\partial S}{\partial E}\right)_{T,H,\Theta} & \left(\dfrac{\partial S}{\partial H}\right)_{T,E,\Theta} & \left(\dfrac{\partial S}{\partial \Theta}\right)_{T,E,H} \\[2ex]
\left(\dfrac{\partial D}{\partial T}\right)_{E,H,\Theta} & \left(\dfrac{\partial D}{\partial E}\right)_{T,H,\Theta} & \left(\dfrac{\partial D}{\partial H}\right)_{T,E,\Theta} & \left(\dfrac{\partial D}{\partial \Theta}\right)_{T,E,H} \\[2ex]
\left(\dfrac{\partial B}{\partial T}\right)_{E,H,\Theta} & \left(\dfrac{\partial B}{\partial E}\right)_{T,H,\Theta} & \left(\dfrac{\partial B}{\partial H}\right)_{T,E,\Theta} & \left(\dfrac{\partial B}{\partial \Theta}\right)_{T,E,H} \\[2ex]
\left(\dfrac{\partial \sigma}{\partial T}\right)_{E,H,\Theta} & \left(\dfrac{\partial \sigma}{\partial E}\right)_{T,H,\Theta} & \left(\dfrac{\partial \sigma}{\partial H}\right)_{T,E,\Theta} & \left(\dfrac{\partial \sigma}{\partial \Theta}\right)_{T,E,H}
\end{bmatrix}
\cdot
\begin{bmatrix}
dT \\
dE \\
dH \\
d\Theta
\end{bmatrix}
$$

(B.19)

Combining Eqs (B.18) and (B.19) results in

$$
\begin{bmatrix}
dS \\
dD \\
dB \\
d\sigma
\end{bmatrix}
=
\begin{bmatrix}
\left(\dfrac{\partial^2 G}{\partial T^2}\right)_{E,H,\Theta} & \left(\dfrac{\partial^2 G}{\partial T\partial E}\right)_{H,\Theta} & \left(\dfrac{\partial^2 G}{\partial T\partial H}\right)_{E,\Theta} & \left(\dfrac{\partial^2 G}{\partial T\partial \Theta}\right)_{E,H} \\[2ex]
\left(\dfrac{\partial^2 G}{\partial E\partial T}\right)_{H,\Theta} & \left(\dfrac{\partial^2 G}{\partial^2 E}\right)_{T,H,\Theta} & \left(\dfrac{\partial^2 G}{\partial E\partial H}\right)_{T,\Theta} & \left(\dfrac{\partial^2 G}{\partial E\partial \Theta}\right)_{T,H} \\[2ex]
\left(\dfrac{\partial^2 G}{\partial H\partial T}\right)_{E,\Theta} & \left(\dfrac{\partial^2 G}{\partial H\partial E}\right)_{T,\Theta} & \left(\dfrac{\partial^2 G}{\partial^2 H}\right)_{T,E,\Theta} & \left(\dfrac{\partial^2 G}{\partial H\partial \Theta}\right)_{T,E} \\[2ex]
\left(\dfrac{\partial^2 G}{\partial \Theta\partial T}\right)_{E,H} & \left(\dfrac{\partial^2 G}{\partial \Theta\partial E}\right)_{T,H} & \left(\dfrac{\partial^2 G}{\partial \Theta\partial H}\right)_{T,E} & \left(\dfrac{\partial^2 G}{\partial^2 \Theta}\right)_{T,E,H}
\end{bmatrix}
\cdot
\begin{bmatrix}
dT \\
dE \\
dH \\
d\Theta
\end{bmatrix}
$$

(B.20)

The second-order derivatives in the diagonal represent properties in the respective domains: mechanical, electrical, magnetic and thermal. All other derivatives represent cross effects. These derivatives are pair-wise equal, since the order of

**Table B.1** Physical Effects Corresponding to the Parameters in Eq. (B.21)

| | | | |
|---|---|---|---|
| Compliancy | Converse Piezoelectricity | Magnetostriction | Thermal expansion |
| Direct piezoelectricity | Permittivity | Magnetodielectric effect | Pyroelectricity |
| Magnetostriction | Magnetodielectric effect | Permeability | Pyromagnetic effect |
| Piezocaloric effect | Electrocaloric effect | Pyromagnetic effect | Heat capacity |

**Table B.2** Symbols, Parameter Names and Units of the Effects in Table B.1

| Symbol | Property | Unit |
|---|---|---|
| $s$ | Compliancy | $m^2 \, N^{-1}$ |
| $\varepsilon$ | Permittivity; dielectric constant | $F \, m^{-1}$ |
| $\mu$ | Permeability | $V \, s \, A^{-1} \, m^{-1}$ |
| $c_p$ | (Specific) Heat capacity | $J \, kg^{-1} \, K^{-1}$ |
| $d$ | Piezoelectric constant | $m \, V^{-1} = C \, N^{-1}$ |
| $\alpha$ | Thermal expansion coefficient | $K^{-1}$ |
| $p$ | Pyroelectric constant | $C \, m^{-2} \, K^{-1}$ |
| $i$ | Pyromagnetic constant | $N \, m^{-1} \, A^{-1} \, K^{-1}$ |
| $m$ | Magnetodielectric constant | $s \, m^{-1}$ |
| $\lambda$ | Magnetostrictive constant | $m \, A^{-1}$ |

differentiation is not relevant. The derivatives in Eq. (B.20) represent material properties; they have been given special symbols. The variables denoting constancy are put as superscripts, to make place for the subscripts denoting orientation.

$$\begin{bmatrix} S \\ D \\ B \\ \sigma \end{bmatrix} = \begin{bmatrix} s^{E,H,\Theta} & d^{H,\Theta} & \lambda^{E,\Theta} & \alpha^{E,H} \\ d^{H,\Theta} & \varepsilon^{T,H,\Theta} & m^{T,\Theta} & p^{T,H} \\ \lambda^{E,\Theta} & m^{T,\Theta} & \mu^{T,E,\Theta} & i^{T,E} \\ \alpha^{E,H} & p^{T,H} & i^{T,E} & \dfrac{\rho}{\Theta} c^{T,E,H} \end{bmatrix} \cdot \begin{bmatrix} T \\ E \\ H \\ \Delta\Theta \end{bmatrix} \quad \text{(B.21)}$$

The 16 associated effects are displayed in Table B.1.

Some of these effects are mentioned in Figure 2.6 (Chapter 6), with the same or other names. The parameters $\varepsilon$, $c_p$ and $s$ correspond to those in Tables A.2, A.5 and A.8 of Appendix A. Table B.2 summarizes the parameters with symbols and units.

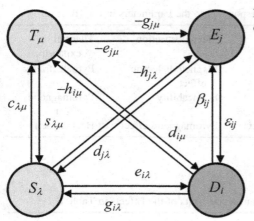

**Figure B.1** Heckmann diagram for two domains.

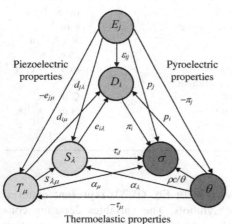

**Figure B.2** Heckmann diagram for the electrical, mechanical and thermal domains.

## B.3 Heckmann Diagrams

The material constants in the mechanical, electrical and thermal domains are defined by

$$\text{mechanical} \quad : \sigma = \left( \frac{\rho}{c\theta} \right) \cdot \theta$$

$$\text{electrical} \quad : S_\lambda = s_{\lambda\mu} \cdot T_\mu \quad \text{or} \quad T_\mu = c_{\lambda\mu} \cdot S_\lambda$$

$$\text{thermal} \quad : D_i = \varepsilon_{ij} \cdot E_j \quad \text{or} \quad E_j = \beta_{ij} \cdot D_i$$

(B.22)

(compare Eqs (A.5), (A.6) and (8.23)).

When extended with the mutual interactions between mechanical and electrical properties (piezoelectricity), four sets of equations can be defined, depending on the choice of the dependent and independent pairs of quantities. These equations are:

$$\begin{cases} D_i = \varepsilon_{ij} \cdot E_j + d_{i\mu} T_\mu \\ S_\lambda = d_{j\lambda} E_j + s_{\lambda\mu} T_\mu \end{cases}$$

$$\begin{cases} E_j = \beta_{ij} D_i - g_{j\mu} T_\mu \\ S_\lambda = g_{i\lambda} D_i + s_{\lambda\mu} T_\mu \end{cases}$$

$$\begin{cases} D_i = \varepsilon_{ij} \cdot E_j + e_{i\lambda} S_\lambda \\ T_\mu = - e_{j\mu} E_j + c_{\lambda\mu} S_\lambda \end{cases}$$

$$\begin{cases} E_j = \beta_{ij} D_i - h_{j\lambda} S_\lambda \\ T_\mu = - h_{i\mu} D_i + c_{\lambda\mu} S_\lambda \end{cases} \tag{B.23}$$

representing in total eight material properties: the reciprocal pairs $s$-$c$ and $\varepsilon$-$\beta$, and four piezoelectric parameters (compare Table 8.1 in Chapter 8). These relations are visualized in the Heckmann diagram for the electrical and mechanical domains (Figure B.1). The circles represent the state variables, and the arrows represent the material properties.

When the thermal domain is also included sets of three equations can be defined. One of them is the set given in Chapter 2, by Eq. (2.10), yielding the most practical material properties. Figure B.2 shows the Heckmann diagram for three domains, where the double arrows are left out for clarity.

# Appendix C
# Basic Interface Circuits

In this appendix we review the transfer functions of some basic electronic circuits, to facilitate the understanding of the interface circuits as used throughout the text. All circuits are built up with operational amplifiers. Circuits and associated transfer functions are given for the current-to-voltage converter, inverting and non-inverting amplifiers, differential amplifier, integrator, differentiator and some analogue filter circuits. In first instance the operational amplifier is considered to have ideal characteristics. Where appropriate the consequences of a deviation from the ideal behaviour are given, as well as measures to reduce such unwanted effects.

## C.1 Operational Amplifier

The symbol for an operational amplifier is given in Figure C.1A. Usually the power supply connections are left out (which will be done hereafter). The model in Figure C.1B shows the major deviations from the ideal behaviour.

The meaning of these circuit elements is as follows:

$A$: voltage gain (frequency dependent); $A_0$: low-frequency voltage gain
$V_{off}$: input offset voltage
$I_b^+$ and $I_b^-$: input bias currents through the plus and minus terminals
$I_{off} = |I_b^+ - I_b^-|$: input offset current
$Z_c^-$ and $Z_c^+$: common input impedances from either input to ground
$Z_d$: differential input impedance (between the two input terminals)
$Z_o$: output impedance

Further the common mode rejection ratio (CMRR) accounts for the suppression of pure common mode signals relative to differential mode signals. Finally amplifier noise is specified in terms of (spectral) voltage noise and current noise. For white noise these error sources are specified in terms of $V/\sqrt{Hz}$ and $A/\sqrt{Hz}$, respectively. In Figure C.1 the noise voltage can be modelled with a voltage source in series with the offset voltage and the noise current by a current source in parallel to the bias current. For an ideal operational amplifier $A$, $Z_c$, $Z_d$ and $H$ are infinite, and $V_{off}$, $I_b$ and $Z_o$ are all zero. Any deviation of these ideal values is reflected in the transfer properties of interface circuits built up with operational amplifiers.

We will consider only a limited set of specifications. Table C.1 shows typical values for these specifications, for three types of operational amplifiers: types I and

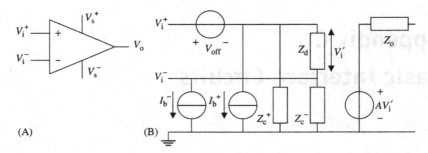

**Figure C.1** (A) Symbol of an operational amplifier, (B) modelling of some deviations from the ideal behaviour.

**Table C.1** Selected Specifications of General Purpose Operational Amplifiers

| Parameter | Type I | Type II | Type III | Unit |
|---|---|---|---|---|
| $V_{off}$ | 1 | 0.5 | 0.5 | mV |
| tc of $V_{off}$ | 20 | 2–7 | 10 | $\mu$V/K |
| $I_b$ | $80 \times 10^{-9}$ | $10-50 \times 10^{-12}$ | $2 \times 10^{-12}$ | A |
| $I_{off}$ | $20 \times 10^{-9}$ | $10-20 \times 10^{-12}$ | $2 \times 10^{-12}$ | A |
| $A_0$ | $2 \times 10^5$ | $2 \times 10^5$ | 1500 | – |
| $R_i$ | $2 \times 10^6$ | $10^{12}$ | $10^{12}$ | $\Omega$ |
| CMRR | 90 | 80–100 | 100 | dB |
| SVRR | 96 | 80–100 | 70 | dB |
| $f_t$ | 1 | 1–4 | 300 | MHz |
| Slew rate | 0.5 | 3–15 | 400 | V/$\mu$s |

II are general purpose low-cost operational amplifiers, one with bipolar transistors at the input side (type I) and the other with JFETs at the input (type II). Type III is designed for high-frequency applications, with MOSFET transistors at the input. Also included are the temperature coefficient (tc) of the offset voltage, the CMRR, the supply voltage rejection ratio (SVRR), the slew rate, the DC gain $A_0$ and the unity gain bandwidth $f_t$.

In the analysis that follows we distinguish two types of errors: additive and multiplicative errors. Offset and bias currents cause additive errors. Multiplicative errors are caused by the finite gain and the first-order frequency dependence of the gain; in the analysis of the frequency dependence it is assumed that the DC loop gain is much larger than unity: $A_0\beta \gg 1$.

For a number of interface circuits we give, without any further derivation, their performance in terms of transfer function and input resistance, taking into account the non-ideal behaviour of the operational amplifier. Knowing the input impedance of an interface circuit, the error due to loading the sensor can easily be estimated.

## C.2 Current-to-Voltage Converter

Most AD converter types require an input voltage with a range from 0 to the ADC's reference voltage. When the ADC should convert the output of a sensor with current output (such as a photo detector) that current must first be converted into a proper voltage that matches the ADC's input range. The basic configuration that accomplishes this task is given in Figure C.2A.

Assuming the operational amplifier has ideal properties, the transfer of this I–V converter is simply $V_o = -I_i R$ and the input resistance is zero. The bias current of the operational amplifier sets a lower limit to the input range, the supply voltages an upper limit. The transfer including *additive errors* amounts:

$$V_o = -I_i R + V_{off} + I_{b1} R \tag{C.1}$$

To reduce the effect of bias current a current compensation resistance with value $R$ is inserted in series with the non-inverting input terminal (Figure C.2B). The transfer becomes:

$$V_o = -I_i R + V_{off} + I_{off} \tag{C.2}$$

which shows that current compensation reduces the error due to the bias current by about a factor of 5 (see Table C.1). The transfer including *multiplicative errors* due to finite gain $A_0$ is:

$$V_o = -I_R \frac{A_0}{1 + A_0} \approx -I_i R \left( 1 - \frac{1}{A_0} \right) \tag{C.3}$$

which means that the scale error is about the inverse of the DC gain of the amplifier. The first-order approximation of the transfer of an operational amplifier is:

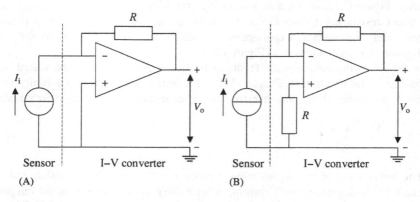

**Figure C.2** Current-to-voltage converter: (A) without and (B) with bias current compensation.

$$A(\omega) = \frac{A_0}{1 + j\omega\tau_A} \qquad (C.4)$$

with $\tau_A$ the first-order time constant of the amplifier. Amplifiers with this transfer have a constant gain-bandwidth product: in feedback mode, the overall gain decreases as much as the bandwidth increases. At unity gain the bandwidth equals the 'unity gain bandwidth' which is the value specified by the manufacturer, as in Table C.1. Apparently $f_t = \omega_t/2\pi = A_0/2\pi\tau_A$. Taking this into account the transfer of the I—V converter is:

$$V_o = -I_i R \frac{1}{1 + j\omega\tau_A/A_0} \qquad (C.5)$$

Note that the bandwidth of the converter equals the unity gain bandwidth of the operational amplifier. The input resistance in terms of input admittance amounts is:

$$G_{in} = \frac{1}{R_i} + \frac{A_0}{R} \qquad (C.6)$$

In general $1/R_i \ll A_0/R$ so the input impedance $R_{in} = 1/G_{in}$ is about $R/A_0$, in most cases a sufficiently low value to guarantee that the whole input current flows through the feedback resistor.

## C.3  Non-Inverting Amplifier

With a voltage amplifier the output voltage range of a sensor can be matched to the input range of an ADC. Voltage amplification is accomplished by either an inverting or a non-inverting amplifier configuration, depending on the required signal polarity. Figure C.3A shows a non-inverting amplifier with gain $1 + R_2/R_1$ and infinite input resistance. Hence the gain can be arbitrarily chosen by the two resistances, and the sensor experiences no electrical load. For $R_2/R_1 = 0$, the configuration is called a buffer (Figure C.3b).

In the following formulas the feedback portion $R_1/(R_1 + R_2)$ is abbreviated by $\beta$ according to the notation in Chapter 3. For the buffer amplifier, $\beta = 1$ (unity feedback). The transfer of the non-inverting amplifier including additive errors equals:

$$V_o = \frac{V_i}{\beta} + \frac{V_{off}}{\beta} + I_b^- R_2 \qquad (C.7)$$

The lower limit of the signal voltage range is set mainly by the offset voltage and to a lesser degree the bias current. After nullifying the offset it is the temperature dependency of the offset that limits the range. The transfer shows a scale error (multiplicative error) due to the finite gain:

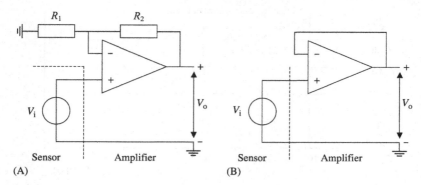

**Figure C.3** (A) Non-inverting amplifier, (B) buffer.

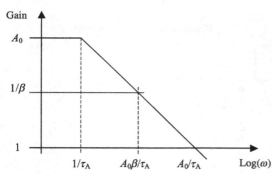

**Figure C.4** Frequency characteristic of a non-inverting amplifier: asymptotic approximation according to Eq. (C.9).

$$V_o = \frac{V_i}{\beta} \frac{A_0 \beta}{1 + A_0 \beta} \approx \frac{V_i}{\beta} \left(1 - \frac{1}{A_0 \beta}\right) \tag{C.8}$$

The scale error is the inverse of the loop gain $A_0\beta$. Taking into account the frequency-dependent gain as in Eq. (C.4), the transfer of the non-inverting amplifier is:

$$V_o = \frac{V_i}{\beta} \frac{1}{1 + j\omega\tau_A/A_0\beta} \tag{C.9}$$

Obviously the bandwidth is proportional to $A_0\beta$, and the transfer is inversely proportional to $A_0\beta$, hence the product of gain and bandwidth has a fixed value: the gain-bandwidth product. This relationship is shown in Figure C.4A.

*Numerical example.* $A_0 = 10^4$, $\beta = 0.01$, so $A_0\beta = 100$; unity gain bandwidth $A_0/\tau_A = 10^6$ rad/s, so the operational amplifier's cut-off frequency is $1/\tau_A = 100$ rad/s and the non-inverting amplifier bandwidth is $10^4$ rad/s.

Figure C.5 shows a simulation[1] of the transfer characteristic using a general purpose, low-cost operational amplifier for three gain factors: 1 (buffer), 100 and

[1] Type μA741 (original design by Fairchild, 1968); simulated by PSpice (OrCAD Inc.).

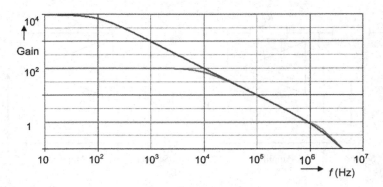

**Figure C.5** Simulation with a low-cost operational amplifier.

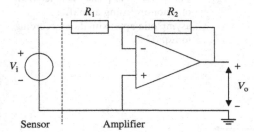

**Figure C.6** Inverting amplifier.

Sensor    Amplifier

10,000. Clearly the product of gain and bandwidth is constant. At higher frequencies the second-order behaviour of the operational amplifier becomes apparent.

Finally the input resistance of the non-inverting amplifier is found to be:

$$R_{in} = R_i(1 + A_0\beta) + \beta R_2 \approx A_0\beta R_i \tag{C.10}$$

which is sufficiently high in most applications.

## C.4 Inverting Amplifier

Figure C.6 shows the inverting amplifier configuration. Its voltage transfer is $-R_2/R_1$ and the input resistance amounts $R_1$, a much lower value than in the case of the non-inverting amplifier.

The transfer including additive errors amounts is:

$$V_o = -V_i\frac{R_2}{R_1} + V_{off}\left(1 + \frac{R_2}{R_1}\right) + I_b^- R_2 \tag{C.11}$$

The contribution of the bias current $I_b^-$ can be reduced by inserting a resistance $R_3$ in series with the non-inverting input terminal, with a value equal to $R_1//R_2$, that is the parallel combination of $R_1$ and $R_2$. With this bias current compensation, the term $I_b^-$ in Eq. (C.11) is replaced by $I_{off}$.

The transfer including multiplicative errors due to a finite gain is:

$$\frac{V_o}{V_i} = -\frac{R_2}{R_1}\frac{A_0\beta}{1+A_0\beta} \approx -\frac{R_2}{R_1}\left(1-\frac{1}{A_0\beta}\right) \tag{C.12}$$

which, again, introduces a scale error equal to the inverse of the loop gain. The frequency-dependent gain is given by

$$\frac{V_o}{V_i} = -\frac{R_2}{R_1}\frac{1}{1+j\omega\tau_A/A_0\beta} \tag{C.13}$$

Here, too, the product of gain and bandwidth is constant. The frequency characteristics are similar to those in Figure C.4. The input resistance becomes:

$$R_{in} = R_1\left(1 + \frac{R_2}{A_0R_i}\right) \tag{C.14}$$

a value that is almost equal to $R_1$. Note that this value can be rather low, resulting in an unfavourable load on the transducer.

## C.5  Comparator and Schmitttrigger

### C.5.1  Comparator

A voltage comparator (or short comparator) responds to a change in the polarity of an applied differential voltage. The circuit has two inputs and one output (Figure C.7A). The output has just two levels: high or low, all depending on the polarity of the voltage between the input terminals. The comparator is frequently used to determine the polarity in relation to a reference voltage.

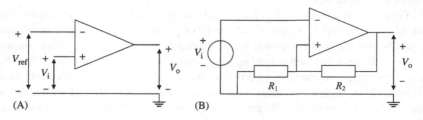

**Figure C.7** (A) Comparator, (B) Schmitttrigger.

**Table C.2** Specifications for Two Types of Comparators: Fast (Type I) and Accurate (Type II)

|                        |                 | Type I                       | Type II                   |
| ---------------------- | --------------- | ---------------------------- | ------------------------- |
| Voltage gain           | $A$             | —                            | $2 \times 10^5$           |
| Voltage input offset   | $V_{\text{off}}$ | $\pm 2$ mV $\pm 8\mu$V/K     | $\pm 1$ mV, max. $\pm 4$ mV |
| Input bias current     | $I_{\text{bias}}$ | $5\,\mu$A                   | 25 nA, max. 300 nA        |
| Input offset current   | $I_{\text{off}}$ | $0.5\,\mu$A $\pm 7$ nA/K     | 3 nA, max. 100 nA         |
| Input resistance       | $R_i$           | 17 k$\Omega$                 | —                         |
| Output resistance      | $R_o$           | 100 $\Omega$                 | 80 $\Omega$               |
| Response time          | $t_r$           | 2 ns                         | 1.3 $\mu$s                |
| Output voltage, high   | $v_{o,h}$       | 3 V                          | —                         |
| Output voltage, low    | $v_{o,l}$       | 0.25 V                       | —                         |

It is possible to use an operational amplifier without feedback as a comparator. The high gain makes the output either maximally positive or maximally negative, depending on the input signal. However an operational amplifier is rather slow, in particular when it has to return from the saturation state. Table C.2 gives the specifications for two different types of comparators, a fast type and an accurate type.

Purpose-designed comparators have a much faster recovery time with response times as low as 10 ns. They have an output level that is compatible with the levels used in digital electronics (0 and +5 V). Their other properties correspond to a normal operational amplifier and the circuit symbol resembles that of the operational amplifier.

## C.5.2 Schmitttrigger

A Schmitttrigger can be conceived as a comparator with hysteresis. It is used to reduce irregular comparator output changes caused by noise in the input signal. A simple Schmitttrigger consists of an operational amplifier with positive feedback (Figure C.7B). The output switches from low to high as soon as $V_i$ exceeds the upper reference level $V_{\text{ref1}}$, and from high to low as soon as $V_i$ drops below the lower level $V_{\text{ref2}}$. For proper operation the hysteresis interval $V_{\text{ref1}} - V_{\text{ref2}}$ must exceed the noise amplitude. However a large hysteresis will lead to huge timing errors in the output signal.

In Figure C.7B a fraction $\beta$ of the output voltage is fed back to the non-inverting input: $\beta = R_1/(R_1 + R_2)$. Suppose that the most positive output voltage is $E^+$ (usually just below the positive power supply voltage) and the most negative output is $E^-$. The voltage at the non-inverting input will be either $\beta E^+$ or $\beta E^-$. When $V_i$ is below the voltage on the non-inverting input, then $V_o$ equals $E^+$ (because of the high gain). This remains a stable situation as long as $V_i < \beta E^+$. If $V_i$ reaches the value $\beta E^+$, the output will decrease sharply and so will the non-inverting input voltage. The voltage difference between both input terminals decreases much faster than $V_i$ increases so, within a very short period of time, the output becomes maximally negative $E^-$. As long as $V_i > \beta E^-$, the output remains $V_o = E^-$, a new stable state.

The comparator levels of the Schmitttrigger are apparently $\beta E^+$ and $\beta E^-$. In conjunction with the positive feedback even a rather slow operational amplifier can have a fast response time. Evidently a special comparator circuit is even better. The switching levels can be adjusted by connecting $R_1$ to an adjustable reference voltage source $V_{ref}$ instead to ground. Both levels shift by a factor $V_{ref}R_1/(R_1 + R_2)$.

## C.6  Integrator and Differentiator

Velocity information can be derived from an acceleration sensor (for instance a piezoelectric transducer) and displacement from velocity (for instance from an induction-type sensor) by integrating the sensor signals. The reverse is also possible, by differentiation, but this approach is not recommended because of an increased noise level. Basic integrator and differentiator circuits are shown in Figure C.8. Ideally the transfers in time and frequency domains are given by:

$$V_o = -\frac{1}{RC}\int V_i \, dt \quad V_o = -RC\frac{dV_i}{dt} \tag{C.15}$$

$$V_o = -\frac{1}{j\omega RC}V_i \quad V_o = -j\omega RCV_i \tag{C.16}$$

However these circuits will not work properly. The integrator runs into overload because the constant offset voltage and bias current will be integrated too. This is prevented by limiting the integration range, simply by adding a feedback resistor in parallel to the capacitance $C$. The differentiator is unstable because of the (parasitic) second-order behaviour of the operational amplifier. This too is prevented by adding a resistor, in series with the capacitance $C$, resulting in a limitation of the differentiation range.

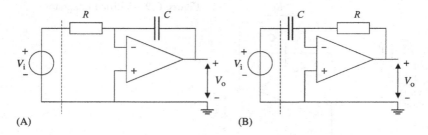

**Figure C.8** (A) Basic integrator and (B) differentiator are unable to operate well.

## C.6.1 Integrator

Figure C.9 shows the modified or 'tamed' integrator. Taking into account the finite gain of the operational amplifier $A$, the transfer function of this integrator is:

$$\frac{V_o}{V_i} = -\frac{A}{1 + (1+A)(R/R_f) + j\omega(1+A)RC} \tag{C.17}$$

Assuming a frequency-independent amplifier gain: $A = A_0$, the transfer function becomes:

$$\frac{V_o}{V_i} = -\frac{A_0}{1 + (1+A_0)(R/R_f) + j\omega(1+A_0)RC} \approx -\frac{R}{R_f}\frac{1}{1+j\omega R_f C} \tag{C.18}$$

The condition for the last approximation is valid for $A_0 \gg 1$ (which is always the case) and $A_0 R \gg R_f$ (which is usually the case). The inclusion of resistance $R_f$ limits the LF transfer to $R_f/R$, whereas the lower limit of the integration range is set to $f_L = 1/(2\pi R_f C)$.

*Numerical example.* Assume a lower limit of the integration range of 100 Hz and a DC gain of 100 are required. The component values should satisfy $R_f/R = 10^2$ and $R_f C = 1/200\pi$, accomplished by, for instance $R_f = 100R = 100\ \mathrm{k}\Omega$ and $C = 16$ nF. Figure C.10 shows the frequency characteristic for this design, based on a simulation with an ideal and a real operational amplifier. Clearly the higher limit of the integration range is set by the frequency-dependent gain of the operational amplifier.

## C.6.2 Differentiator

The transfer of the basic differentiator, taking into account the finite gain of the operational amplifier, is given by:

$$\frac{V_o}{V_i} = -j\omega RC \cdot \frac{A}{1 + A + j\omega RC} \tag{C.19}$$

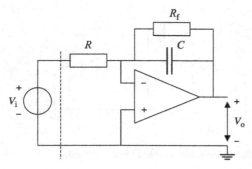

**Figure C.9** Modified integrator.

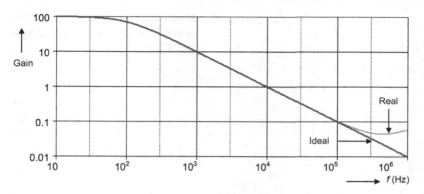

**Figure C.10** Frequency characteristic of the modified integrator according to Eq. (C.18) (ideal) and according to a simulation with a real amplifier (real).

For $A = A_0$ and $A_0 \gg 1$ this is:

$$\frac{V_o}{V_i} = -j\omega RC \cdot \frac{A_0}{1 + A_0 + j\omega RC} = -A_0 \frac{j\omega RC / A_0}{1 + j\omega RC / A_0} \qquad (\text{C.20})$$

This corresponds to a first-order high-pass characteristic with cut-off frequency at $\omega = A_0/RC$. The differentiation range is limited to this frequency. For much lower frequencies the transfer function is just $-j\omega RC$ and corresponds to that of an ideal differentiator.

However when the frequency dependence of the operational amplifier gain according to Eq. (C.4) is taken into account, the transfer function becomes:

$$\frac{V_o}{V_i} = -j\omega RC \cdot \frac{A_0}{A_0 + j\omega(\tau_A + RC) - \omega^2 \tau_A RC} \qquad (\text{C.21})$$

This characteristic shows a sharp peak at $\omega^2 \tau_A \tau / A_0 \approx 1$, for which value the transfer is about $A_0 RC$.

*Numerical example.* The required gain is 0.1 at 1 rad/s. So the component values $R$ and $C$ should satisfy $1/RC = 10$ rad/s. Suppose the amplifier properties are: $A_0 = 10^5$, unity gain bandwidth $10^5$ rad/s. Figure C.11 presents the calculated characteristic for this example. The transfer shows a peak of $10^4$ at the frequency $10^3$ rad/s.

Figure C.12 shows a simulation for a general purpose operational amplifier with frequency compensation (i.e. it behaves almost as a first-order system). Component values in this simulation are $R = 100$ k$\Omega$, $C = 1$ μF. Differences compared to Figure C.11 are due to different amplifier parameters $A_0$ and $\tau_A$ of the real amplifier.

If the operational amplifier has a second-order cut-off frequency (and most do have), instability may easily occur. Therefore the differentiating range should be limited to some upper frequency, well below the cut-off frequency of the

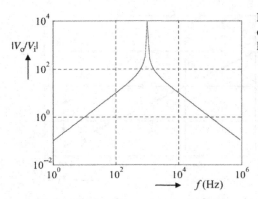

**Figure C.11** Frequency characteristic of the basic differentiator according to Eq. (C.21).

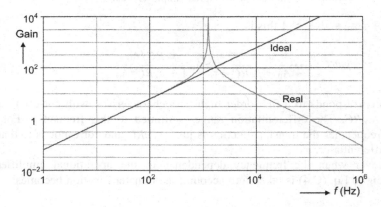

**Figure C.12** Simulation with ideal and a general purpose operational amplifier.

operational amplifier. This is accomplished by inserting a resistor $R_s$ in series with the capacitor $C$, as shown in Figure C.13.

With the 'taming resistor' $R_s$ the transfer becomes:

$$\frac{V_o}{V_i} = -j\omega\tau \cdot \frac{A}{1 + A + j\omega\{RC + (1 + A)R_sC\}} \tag{C.22}$$

Taking into consideration the finite and frequency-dependent gain of the operational amplifier according to Eq. (C.4) and assuming $A_0 \gg 1$, the transfer is modified to:

$$\frac{V_o}{V_i} = -j\omega RC \cdot \frac{1}{1 + j\omega\{R_sC + (RC + \tau_A)/A_0\} - \omega^2\tau_A(R_sC + RC)/A_0} \tag{C.23}$$

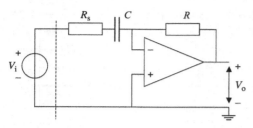

**Figure C.13** Modified or 'tamed' differentiator.

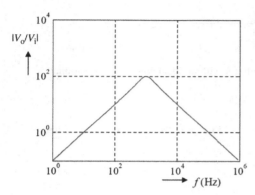

**Figure C.14** Frequency characteristic of the compensated differentiator according to Eq. (C.24).

For $R \gg R_s$ and $R_s C \gg (RC + \tau_A)/A_0$ this can be approximated by:

$$\frac{V_o}{V_i} = -j\omega RC \frac{1}{1 + j\omega R_s C - \omega^2 \tau_A RC / A_0} \qquad (C.24)$$

*Numerical example.* Same values as for the uncompensated differentiator and $R_s = R/100$. The transfer is calculated from Eq. (C.24) and displayed in Figure C.14. It shows that the peak is reduced by inserting resistance $R_s$. The differentiation range runs from very low frequencies up to about 1000 rad/s.

The simulation in Figure C.15 shows the effect of a taming resistor, here 100 Ω. Note that the sharp peak has completely disappeared. Further note the big difference between an ideal operational amplifier and a real type.

## C.7  Filters

Removing unwanted signal components (noise, interference, intermodulation products, etc.) can be accomplished by (analogue) filters. Simple filters consist of passive components only. However, loading such filters (at both input and output) may easily affect the performance. In such cases active filters (with operational amplifiers) are preferred, providing independent choices for various filter parameters (input and output impedance, cut-off frequency, gain). The first-order

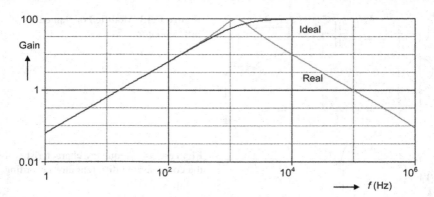

**Figure C.15** Simulation of a tamed differentiator with an ideal and a real operational amplifier.

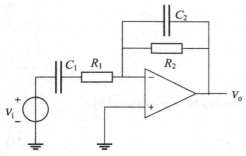

**Figure C.16** Active band-pass filter.

integrator and differentiator of the preceding section are essentially a low-pass and a high-pass filter, respectively. Figure C.16 shows an active band-pass filter.

Assuming ideal operational amplifier characteristics the transfer function is:

$$\frac{V_o}{V_i} = -\frac{Z_2}{Z_1} = -\frac{j\omega R_2 C_1}{(1 + j\omega R_1 C_1)(1 + j\omega R_2 C_2)} = -\frac{R_2}{R_1}\frac{j\omega R_1 C_1}{1 + j\omega R_1 C_1}\frac{1}{1 + j\omega R_2 C_2}$$

(C.25)

so a combination of gain $(-R_2/R_1)$, low-pass filter $(f_l = 1/2\pi R_2 C_2)$ and high-pass filter $(f_h = 1/2\pi R_1 C_1)$. Figure C.17 shows the transfer function (frequency characteristic) obtained by simulation with an ideal and a low-cost operational amplifier. In this design, the pass-band gain is set to 100 and the high-pass and low-pass frequencies are set at 100 Hz and 1 kHz, respectively, a combination that can be obtained with the component values $R_1 = 1$ k$\Omega$, $R_2 = 100$ k$\Omega$, $C_1 = 1600$ nF and $C_2 = 1.6$ nF. Both curves are almost identical, since the unity gain bandwidth of the (real) operational amplifier is much higher than the filter gain over the whole frequency range.

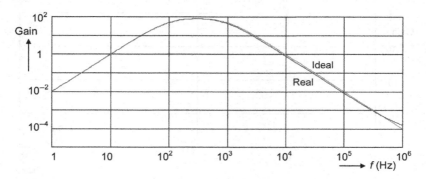

**Figure C.17** Transfer characteristic of the band-pass filter from Figure C.16.

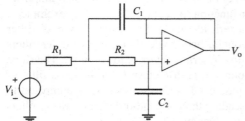

**Figure C.18** Sallen-Key low-pass filter.

Figure C.18 shows a second-order low-pass filter of the 'Sallen-Key' configuration.

Its transfer function equals:

$$\frac{V_o}{V_i} = \frac{1}{1 + j\omega(R_1 + R_2)C_2 - \omega^2 R_1 R_2 C_1 C_2} \tag{C.26}$$

Under particular conditions for the resistance and capacitance values this filter has a Butterworth characteristic, which means a maximally flat amplitude characteristic in the pass-band. The general transfer function of a Butterworth filter of order $2n$ is:

$$\left|\frac{V_o}{V_i}\right| = \frac{1}{\sqrt{1 + (\omega/\omega_c)^{2n}}} \tag{C.27}$$

For $n = 1$ the Butterworth condition is $C_1/C_2 = (R_1 + R_2)^2/2R_1R_2$. When $R_1 = R_2 = R$, then $C_1 = 2C_2$. The cut-off frequency is $f_c = 1/2\pi RC_1$. Figure C.19 depicts a simulation of such a filter, with a cut-off frequency of 1 kHz (by making $R_1 = 1$ k$\Omega$ and $C_1 = 160$ nF).

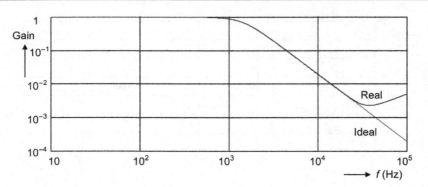

**Figure C.19** Transfer characteristic of the Butterworth filter from Figure C.18.

Here, the influence of the finite unity gain bandwidth of the operational amplifier is clearly visible and limits a proper filter function to about 20 kHz in this example.

The slope of the stop-band characteristic of a first-order low-pass filter is $-6$ dB/octave or $-20$ dB/decade. To obtain better selectivity, higher order filters can be obtained by simply cascading (putting in series) filters of lower order. The slope of the characteristic in the stop-band amounts $-6n$ dB/octave for a low-pass filter of order $n$. Note that the attenuation in the pass-band is also $n$ times more compared to that of a first-order filter. For instance, cascading three second-order Butterworth filters of the kind shown in Figure C.18 results in a slope of $-36$ dB/octave.

Printed in the United States
By Bookmasters

Printed in the United States
By Bookmasters